DNA Fingerprinting: Approaches and Applications

Edited by Terry Burke
         Gaudenz Dolf
         Alec J. Jeffreys
         Roger Wolff

Birkhäuser Verlag
Basel · Boston · Berlin

Editors' addresses:

Dr. Terry Burke
Dept. Zoology
University of Leicester
Leicester LE1 7RH
England

Dr. Gaudenz Dolf
Universität Bern
Institut für Tierzucht
3012 Bern
Switzerland

Prof. Alec J. Jeffreys
Dept. Genetics
University of Leicester
Leicester LE1 7RH
England

Dr. Roger Wolff
UCSF-HSE 1556
Box # 0554
San Francisco
CA 94143-0554
USA

Library of Congress Cataloging-in-Publication Data
DNA fingerprinting: approaches and applications/edited by Terry Burke . . . [et al.].
  p.   cm.—(Experientia. Supplementum; vol. 58)
  Selected papers presented at the First International Symposium on DNA Fingerprinting, held in Bern, Oct. 1–3, 1990.
  Includes bibliographical references and index.
  ISBN 3-7643-2562-3 (acid-free paper)—ISBN 0-8176-2562-3 (acid-free paper)
  1. DNA fingerprints—Congresses. 2. Molecular genetics—Technique—Congresses.
I. Burke, Terry, 1957–   .
II. International Symposium on DNA Fingerprinting (1st: 1990: Bern, Switzerland)
RA1057.55.D635   1991
574.87'328—dc20

Deutsche Bibliothek Cataloging-in-Publication Data
DNA fingerprinting: approaches and applications/ed. by Terry Burke . . . – Basel; Boston; Berlin . . . Birkhäuser, 1991
  (Experientia: Supplementum; 58)
  ISBN 3-7643-2562-3 (Berlin . . .)
  ISBN 0-8716-2562-3 (Boston)
NE: Burke, Terry [Hrsg.]; Experientia/Supplementum

© 1991 Birkhäuser Verlag
     P.O. Box 133
     4010 Basel
     Switzerland

Printed in Germany on acid-free paper

ISBN 3-7643-2562-3
ISBN 0-8176-2562-3

Contents

vi

Population Genetics and Evolutionary Biology

Economically-important Animals and Plants

Implementation of DNA Typing

Preface

Although DNA fingerprinting is a very young branch of molecular
genetics, being barely six years old, its recent impact on science,
law and politics has been dramatic. The application of DNA fingerprinting to forensic and legal medicine has guaranteed a high public
profile for this technology, and indeed, scarcely a week goes by without the press reporting yet another crime successfully solved by molecular genetics. Less spectacularly, but equally importantly, DNA typing
methods are steadily diffusing into an ever wider set of applications
and research fields, ranging from medicine through to conservation
biology.

To date, two DNA fingerprinting workshops have been held in the
UK, one in 1988 organised by Terry Burke at the University of
Leicester, and the second in 1989 at the University of Nottingham,
co-ordinated by David Parkin. In parallel with these workshops, which
have provided an important focus for researchers, Bill Amos and
Josephine Pemberton in Cambridge have established an informal
newsletter "Fingerprint News" which is playing a major role as a
forum for DNA fingerprinters. By 1989, it was clear that the field had
broadened sufficiently to warrant a full international meeting. As a
result, Gaudenz Dolf took on the task of organising the first, of what I
hope will be many, International Symposium of DNA Fingerprinting
held at Bern during 1st–3rd October 1990. The success of the meeting
can be judged from the remarkable attendance, with 270 delegates
from no less than 30 countries. This volume provides a compilation of
some of the many presentations given at the symposium

We have tried to organise the papers according to subject area,
though in many cases the boundaries are blurred and the classification
of some papers is rather arbitrary. Contributors in the first section
discuss the types of DNA typing systems currently available, as well as
the molecular genetics of the highly variable loci which underpin DNA
fingerprinting and some of their medical applications. In the next
section, the population genetics and evolutionary biology of hypervariable DNA are explored. Next, contributors interested in animal and
plant DNA typing describe some of the roles that DNA fingerprinting
is playing in analysing the population genetics and mating systems of
natural populations and in developing DNA markers to assist animal
and plant breeders. Finally, the large scale implementation of DNA

typing systems is discussed, particularly in the context of forensic and legal medicine.

We trust this volume will provide a useful broad survey of current research into DNA fingerprinting as we enter the 1990s.

Alec J. Jeffreys,
Leicester

February 1991

DNA Fingerprinting: Approaches and Applications
ed. by T. Burke, G. Dolf, A. J. Jeffreys & R. Wolff

Principles and Recent Advances in Human DNA Fingerprinting

A. J. Jeffreys, N. J. Royle, I. Patel, J. A. L. Armour, A. MacLeod, A. Collick, I. C. Gray, R. Neumann, M. Gibbs, M. Crosier, M. Hill, E. Signer and D. Monckton.

Department of Genetics, University of Leicester, University Road, Leicester LE1 7RH, Great Britain

Summary
Since 1985, DNA typing systems have played an increasingly important role in many aspects of human genetics, most notably in forensic and legal medicine. This article reviews the development of multilocus and single locus minisatellite DNA probes, and more recently the use of PCR to amplify hypervariable DNA loci, as well as discussing the biological properties of the unstable regions of DNA which form the basis of almost all DNA fingerprinting systems.

Introduction

In early 1985, we described the first development of multilocus DNA fingerprints and speculated that these individual-specific DNA patterns might provide a powerful method for individual identification and paternity testing (Jeffreys *et al.*, 1985b). At the time, we suspected that the implementation of these applications would be protracted, and that major legal problems would be encountered as DNA evidence proceeded from the research laboratory to the court room. Subsequent history showed that we were unduly pessimistic. By April 1985 the first case, involving a U.K. immigration dispute, had been satisfactorily resolved by DNA fingerprinting (Jeffreys *et al.*, 1985a). Shortly thereafter, DNA evidence in a paternity dispute was admitted in a U.K. civil court. DNA typing in criminal investigations saw its debut in October 1986 with the Enderby murder case, an investigation which led to the first instance of the release of a prime suspect proved innocent by DNA evidence (Gill and Werrett, 1987; Wong *et al.*, 1987; see Wambaugh, 1989). By 1987, DNA typing results had been admitted in evidence in criminal courts in the U.K. and U.S.A., and in 1988 the U.K. Home Office and Foreign and Commonwealth Office had ratified the use of DNA fingerprinting for the resolution of immigration disputes which hinge upon disputed family relationships (Home Office, 1988). 1989 saw the first major attack, in the U.S.A., on the procedural and scientific validity of DNA typing in forensics (see Lander, 1989), resulting in a major independent review carried out by the U.S. Congress Office of Technology Assessment (1990) which concluded that DNA-based identification was scientifically valid provided that appropriate technology, quality control and quality assurance procedures were implemented. Today (September 1990), DNA typing systems are in place in public and commercial forensic laboratories in at least 15 different countries, with many other countries actively considering DNA analysis in forensic and legal medicine.

Together with this extraordinarily rapid spread of DNA typing, the last few years have seen major developments in the underpinning technology and our understanding of the genetics of hypervariable DNA. We will review the various DNA test systems now available and will discuss the biological properties of the highly variable DNA regions which provide the basis for DNA typing.

Multilocus DNA Fingerprinting

The most variable loci discovered in the human genome consist of tandemly-repeated minisatellites, otherwise known as VNTR (variable number of tandem repeats) loci (Nakamura *et al.*, 1987), and provide the basis for most currently used DNA typing systems. Surprisingly, there are clear DNA sequence similarities amongst the tandem repeat sequences of many human minisatellites, which define the minisatellite "core" sequence (Jeffreys *et al.*, 1985b). The core is presumably involved in the generation of minisatellites and/or maintenance of variability in allelic repeat unit copy number at these loci, possibly by promoting unequal crossing over between tandem repeats. Direct evidence for the recombinational proficiency of minisatellites has come from studies of inter-plasmid recombination in transfected mammalian cells (Wahls *et al.*, 1990). Recently, we have discovered a DNA binding protein which interacts with high affinity with tandem repeats of the minisatellite core sequence (Collick and Jeffreys, 1990) which may provide a route to unravelling the biochemical processes which operate at these extraordinary VNTR loci.

Whatever the role of the minisatellite core sequence, its discovery immediately suggested that DNA hybridization probes comprised of tandem repeats of the core sequence should detect multiple variable human DNA fragments by Southern blot hybridization, to produce an individual-specific DNA "fingerprint". We have developed two such multilocus probes, termed 33.6 and 33.15, which vary in the length and precise sequence of the core repeat and which are routinely used in casework analysis, primarily in parentage testing (Jeffreys *et al.*, 1985b, c). Each probe detects typically 17 variable DNA fragments per individual in the size range 3.5–20+ kb, plus many smaller DNA fragments which are too complex to resolve electrophoretically and which are not used in the statistical evaluation of casework data. Extensive pedigree analysis shows that 33.6 and 33.15 detect independent sets of variable DNA fragments, with only approximately 1% of fragments co-detected by both probes (Jeffreys et al., 1986; A. J. Jeffreys, M. Turner and P. Debenham, manuscript submitted), and therefore provide independent DNA fingerprints. Incidentally, these probes cross-hybridize to variable loci in a wide range of animal, bird and plant species, though the many applications of non-human DNA fingerprints in breeding, population genetics and conservation biology will not be discussed here.

Following the development of multilocus probes 33.6 and 33.15, many other tandem-repeated DNA fingerprinting probes have been reported (see for example Ali et al., 1986; Vassart et al., 1987; Fowler et al., 1988). Indeed, Vergnaud (1989) has shown that almost any tandem-repeated sequence can, to some extent at least, detect multiple variable human DNA fragments. However, the most effective multilocus probes all tend to be G-rich and similar in sequence to the original minisatellite core sequence. Most importantly, little research has been done on establishing the degree of independence of the sets of loci detected by different multilocus probes (other than 33.6 and 33.15), and indeed there is now clear evidence that significant overlap between different probes may exist (Armour et al., 1990; Julier et al., 1990). It is still unclear what molecular features are essential to an efficient multilocus probe, although the idea of the minisatellite core sequence remains as the only conceptual basis for understanding probe effectiveness.

Properties of Human DNA Fingerprints

The statistical evaluation of multilocus DNA fingerprint evidence, whether in comparing a forensic specimen with a criminal suspect or in determining paternity, rests on a single parameter, namely the proportion of bands x which on average are apparently shared between unrelated people (Jeffreys et al., 1985c). x has been estimated at 0.14 for probe 33.6 and 33.15 from extensive Caucasian casework, with no evidence for significant shifts in the value of x between different ethnic groups. A conservative value of x of 0.25 is deliberately used in casework evaluation, to prevent over-interpretation with respect to the defendant and to allow for reduction in variability due to inbreeding ($x = 0.25$ is equivalent to an assumption of first cousin relationship in an outbred Caucasian population). Assuming statistical independence of bands, then the chance that n bands in individual A would all be matched by bands of similar electrophoretic mobility in B is given by x^n.

This probability is not that of full identity, since (a) individual B may have additional bands not present in A, (b) non-scored bands smaller than 3.5 kb would also have to correspond between A and B, (c) for identity, bands would have to match in precise relative position and relative intensity, a criterion far more stringent than used in determining the band-sharing coefficient x. The statistical estimator x^n is therefore extremely conservative. Note incidentally that the individual specificity of multilocus DNA fingerprints is derived, not from the (modest) statistical weight attributed to a given band, but from the large number of bands scored.

Statistical evaluation of DNA fingerprints, whether using the estimator x^n or more formal Bayesian analyses (Hill, 1986; Brookfield, 1989; Evett et al., 1989a, b), makes a number of critical but testable assumptions. First, the band sharing frequency x is assumed to be constant for all bands (or more correctly molecular weight intervals on the DNA fingerprint). In fact, x decreases with increased DNA fragment size (Jeffreys et al., 1985c), and algebraic considerations show that this heterogeneity in x results in x^n being a conservative match estimate. Second, all bands are assumed to be statistically independent. Non-independence could in principle arise through linkage or allelism between bands, and through inbreeding leading to relatively homogeneous subpopulations within the global population from which x was estimated. Extensive analyses of pedigrees and of loci cloned from DNA fingerprints shows widespread dispersal of the variable DNA fragments around the human genome, and enormous allelic length variability, resulting in very low levels of allelism and linkage between human DNA fingerprint fragments, and therefore substantial independence of bands (Jeffreys et al., 1986; A. J. Jeffreys, M. Turner and P. Debenham, manuscript submitted). This is not necessarily true for all species, and indeed there is clear evidence of major linkage and, probably, linkage disequilibrium between different DNA fragments detected in mice (Jeffreys et al., 1987) and cattle (unpub. data). In man, the sole residual concern is therefore that of inbreeding. While non-random mating obviously occurs, the relevant question is whether human inbreeding in practice could result in a significant increase in the band sharing frequency within a local community. This is unlikely. The band sharing frequency x is largely dictated by gel electrophoretic resolution, with many of the underlying loci showing far higher levels of variability. As a result, many different variable loci can, and do, contribute bands to a given gel interval (Wong et al., 1987), thereby providing a buffer against shifts in allele frequency at individual contributing loci due to inbreeding. In practice, the bands sharing frequency x in Caucasians remains the same if comparisons are restricted to husband-wife pairs (who by definition are representatives of local breeding communities), with as yet no evidence for a subset of consanguineous marriages which might signal significant inbreeding (A. J. Jeffreys, M. Turner and P. Debenham, manuscript submitted). In only one instance, involving a highly inbred small community from the Gaza strip where consanguineous marriages are a cultural norm, was the band-sharing coefficient between "unrelated" individuals significantly increased (R. J. Bellamy, C. F. Inglehearn, I. K. Jalili, A. J. Jeffreys and S. S. Bhattacharya, manuscript submitted). However the magnitude of the increase in x (approximately 0.2 per probe) was modest and still resulted in DNA fingerprints which were essentially completely individual-specific.

Extension of multilocus DNA fingerprints to paternity disputes raises one other concern, namely the effect of germline mutation of DNA fingerprint bands on the efficiency of parentage estimation. Such mutation will generate one or more offspring bands which cannot be attributable to either genuine parent, providing evidence which could be interpreted as an exclusion. The incidence of mutation at these highly unstable loci is not insignificant; using both 33.6 and 33.15, 27% of offspring show one mutant band, 1.2% show two mutations and an estimated $< 0.3\%$ show three unassignable bands. To determine whether this level of mutation significantly blurs the discrimination of fathers from non-fathers, we have determined the proportion of non-maternal bands in a child which cannot be attributed to the alleged father. For a sample of 1419 true fathers, this proportion was, as expected, low (mode $= 0$, range $= 0 - 0.18$). For a corresponding sample of 283 falsely accused non-fathers, this proportion was much higher (mode 0.77, range $0.43 - 1$). Thus this proportion provides a single statistic which efficiently distinguishes fathers from non-fathers, even in the presence of mutation (A. J. Jeffreys, M. Turner and P. Debenham, manuscript submitted.) (In practice paternity disputes, particularly those showing mutant bands, are further tested using single locus minisatellite probes, see below). For a given paternity case, statistical evaluation uses the x^n statistic or Bayesian analysis combined with the empirically observed frequency of new mutant bands to determine an appropriate paternity index.

Multilocus human DNA fingerprinting is thus supported by a substantial body of genetic and population data, and is now routinely used in paternity and immigration testing. These probes have proved particularly useful in the latter disputes, where no prior assumptions about claimed family relationships (for example between the UK sponsor, his alleged wife and alleged children) may be made, and where one is sometimes required to prove for example that a man is the family's father, rather than the *brother* of the true father. All of these analyses use pristine DNA obtained from fresh blood and do not involve inter-blot comparisons, thereby minimizing problems in data production and interpretation. In contrast, DNA fingerprints have proved less powerful in routine forensic testing, due to the relative lack of probe

6

sensitivity, problems in interpreting incomplete patterns resulting from
partial DNA degradation and/or poor DNA recovery, inter-blot comparison problems due to subtle variations in hybridization stringency
and in the resulting patterns, the (not insurmountable) difficulty of data
banking the complex patterns, and the inability to detect mixed DNA
samples originating from more than one person. Nevertheless, DNA
fingerprints have been used successfully in a number of criminal investigations (see for example Gill and Werrett, 1987), and provide rich
patterns in which electrophoretic band shift, which has bedevilled some
single locus probe analyses (Lander, 1989), is readily identifiable and
correctable.

Single Locus Minisatellite Probes

DNA fingerprints provide DNA phenotypes, not genotypes, in which
information on loci and alleles is unavailable. In contrast, cloned
human minisatellites can produce locus-specific DNA hybridization
patterns from which genotypic information can be deduced, of critical
importance in for example linkage analysis. Hundreds of cloned minisatellites have now been isolated, either by chance, by screening human
genomic libraries for clones which detect hypervariable loci (Knowlton
et al., 1986), or by hybridization screening of libraries using oligonucleotides based on known VNTR sequences (Nakamura *et al.*, 1987, 1988),
or by selective cloning of DNA fingerprint fragments into λ bacteriophage vectors (Wong *et al.*, 1986, 1987) or more efficiently into a
charomid vector followed by ordered-array library screening with a
range of multilocus probes (Armour *et al.*, 1990). This latter approach
has proved to be very effective both on humans and on other mammalian and avian species.
  Many of these cloned minisatellites have been localised in the human
genome (and indeed provide critically informative landmark loci in
human linkage maps). They have been found on essentially every
human chromosome, including the X chromosome and the X-Y pairing
region, though as yet no Y-specific minisatellites have been described.
Minisatellites are however not randomly distributed in the genome but
instead preferentially localize near the ends of human chromosomes
(Royle *et al.*, 1988). DNA sequence analysis of these proterminal
minisatellites and their immediate environs shows that they are frequently closely linked to other hypervariable loci (sometimes within a
few base pairs) and to dispersed repeat elements such as Alu, L1 and
proretroviral LTRs (Armour *et al.*, 1989b). In two instances, a human
minisatellite has been shown to have evolved by tandem-repeat amplification from within such a dispersed element. The clustering of minisatellites in these proterminal chromosomal regions is intriguing,

particularly in view of the role of these regions in chromosome pairing and recombination. However, there is no evidence for the direct involvement of minisatellites in these aspects of chromosome mechanics, and instead it seems likely that their presence in these regions may reflect the existence of relatively unstable DNA domains adept at accumulating elements such as minisatellites and retroposons. Finally, there is no evidence for expression, function or coding potential of human minisatellites, with the sole exception of the hypervariable MUC1 locus which, remarkably, encodes a highly polymorphic mucin (Swallow *et al.*, 1987). It is therefore extremely unlikely that genotypic data gleaned from the minisatellites used in forensics will ever provide phenotypic information, for example on disease liability. This is of course critical if the public is to accept the widespread use of DNA typing in civil and criminal investigations.

Given the large number of cloned minisatellites, it is possible to select a combination of probes appropriate for forensic analysis. The probes should be unlinked to minimize the risk of allelic association (linkage disequilibrium) between different loci. Each probe must be locus-specific and detect, by Southern blot hybridization analysis, a hypervariable locus with one band per allele, giving two-band patterns (heterozygotes) or one-band patterns (presumptive homozygotes or heterozygotes with alleles of similar size). We have established a set of 5 hypervariable loci conforming to these requirements and all typable on HinfI blots (Wong *et al.*, 1987; Smith *et al.*, 1990). The FBI are currently evaluating a corresponding set of loci, using HaeIII as a standard restriction enzyme. Our opinion is that HinfI is a more appropriate enzyme for minisatellite typing. First, most minisatellites are GC-rich and are relatively susceptible to HaeIII cleavage (recognition sequence GGCC). If HaeIII cleaves every repeat unit, the minisatellite is destroyed and rendered untypable. On other occasions, only a few of the repeat units are susceptible to cleavage, due to the presence of variant repeat units containing a HaeIII site dispersed along the tandem repeat array. Such internal site(s) result in an allele producing, not one, but two or more cosegregating DNA fragments to create a complex hybridization profile. At some loci, these internally-cleavable alleles can be relatively rare, and could result in the occasional forensic case where profile interpretation could be seriously compromised. The other problem with HaeIII is the lack of an appropriate control to ensure complete digestion. In contrast, the minisatellite clone pMS51 (Armour *et al.*, 1989b) detects a variable locus flanked by a HinfI site remarkably resistant to digestion; pMS51 can therefore be used as an internal quality control in forensic casework to ensure complete HinfI digestion of genomic DNA.

The single locus minisatellite probes chosen for forensic casework all show extraordinary levels of allelic variability. The most variable and informative locus described to date is MS1, with alleles ranging from 1 to 23 kb long, and with $> 99\%$ of individuals showing two resolvable alleles by Southern blot hybridization. The repeat unit is 9 bp long, yielding in principle 2400 different allelic length states. Genomic DNA mixing experiments (Wong *et al.*, 1987) and determination of allele length frequency distributions at MS1 in human populations (Smith *et al.*, 1990) show that there are no common alleles at this locus, and theoretical considerations suggest that most or all of the 2400 possible allelic length states exist in human populations. Of course, not all alleles can be electrophoretically resolved, resulting in the allele length frequency distributions being quasi-continuous. In contrast, loci with lower variability ($< 96\%$ heterozygosity) tend to show a more limited number of distinct alleles with real and measurable population frequencies (Wong *et al.*, 1987; Smith *et al.*, 1990); allele length frequency distributions at such loci tend to be discontinuous or "spiky", rather than smooth, with the result that small errors in allele sizing can result in large errors in allele frequency estimates. Such loci will also tend to be more vulnerable to genetic drift and inbreeding effects (see below), and are best avoided for forensic use, if possible.

Single locus minisatellite probes provide a very powerful tool for forensic analysis provided that DNA of sufficient quality and quantity can be recovered from the forensic specimen. Detection based on ^{32}P-label or enhanced chemiluminescence is sensitive, the limits being approximately 10 ng genomic DNA. Mixed DNA samples (e.g. semen-bearing vaginal swabs, blood from more than one victim) can be readily identified. The individual-specificity achievable amongst unrelated people with a battery of four sequential single locus probe tests is comparable to that achievable using one multilocus probe. This follows, not from the large number of bands scored as with multilocus probes, but from the low population frequency of each of the single locus probe alleles. However, these probes are relatively poor at discriminating between close relatives. For example, a single probe has, at best, only a 75% chance of distinguishing two siblings ($< 99.6\%$ for 4 probes). For this reason, we prefer the term "DNA profiling", rather than DNA fingerprinting, to describe single locus probe analysis.

Evaluation of Single Locus Probe Profiles

The statistical evaluation of DNA profile evidence requires knowledge of allele frequencies and assumptions about population structuring. Suppose that a single locus probe yields two alleles a and b, with frequencies q_a and q_b measured in the reference population. Suppose further that the forensic profile is indistinguishable from that of the suspect (allowing for the occasional instance of minor electrophoretic band shift, within an acceptable range established from extensive casework experience). The probability of chance "match", under the assumption that alleles associate at random in the population, is given under the Hardy-Weinberg equilibrium by $2q_aq_b$. However, the sizes of the alleles are not known with absolute precision, either in the casework samples or in the reference database, and thus some form of allele pooling or "binning" is required to estimate appropriate values of q_a and q_b and to correct these values for sampling errors arising from the finite size of any population database. It is important that allele pooling is conservative relative to the criteria used to declare a forensic match, such that the statistical weight of the evidence is biased in favour of the defendant (equivalent to the use of a conservative band-sharing frequency in the evaluation of multilocus DNA fingerprints). It is also important to appreciate that there is not necessarily a sharp distinction between "match" and exclusion in forensic analyses, and that severe cases of electrophoretic band shift could yield results that would have to be declared inconclusive. The obvious solution to this problem is to identify and eliminate the causes of such shifts (which in our experience have very rarely proved to be a significant problem).

The only biological problem to emerge from the recent debate is whether appropriate ethnic reference databases are used and whether the assumption of Hardy-Weinberg equilibrium is valid. This problem resolves into two questions. First, do significant differences in allele length frequency distributions exist between different ethnic groups? Second, do localized inbred sub-populations exist within a given population in which the chance of allelic identity is far higher than estimated from allelic frequencies derived from the population as a whole? The first question can be resolved by establishing databases from different ethnic groups. For the most variable loci, such as MS1, very similar allele length frequency distributions are found even amongst radically different ethnic groups (P. Debenham, pers. commun.). As the variability is lowered, evidence for minor ethnic divergence becomes apparent (Balazs *et al.*, 1989), and with low variability probes (e.g. MS8 with a heterozygosity of 85% and only four major alleles) significant frequency shifts are seen, particularly of the common alleles (Flint *et al.*, 1989). This progressive increase in inter-population divergence with decreased variability is to be expected in view of the opposing force of genetic drift, leading to shifts in allele frequency and eventual allele fixation or extinction, and recurrent mutation, pumping new length alleles into the population. As discussed below, the most variable loci have the highest mutation rate, and at MS1 this rate is so high that drift will be effectively counteracted by mutation, re-establishing variability and preventing any allele from attaining significant population frequencies. This is illustrated by the work of Flint *et al.* (1989) who showed a significant drop in heterozygosity in Polynesians versus Melanesians (presumably as the result of a population bottleneck) which could be detected using moderately variable minisatellite loci but *not* with the ultravariable locus MS1. Although more work is needed on this problem, the current impression is that, by using ultravariable loci, problems of ethnic group divergence and identification of an appropriate reference population database should be minimized.

The second population problem concerns the possible existence of local inbred communities within large populations. Lander (1989) has reported finding a significant excess of "homozygotes" in some population databases, implying departure from Hardy-Weinberg equilibrium and therefore non-random association of alleles in individuals, consistent with significant population structuring. There are, however, two objections to this argument. First, any difference in the criteria used to bin alleles of similar size and the criterion used to define "homozygotes" (single band individuals) will lead to spurious apparent departures from Hardy-Weinberg equilibrium. Second, the assumption that single band individuals are homozygous is not necessarily correct. Instead, such individuals may be heterozygous and contain a second very small allele which has either been electrophoresised off the gel or hybridizes too poorly to be detected. Alternatively, true null alleles may exist at some loci (Armour *et al.*, 1990; Wong *et al.*, 1990). Indeed 90% of the alleles at one hypervariable locus recently isolated in our laboratory are null, and two-band individuals are rarely seen (unpubl. data). Thus, failure to detect all alleles overinflates estimates of homozygosity and renders Hardy-Weinberg tests on population databases invalid.

To circumvent these problems, we have recently investigated the frequency of allele matching between husband-wife pairs, using a battery of ultravariable single locus probes (unpubl. data). If inbreeding occurs in local communities, then true husband-wife pairs (who are representatives of such local communities) should show a greater level of allele sharing than randomized male-female pairs from the overall population. In practice, there is some evidence for enhanced allele sharing in the ethnic groups surveyed, though the effect is small (approximately 1–2% at the borderline of statistical significance) and would have little effect on the statistical evaluation of single locus probe profiles.

Amplification of Minisatellites by PCR

Single locus minisatellite probes can be used to type as little as 10 ng human genomic DNA, corresponding to 1700 diploid cells. To improve sensitivity, we and others have shown that it is possible to amplify minisatellites by the polymerase chain reaction (PCR; Saiki *et al.*, 1988), using amplimers designed from the unique sequence DNA flanking the minisatellite tandem repeat arrays (Jeffreys *et al.*, 1988a, 1990; Boerwin-kle *et al.*, 1989; Horn *et al.*, 1989). Sub-nanogram amounts of genomic DNA can be readily typed using this approach, and multiple hypervari-able loci can be simultaneously amplified using appropriate combina-tions of amplimers. Alleles up to 10 kb long can be faithfully amplified and detected by Southern blot hybridization with the appropriate tandem-repeat probes. By limiting the number of cycles to the exponen-tial phase of PCR, the yield of product becomes proportional to human DNA input, an important consideration in forensic analysis. If the PCR cycle number is increased, alleles up to 6 kb long can be directly visualised on ethidium bromide stained agarose gels, without any need for blotting/hybridization. However, at such high cycle numbers, the relationship between DNA input and product yield can be lost, and collapse of these tandem-repeat alleles to give complex profiles of spurious minisatellite products can occur, particularly with large and inefficiently-amplified alleles. Also, spurious DNA products may arise through mispriming elsewhere in the genome. PCR typing of minisatel-lites can also be extended with good efficiency and fidelity to the single molecule/single cell level (Jeffreys *et al.*, 1988, 1990). While such sensi-tivity dramatically increases the potential range of forensic analysis to hair root, saliva, urine and skeletal remains (see below), it also brings in formidable problems of sample contamination, both by carry-over of previous PCR products and by inadvertent contamination of eviden-tiary material with extraneous human cells (remember, for example, that saliva and nasal mucus can contain hundreds of cells per mi-crolitre!).

Internal Variation in Minisatellites

PCR has also allowed a further level of minisatellite variability to be explored, namely subtle DNA sequence variation between repeat units in an allele (Jeffreys *et al.*, 1990). Such minisatellite variant repeats (MVRs) commonly occur in an intermingled fashion along a minisatel-lite allele. If a minisatellite contains, for example, two types of variant repeats, A and B, then it is possible to chart the positions of these variants along a PCR-amplified allele, to give a binary code or "se-quence" of repeat units along an allele, for example AABBABBBAA-BABBABAAAB for a 20 repeat unit allele. This MVR mapping techique, though at present laborious, gives an absolute value for allele length (number of repeat units) and an unambiguous binary coding of any allele, providing in principle a powerful solution to the problem of identifying allele matches in forensic analyses. Furthermore, MVR mapping is capable of discriminating between vast numbers of different allelic states ($> 10^{70}$ states for minisatellite MS32, cf. approximately 100 states resolvable by conventional Southern blot analysis). We have applied this system to the hypervariable minisatellite MS32, to explore the true level of variability at this locus. Several remarkable features have emerged from this analysis. First, alleles of identical length in unrelated individuals seldom show identical MVR maps, indicating that the true level of allelic variability at MS32 is far greater than that suggested by conventional Southern blot analysis. Second, alleles of different length frequently show strong similarities or identities over part of the MVR map, implying recent divergence from a common ancestral allele. Curiously, MVR map similarity between alleles is usually restricted to the beginning (5′ end) of the tandem repeat array, with the 3′ ends showing far greater interallelic variability. This gradient of variability almost certainly indicates that mutation events which alter the length of MS32 alleles and reshuffle the pattern of MVRs are preferentially confined to the 3′ ends of alleles, implying the existence of some flanking DNA element(s) which modulate the location of length change mutation events. In Northern Europeans, almost all MS32 alleles can be classified into just three types of 5′ MVR map (haplotypes). In contrast, other ethnic groups (Pakistanis, Bangladeshis and particularly Africans/Afro-Caribbeans) show a much greater range of 5′ haplotypes (including the three European haplotypes) (unpubl. data). This curious result suggests a major recent elimination of haplotypic variability in Europeans, possibly reflecting a prehistoric population bottleneck. Despite this loss of *haplotypic* variability, allele *length* variability and allele length frequency distributions in these ethnic groups are similar, providing further evidence for the earlier assertion that recurrent mutation will counteract genetic drift of allele length classes at the most variable and unstable minisatellite loci.

PCR has also allowed us to explore the more distant evolutionary origins of contemporary minisatellite alleles in human populations, by amplifying and analysing the homologous loci from primates (I. C. Gray and A. J. Jeffreys, manuscript submitted). Analysis of the ultravariable loci MS32 and MS1 has shown that hypervariability is a remarkably transient evolutionary phenomenon attained within the last few million years in humans, following the divergence of man from the great apes. In the latter species, these loci are largely monomorphic with very low repeat unit copy number, a progenitor state for hypervariable loci which appears to be evolutionarily stable over periods of tens of millions of years.

Minisatellite Mutation

Assuming that minisatellites are without phenotypic effect, a reasonable assumption, high allelic length variability at these loci must reflect high rates of *de novo* mutation producing new length alleles. Direct measurement of minisatellite mutation rates, both in the germline and in somatic tissue, is not only important for unravelling the molecular processes which generate variability at these loci, but is also of direct relevance to forensic and legal applications of single locus probes. Germline mutation will produce apparent exclusions in paternity testing, and somatic mutation could in principle produce divergence in DNA phenotypes between different tissues (e.g. blood and sperm) in the *same* individual.

Germline length change mutation rates at human loci have been directly measured in human pedigrees (Jeffreys *et al.*, 1988a). As expected from the neutral mutation – random drift hypothesis, mutation rate increases with variability and becomes significant above approximately 96% heterozygosity. For the most variable human minisatellite, MS1, the mutation rate is an extraordinary 0.05 per gamete. Offspring are as likely to inherit a mutant allele from their mother as from the father, and mutant alleles can be, with equal frequency, either larger or smaller than their progenitor allele. Mutation events can sometimes result in the gain or loss of kilobases of the tandem repeat array, though most mutation events are small, involving the gain or loss of only a few repeat units. Mutation rates of approximately 10^{-2} per gamete do not significantly interefere with the use of these probes in paternity analysis, provided that mutation rates are known and can be incorporated into statistical likelihood ratio analyses of paternity against non-paternity. Less variable loci with unknown mutation rates will, in contrast, generate the occasional paternity case where "exclusion" with one probe, but inclusion with remaining probes, will lead to an inconclusive result where the relative likelihood of paternity with mutation against non-paternity cannot be determined, except by indirect estimation of mutation rates from locus heterozygosity (Jeffreys *et al.*, 1988a).

In pedigree analysis, mutant offspring provide indirect information on the incidence of mutant parental gametes. An alternative strategy to estimating mutation rates is to determine the density of new mutant minisatellite molecules directly in gametic (sperm) DNA. This can be achieved by fractionation of sperm DNA by gel electrophoresis to remove progenitor alleles, followed by single molecule PCR analysis to count directly the number of new mutant molecules of abnormal length (Jeffreys *et al.*, 1990). This approach is exquisitely sensitive, being capable of detecting mutation events as infrequent as 10^{-7} per gamete, and yields an estimate of MS32 minisatellite mutation rate comparable with that obtained by pedigree analysis. This approach could in principle be used to estimate mutation rates at less variable loci where pedigree analysis is not feasible.

Minisatellite mutation is not restricted to the germline, but also occurs somatically, as shown by the analysis of clonal tumour cell populations and lymphoblastoid cell lines (Armour *et al.*, 1989a) and by single molecule PCR analysis of normal somatic DNA (Jeffreys *et al.*, 1990). As a result, any tissue will contain a majority of cells with the two progenitor minisatellite alleles, plus a diversity of cells containing various mutant new length alleles. While the proportion of cells harboring new mutant alleles can be significant, the heterogeneity in new mutant allele length will prevent their detection by conventional Southern blot hybridization. This however will not necessarily be true for PCR analyses operating at, or close to, the single DNA molecule level, a consideration of relevance to PCR analysis of minute forensic specimens. Also, it will not be true if a mutation occurs in a very early stem cell lineage, which will create a tissue either mosaic for original non-mutant cells plus cells descended from the same mutant progenitor cell (creating a tissue with *three* alleles) or tissue homogeneously composed of mutant cells. Such a process could result in the divergence of single locus profiles between different tissues of the same individual (e.g. blood and sperm), of obvious concern in forensic analysis. To date, we have seen no evidence for early stem cell events in man, through very low-level mosaicism has been detected in blood and sperm DNA (Jeffreys *et al.*, 1990). In contrast, two highly variable and unstable minisatellites have been identified in mice at which a significant proportion of new mutations occur very early in development, within the first few cell divisions following fertilization, to produce 3-allele mice globally mosaic for similar numbers of original and new mutant cells. This process also creates mice with 3 alleles in the germline which segregate in a non-mendelian fashion into offspring, and can also generate differences in DNA profiles between for example embryonic and extra-embryonic tissues of the same individual (Kelly *et al.*, 1989; unpublished data). It still remains to be seen whether this process is unique to mice, or can occasionally occur in man.

Investigation of new mutant minisatellite alleles also sheds light on the molecular processes generating these new length mutants. By analysing genetic markers closely flanking the new mutant allele (Wolff *et al.*, 1989, 1990), or more effectively by carrying out MVR mapping comparisons of new mutant alleles with their progenitor alleles (Jeffreys *et al.*, 1990), it has become clear that, for the loci studied, most or all mutation events in the germline occur before meiosis and seldom if ever involve unequal crossing over between the tandem-repeat arrays of different alleles in an individual. Instead, the predominant mutation processes appear to be unequal exchange between sister chromatids and, possibly, DNA polymerase slippage at these tandem repeats at DNA replication forks. The interesting implication of these findings is that minisatellites are largely uninfluenced by the fact that they exist in diploid individuals, and instead evolve primarily along haploid chromosome lineages.

Forensic Analysis of Degraded DNA

Frequently, the DNA recovered from forensic specimens is too degraded to allow single locus probe analysis, either by Southern blotting procedures or by PCR, since these minisatellite loci usually have alleles in the kilobase size range. The recent development of PCR-typable DNA marker systems based on much shorter segments of human DNA provides a solution to this problem. Saiki *et al.* (1986) and Higuchi and Blake (1989) have developed a marker system based on the polymorphic HLA-DQα locus, involving PCR amplification of a short DNA segment followed by allele classification by dot-blot hybridization with a range of allele-specific oligonucleotide probes. This system can currently distinguish 6 alleles and thus 21 different genotypes. However, only one locus is analysed, allowing exclusion but not definitive inclusion in casework analysis. Another potentially very useful set of markers have been developed from short simple sequence regions or "microsatellites", particularly regions comprised of tandem repeats of the dinucleotide CA (Litt and Luty, 1989; Tautz, 1989; Weber and May, 1989). Such loci are frequently polymorphic in CA repeat copy number, and provide PCR-typable markers with alleles in the 70–200 bp range. However, length variability is limited, with at most only 30 CA repeats being found at the large number of loci characterized (Weber, 1990). The forensic informativeness of these loci is correspondingly low, with < 90% heterozygosity, small numbers of alleles and "spiky" allele frequency distributions potentially vulnerable to inbreeding and ethnic group divergence effects. However, all allele length states, including alleles which differ by a single CA repeat, can be resolved, either by DNA sequencing gel analysis (Litt and Luty, 1989; Weber and May, 1989) or by agarose gel electrophoresis (A. J. J., unpubl. data), although in the former method, rather complex phenotypes consisting of several bands per allele are produced, apparently as a by-product of the PCR reaction. The third class of marker system is based on DNA sequence analysis of the highly variable control region of mitochondrial DNA (Greenberg *et al.*, 1983), which should prove to be particularly useful in minute degraded DNA samples where the yield of nuclear DNA is too small for typing; in such samples, multicopy mitochondrial DNA may still survive in PCR-typable amounts. However, mitochondrial DNA is strictly maternally inherited and can give no information in paternity analyses.

One area of considerable forensic and anthropological interest is the possibility of typing DNA from skeletal remains. Mitochondrial DNA has been shown to be amplifiable from surprisingly ancient bones (Hagelberg *et al.*, 1990), though it is difficult to be certain that the DNA amplified has not arisen through adventitious contamination. We have recently extended this analysis to nuclear DNA markers in skeletal remains exhumed several years after interment (E. Hagelberg and A. J.

J., unpub. data). As expected, most of the DNA recovered (usually > 99%) is of non-human origin, and presumably arises from bacteria and fungi in the remains. The human DNA component, detected by hybridization with an Alu probe, is generally severely degraded. Despite this degradation, contamination with non-human DNA and the presence of PCR inhibitors in skeletal DNA extracts, typing of nuclear microsatellite markers has proved possible. Furthermore, in one case analysed to date, the bone DNA typing information has been shown to be authentic. This case involved the skeletal remains of a murder victim provisionally identified from facial reconstructions and dental records. Comparison of bone DNA with the parents of the putative victim showed a complete parent-offspring match with each of six microsatellite markers used, not only establishing the identification of the murder victim (with > 99.9% certainty) but also providing strong evidence that the human DNA recovered from the bones was authentic and not a contamination.

Future Perspectives

The scientific and legal framework of DNA typing in forensic and legal medicine has now been firmly established. There are four major areas of future development. First, using the extreme sensitivity of PCR, to what extent can DNA testing be extended from the traditional forensic samples such as blood, semen stains and vaginal swabs to more esoteric samples such as saliva traces (for example on blackmail letters) and skeletal remains? Second, what improvements in DNA marker systems and marker detection are possible? An ideal marker would perhaps have the following features: (*a*) alleles limited in size to 100–500 bp, such that all alleles can be efficiently amplified by PCR, even in degraded DNA, (*b*) all allelic length states resolvable, such that precise allelic classification is possible, (*c*) large numbers of alleles, with no common allele, (*d*) a quantifiable and sufficiently high mutation rate (perhaps in the range 10^{-2}–10^{-3} per gamete) to ensure insulation against genetic drift effects. No such marker has yet been identified, and may actually not exist in the human genome. In terms of DNA marker detection, highly sensitive non-isotopic probes detected by enhanced chemiluminescence should soon replace radioisotope-based detection systems. Many aspects of PCR-based typing should be amenable to automation, particularly using automated gel electrophoretic or microcapillary electrophoretic analysis of fluorescence-tagged PCR products. The third area of development concerns databasing, of criminal offenders or potentially entire populations, which raises considerable problems, not only of a social and legal nature but also in relationship to standardisation of DNA markers and analytical procedures. The final area relates to the potential of human molecular genetics to deliver DNA markers capable of giving phenotypic information from forensic DNA samples where there is no suspect; such information might include sex (already testable), ethnicity, and visual appearance (e.g. hair colour, eye colour, stature). The latter characters, in particular, present a far-from-trivial problem and will require a profound revolution in our ability to dissect the molecular genetics of quantitative characters in man; indeed such DNA analyses may eventually prove in practice to be impossible.

Acknowledgements
We are grateful to Sarah Laband for preparing this paper. A. J. J. is a Lister Institute Research Fellow. Research was supported by grants to A. J. J. from the Medical Research Council and the Wolfson Foundation. The minisatellite probes described in this paper are the subject of Patent Applications, and commercial enquiries should be addressed to Cellmark Diagnostics, 8 Blacklands way, Abingdon Business Park, Abingdon, Oxfordshire, OX14 1DY, UK.

References

Ali, S., Muller, C. R. and Epplen, J. T. (1986) DNA finger printing by oligonucleotide probes specific for simple repeats. *Hum. Genet.* 74: 239–243.

Armour, J. A. L., Patel, I., Thein, S. L., Fey, M. F. and Jeffreys, A. J. (1989a) Analysis of somatic mutations at human minisatellite loci in tumours and cell lines. *Genomics* 4: 328–334.

Armour, J. A. L., Wong, Z., Wilson, V., Royle, N. J. and Jeffreys, A. J. (1989b) Sequences flanking the repeat arrays of human minisatellites: association with tandem and dispersed repeat elements. *Nucleic Acids Res.* 17: 4925–4935.

Armour, J. A. L., Povey, S., Jeremiah, S. and Jeffreys, A. J. (1990) Systematic cloning of human minisatellites from ordered array charomid libraries. *Genomics* 8: in press.

Balazs, I., Baird, M., Clyne, M. and Meade, E. (1989) Human population genetic studies of five hypervariable DNA loci. *Am. J. Hum. Genet.* 44: 182–190.

Boerwinkle, E., Xiong, W., Fourest, E. and Chan, L. (1989) Rapid typing of tandemly repeated hypervariable loci by the polymerase chain reaction: application to the apolipoprotein B 3′ hypervariable region. *Proc. Nat. Acad. Sci. USA* 86: 212–216.

Brookfield, J. F. Y. (1989) Analysis of DNA fingerprinting data in cases of disputed paternity. *IMA J. of Mathematics Applied in Medicine and Biology* 6: 111–131.

Collick, A. and Jeffreys, A. J. (1990) Detection of a novel minisatellite-specific DNA-binding protein. *Nucleic Acids Res.* 18: 625–629.

Evett, I. W., Werrett, D. J. and Buckleton, J. S. (1989a) Paternity calculations from DNA multilocus profiles. *J. Forensic Sci. Soc.* 29: 249–254.

Evett, I. W., Werrett, D. J. and Smith, A. F. M. (1989b) Probabilistic analysis of DNA profiles. *J. Forensic Sci. Soc.* 29: 191–196.

Flint, J., Boyce, A. J., Martinson, J. J. and Clegg, J. B. (1989) Population bottlenecks in Polynesia revealed by minisatellites. *Hum. Genet.* 83: 257–263.

Fowler, S. J., Gill, P., Werrett, D. J. and Higgs, D. R. (1988) Individual specific DNA fingerprints from a hypervariable region probe: alpha-globin 3′ HVR. *Hum. Genet.* 79: 142–146.

Gill, P. and Werrett, D. J. (1987) Exclusion of a man charged with murder by DNA fingerprinting. *Forensic Science International* 35: 145–148.

Greenberg, B. D., Newbold, J. E. and Sugino, A. (1983) Intraspecific nucleotide-sequence variability surrounding the origin of replication in human mitochondrial DNA. *Gene* 21: 33–49.

Hagelberg, E., Sykes, B. and Hedges, R. (1990) Ancient bone DNA amplified. *Nature* 342: 485.

18

Higuchi, R. and Blake, E. T. (1989) Applications of the polymerase chain reaction in forensic science. In Banbury Report 32: DNA Technology and Forensic Science (eds. J. Ballantyne, G. Sensabaugh and J. Witkowski; Cold Spring Harbor Laboratory Press, 1989) pp. 265–281.

Hill, W. G. (1986) DNA fingerprint analysis in immigration test-cases. *Nature* 322: 290–291.

Home Office (1988) DNA profiling in immigration casework. Report of a pilot trial by the Home Office and Foreign and Commonwealth Office. (Home Office, London).

Horn, G. T., Richards, B. and Klinger, K. W. (1989) Amplification of a highly polymorphic VNTR segment by the polymerase chain reaction. Nucleic Acids Res. 17: 2140.

Jeffreys, A. J., Brookfield, J. F. Y. and Semeonoff, R. (1985a) Positive identification of an immigration test-case using human DNA fingerprints. *Nature* 317: 818–819.

Jeffreys, A. J., Wilson, V. and Thein, S. L. (1985b) Hypervariable 'minisatellite' regions in human DNA. *Nature* 314: 67–73.

Jeffreys, A. J., Wilson, V. and Thein, S. L. (1985c) Individual-specific 'fingerprints' of human DNA. *Nature* 316: 76–79.

Jeffreys, A. J., Wilson, V., Thein, S. L., Weatherall, D. J. and Ponder, B. A. J. (1986) DNA 'fingerprints' and segregation analysis of multiple markers in human pedigrees. *Am. J. Hum. Genet.* 39: 11–24.

Jeffreys, A. J., Wilson, V., Kelly, R., Taylor, B. A. and Bulfield, G. (1987) Mouse DNA 'fingerprints': analysis of chromosome localization and germ-line stability of hypervariable loci in recombinant inbred strains. *Nucleic Acids Res.* 15: 2823–2836.

Jeffreys, A. J., Royle, N. J., Wilson, V. and Wong, Z. (1988a) Spontaneous mutation rates to new length alleles at tandem-repetitive hypervariable loci in human DNA. *Nature* 332: 278–281.

Jeffreys, A. J., Wilson, V., Neumann, R. and Keyte, J. (1988b) Amplification of human minisatellites by the polymerase chain reaction: towards DNA fingerprinting of single cells. *Nucleic Acids Res.* 16: 10953–10971.

Jeffreys, A. J., Neumann, R. and Wilson, V. (1990) Repeat unit sequence variation in minisatellites: a novel source of DNA polymorphism for studying variation and mutation by single molecule analysis. *Cell* 60: 473–485.

Julier, C., de Gouyon, B., Georges, M., Guenet, J. -L., Nakamura, Y., Avner, P. and Lathrop, G. M. (1990) Minisatellite linkage maps in the mouse by cross-hybridization with human probes containing tandem repeats. *Proc. Nat. Acad. Sci. USA* 87: 4585–4589.

Kelly, R., Bulfield, G., Collick, A., Gibbs, M. and Jeffreys, A. J. (1989) Characterization of a highly unstable mouse minisatellite locus: evidence for somatic mutation during early development. *Genomics* 5: 844–856.

Knowlton, R. G., Brown, V. A., Braman, J. C., Barker, D., Schumm, J. W., Murray, C., Takvorian, T., Ritz, J. and Donnis-Keller, H. (1986) Use of highly polymorphic DNA probes for genotype analysis following bone marrow transplantation. *Blood* 68: 378–385.

Lander, E. S. (1989) DNA fingerprinting on trial. *Nature* 339: 501–505.

Litt, M. and Luty, J. A. (1989) A hypervariable microsatellite revealed by *in vitro* amplification of a dinucleotide repeat within the cardiac muscle actin gene. *Am. J. Hum. Genet.* 44: 397–401.

Nakamura, Y., Leppert, M., O'Connell, P., Wolff, R., Holm, T., Culver, M., Martin, C., Fujimoto, E., Hoff, M., Kumlin, E. and White, R. (1987) Variable number of tandem repeat (VNTR) markers for human gene mapping. *Science* 235: 1616–1622.

Nakamura, Y., Carlson, M., Krapcho, K., Kanamori, M. and White, R. (1988) New approach for isolation of VNTR markers. *Am. J. Hum. Genet.* 43: 854–859.

Royle, N. J., Clarkson, R. E., Wong, Z. and Jeffreys, A. J. (1988) Clustering of hypervariable minisatellites in the proterminal regions of human autosomes. *Genomics* 3: 352–360.

Saiki, R. K., Bugawan, T. L., Horn, G. T., Mullis, K. B. and Erlich, H. A. (1986) Analysis of enzymatically amplified β-globin and HLA-DQα DNA with allele-specific oligonucleotide probes. *Nature* 324: 163–166.

Saiki, R. K., Gelfand, D. H., Stoffel, S., Scharf, S. J., Higuchi, R., Horn, G. T., Mullis, K. B. and Erlich, H. A. (1988) Primer-directed enzymatic amplification of DNA with a thermostable DNA polymerase. *Science* 239: 487–491.

Smith, J. C., Anwar, R., Riley, J., Jenner, D., Markham, A. F. and Jeffreys, A. J. (1990). Highly polymorphic minisatellite sequences: allele frequencies and mutation rates for five locus specific probes in a Caucasian population. *J. For. Sci. Soc.* 30: 19–32.

Swallow, D. M., Gendler, S., Griffith, B., Corney, G., Taylor-Papadimitriou, J. and Bramwell, M. E. (1987) The human tumour-associated epithelium mucins are coded by an expressed hypervariable gene locus PUM. *Nature* 328: 82–84.

Tautz, D. (1989) Hypervariability of simple sequences as a general source for polymorphic DNA markers. *Nucleic Acids Res.* 17: 6463–6471.

US Congress, Office of Technology Assessment (1990) Genetic witness: forensic uses of DNA tests, OTA-BA-438 (Washington DC: US Government Printing Office).

Vassart, G., Georges, M., Monsieur, R., Brocas, H., Lequarre, A. S. and Christophe, D. (1987) A sequence in M13 phage detects hypervariable minisatellites in human and animal DNA. *Science* 235: 683–684.

Vergnaud, G. (1989) Polymers of random short oligonucleotides detect polymorphic loci in the human genome. *Nucleic Acids Res.* 17: 7623–7630.

Wahls, W. P., Wallace, L. J. and Moore, P. D. (1990) Hypervariable minisatellite DNA is a hotspot for homologous recombination in human cells. *Cell* 60: 95–103.

Wambaugh, J. (1989) The Blooding. (William Morrow & Co., Inc., New York, NY).

Weber, J. L. (1990) Informativeness of human (dC-dA)$_n$. (dG-dT)$_n$ polymorphisms. *Genomics* 7: 524–530.

Weber, J. L. and May, P. E. (1989) Abundant class of human DNA polymorphisms which can be typed using the polymerase chain reaction. *Am. J. Hum. Genet.* 44: 388–396.

Wolff, R., Nakamura, Y. and White, R. (1988) Molecular characterization of a spontaneously generated new allele at a VNTR locus: no exchange of flanking DNA sequence. *Genomics* 3: 347–351.

Wolff, R. K., Plaetke, R., Jeffreys, A. J. and White, R. (1989) Unequal crossingover between homologous chromsomes is not the major mechanism involved in the generation of new alleles at VNTR loci. *Genomics* 5: 382–384.

Wolff, R., Nakamura, Y. and White, R. (1988) Molecular characterization of a spontaneously generated new allele at a VNTR locus: no exchange of flanking DNA sequence. *Genomics* 3: 347–351.

Wolff, R. K., Plaetke, R., Jeffreys, A. J. and White, R. (1989) Unequal crossingover between homologous chromosomes is not the major mechanism involved in the generation of new alleles at VNTR loci. *Genomics* 5: 382–384.

Wong, Z., Wilson, V., Jeffreys, A. J. and Thein, S. L. (1986) Cloning a selected fragment from a human DNA 'fingerprint': isolation of an extremely polymorphic minisatellite. *Nucleic Acids Res.* 14: 4605–4616.

Wong, Z., Wilson, V., Patel, I., Povey, S. and Jeffreys, A. J. (1987). Characterization of a panel of highly variable minisatellites cloned from human DNA. *Ann. Hum. Genet.* 51: 269–288.

Wong, Z., Royle, N. J. and Jeffreys, A. J. (1990) A novel human DNA polymorphism resulting from transfer of DNA from chromosome 6 to chromosome 16. *Genomics* 7: 222–234.

DNA Fingerprinting: Approaches and Applications
ed. by T. Burke, G. Dolf, A. J. Jeffreys & R. Wolff
© 1991 Birkhäuser Verlag Basel/Switzerland

Generation of Variability at VNTR Loci in Human DNA

R. Wolff[1,5], Y. Nakamura[2], S. Odelberg[1], R. Shiang[3], and R. White[1,4]

[1]Department of Human Genetics, University of Utah Medical Center, Salt Lake City, Utah 84132, USA; [2]Division of Biochemistry, Cancer Institute, Tokyo 170, Japan; [3]University of Iowa, Iowa City, Iowa 52242, USA; [4]Howard Hughes Medical Institute, University of Utah Medical Center, Salt Lake City, Utah 84132, USA
[5]Current Address: Department of Biochemistry, University of California San Francisco, San Francisco, California 94143–0554, USA

Summary
Our laboratory has constructed linkage maps of the human chromosomes to use as a tool towards the goal of cloning by position the genes responsible for genetic disorders. Construction of the map required the development of polymorphic marker systems in the form of Restriction Fragment Length Polymorphisms (RFLPs). Work by Yusuke Nakamura in the laboratory led to the identification of more than 200 highly informative Variable Number Tandem Repeat (VNTR) markers. The hypervariable nature of these marker loci has allowed individualization at the DNA level. Techniques for individualization have subsequently been adopted by diverse fields including gene mapping, cancer genetics and forensic biology. These markers have also become a resource to test hypotheses as to how the VNTRs generate their intrinsic variability. We have demonstrated that the hypothesis that VNTRs generate their variability by unequal exchange between homologous chromosomes is incorrect (Wolff *et al.*, 1988; Wolff *et al.*, 1989). Our data are consistent with intrachromosomal models such as unequal sister chromatid exchange and replication slippage. Using DNA derived from nonhuman primate species, we have tested hypotheses that try to explain the sequence relationship at dispersed VNTR loci. Our data reveal that VNTR loci are most likely not related by transposition but rather arose independently at multiple loci.

Human Genetic Linkage Maps

By the late 1970s many human diseases were known to contain a genetic component, as determined by inheritance patterns within families. However, identification of a particular genetic defect was nearly impossible without detailed biochemical knowledge of the disease phenotype. In 1980 a new tool was proposed: a genetic linkage map of the human chromosomes based on heritable polymorphisms (RFLPs) in the length of fragments produced by digestion of DNA with restriction endonucleases (Botstein *et al.*, 1980). Genetic maps in diploid organisms require that the mapmaker be able to discriminate between the two homologous chromosomes of an individual at all marker loci, so that he or she can analyze the segregation of those loci in the progeny. When two independent traits consistently cosegregate in the offspring, they are considered to be genetically linked. The map distance between two markers reflects the frequency with which the two traits are separated by meiotic recombination.

The endeavor of creating a linkage map in the human hinged on the ability to detect enough genetic polymorphism distributed throughout the genome. At the time, the existence of RFLPs in humans had only been described in a handful of cases, always near genes that were being studied (Maniatis *et al.*, 1978; Kan *et al.*, 1978). The inherent power of disease-linked RFLPs was demonstrated in 1978 by Kan *et al.*, when they described a positive predictive diagnosis of sickle cell anemia in an individual using an RFLP located near the β-globin gene. A logical extension of this test case would be to identify a specific RFLP allele that consegregates with affection status within a pedigree segregating any genetic disorder. This marker locus would then be considered to be genetically linked to the defect, even in the absence of prior knowledge of that gene's function or chromosomal location. The marker should pave the way for eventual localization of the gene, and permit diagnostic evaluation of an individual's affection status. It should be stressed that both these goals would be achievable without prior knowledge of the biochemical defect responsible for the disorder, or of the identity of the gene involved.

The construction of a linkage map based on RFLPs requires the isolation of markers that identify polymorphic sites, and the collection of DNA from preexisting human families to replace the designed matings, or crosses, commonly used in conventional genetic studies. The information that can be obtained from these analyses depends on the quality and quantity of both family materials and markers. The number of markers needed to cover all chromosomes was determined to be approximately 150, equally spaced about 20 centimorgans (cM) apart, to ensure linkage of every locus to at least one marker (Botstein *et al.*, 1980). Many more than 150 would have to be identified to ensure the 20-cM spacing. The "quality" of the probes is a reflection of their polymorphism information content, or PIC. The information that a marker reveals depends on the allele frequencies found in the test population. Thus, if a marker reveals a two-allele polymorphism, but one allele is common and the other rare, one would need to test more families to identify enough informative meioses to yield significant linkage results. The ideal situation would be to have as many markers as possible that reveal loci that are highly variable in the population.

RFLP Markers

RFLP markers fall into two main categories. Both are able to identify polymorphism upon digestion of genomic DNA with a restriction endonuclease, followed by size-fractionation through agarose gels, and Southern analysis.

Site polymorphisms represent the first category. By the process of mutation, base substitutions occur in DNA. One can scan DNA for such substitutions in two homologous alleles by using restriction enzymes. When a substitution either creates or destroys a recognition site for a particular enzyme, a size difference in the fragments of DNA encompassing the two alleles can been seen. When a site polymorphism identifies two alleles having equal frequency in the population, a maximum theoretical heterozygosity of 50% can be obtained. This means that under the best conditions an average of 50% of the individuals involved in the matings under study will be heterozygous for this locus.

The second type of RFLP is insertion/deletion rearrangement. In this case restriction endonuclease recognition sites are not disrupted; rather the sequence between two sites contains an insertion or deletion that alters the size of the fragment. This type of RFLP can be distinguished from a site polymorphism because it is detectable with any restriction enzyme digest that allows resolution of the size difference between the two alleles.

Highly Polymorphic Markers

The first highly polymorphic locus in humans was identified using arbitrary DNA probe pAW101 (Wyman and White, 1980). The pAW101 polymorphism revealed alleles of eight different sizes in 11 individuals studied. The observed heterozygosity for pAW101 was 0.79, well above the theoretical maximum of 0.50 for a two-allele system. The advantage of this type of marker system over biallelic systems is clear, because the probe can extract more information from a given set of DNA samples.

Other hypervariable regions were subsequently identified at the human insulin gene locus (Bell et al., 1982; Ullrich et al., 1982), the zeta-globin gene and pseudogene (Proudfoot et al., 1982; Goodbourn et al., 1983) and the Ha-ras gene (Capon et al., 1983). Analysis of the nucleotide sequences showed that the structural nature of the polymorphisms in these regions was the presence of a variable number of tandemly repeated oligonucleotides. This type of marker system not only has the advantage of detectability with numerous enzymes, a characteristic of insertion/deletion polymorphisms, but the highly variable amount of DNA that is inserted at the locus produces alleles of specific lengths (Fig. 1).

Attempts to align the oligonucleotide repeats seen at different hypervariable loci revealed partial sequence homology. Both Goodbourn et al. (1983) and Proudfoot et al. (1982) postulated that something inher-

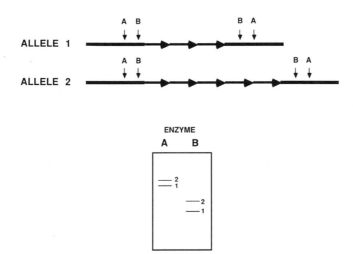

Figure 1. Description of a VNTR. Two alleles at a VNTR locus are diagramed. The patterns shown in the drawing representing a Southern blot of this locus using either restriction enzyme A or B (below). Note that the pattern of the bands is the same regardless of the enzyme used. The difference in the size between allele 1 and allele 2 remains constant regardless of the enzyme used. The only effect resulting from enzyme choice is the resolution between the allele sizes.

ent in the sequence might allow tandem expansion at independent loci dispersed throughout the genome.

Hypervariable Minisatellites

Jeffreys *et al.* (1985) reported finding hypervariable "minisatellite" regions in human DNA. The work was initiated on the premise that the tandem repeats found at the myoglobin locus could be related by transposition to other highly polymorphic loci in the genome. The hypothesis was based on the following observations: 1) The myoglobin gene contains four tandem repeats of a 33-bp sequence within an intron. The tandem repeats are flanked by 9-bp direct repeats, characteristic of a target site duplication generated by a transposition event. 2) The myoglobin repeat unit shows sequence similarity to the repeats at the three other previously sequenced highly variable loci. Low-stringency hybridization to human DNA, using a tandemly repeated core sequence probe, detected many loci, with only the 4- to 20- kb bands being well resolved above a smear. These bands showed extreme polymorphism, to the extent that the hybridization profile provided an individual-specific in DNA "fingerprint". All bands in the offspring were inherited in a Mendelian fashion, and could be traced back to one or the other parent, and subsequently to the grandparents. Because no bands appeared to be sex-specific, mainly autosomal origin was implied. Although the hybridization patterns were complex, no two bands were seen to cosegregate to the offspring; the authors inferred that the core sequence was identifying multiple, dispersed (unlinked) loci.

Variable Number of Tandem Repeat (VNTR) Markers

Yusuke Nakamura in our laboratory extended the Jeffreys finding by isolating more than 200 single-copy VNTR markers (Nakamura *et al.*, 1987). Using five oligonucleotide probes based on the core sequences of the repeats at the myoglobin, insulin, Ha-*ras*, zeta-globin pseudo-gene, and Hepatitis-B X-gene loci, he screened genomic cosmid libraries under conditions of low stringency. Positive cosmids were subsequently used to probe restriction digests of DNA from unrelated human individuals, at high stringency. Highly polymorphic regions were identified in 26% of the cosmids that were prescreened with the oligonucleotide sequence, as compared to 2–4% for unselected cosmids. By aligning the sequences of the oligonucleotides and those seen at six of the resulting VNTRs, Nakamura derived a core sequence. This core sequence, GNNGTGGG, was different from Jeffreys' original core sequence, although the revised core from Wong *et al.* (1987), GGAG-GTGGGCAGGAR (A or G)G (R = purine), does contain the Nakamura core. Using oligonucleotides based on this new core sequence, Nakamura *et al.* (1988a) were able to derive another 34 multiallelic VNTR polymorphisms, with an average heterozygosity of 68%.

This method has led to the efficient (26–33%) isolation of more than 200 VNTR marker loci. These markers are now being widely used in gene-mapping studies (Nakamura *et al.*, 1988b), in forensic biology (Odelberg *et al.*, 1989), and in studies of cancer genetics involving reduction to homozygosity of specific chromosomal regions (Vogelstein *et al.*, 1989).

Mapping studies have confirmed Jeffreys' observation that the minisatellite bands appear to be unlinked. VNTR loci are dispersed throughout the autosomal chromosomes, apparently preferentially near the terminal regions of the autosomes (Nakamura *et al.*, 1988a; Royle *et al.*, 1988).

Molecular Characterization of a Spontaneously Generated New Allele at a VNTR Locus: No Exchange of Flanking DNA Sequence

Jeffreys *et al.* (1988) described a number of spontaneous mutations that generated new length-alleles and reported the generation rates at five independent loci, yet the nature of the generation event remained

elusive. Frequent unequal exchange at meiosis among tandemly repeated DNA sequences would have the effect of conserving the sequence
of the repeats. Jeffreys *et al.* (1985) noted that the core sequence of the
tandem repeats contains strong homology to chi, the recombination
hotspot of *Escherichia coli.* Steinmetz *et al.* (1986) also described a
recombination hotspot in the mouse genome that contains tandem
repeats with homology to the "minisatellite" core sequence of Jeffreys.
If these sequences are recombination hotspots, a high frequency of
crossing over could generate the variability by a concomitant high rate
of unequal crossing over at the repeated sequences. This model allows
flexibility for the generation of new alleles, both larger and smaller. It
also provides an explanation for the apparent homogenization of base
substitutions to other repeat units within the tandem array. A prediction of this model is that flanking markers surrounding the VNTR will
be recombinant following an event that generates a new allele (Fig. 2).

During linkage studies in three-generation Utah pedigrees, a VNTR
marker locus, YNZ22, revealed in one individual an allele not present in
either parent. The father's allele was inherited normally. The new allele
was smaller than either of the maternal alleles (Fig. 3a). To confirm the
biological relationship of the parents to the child, we reviewed the
genotypic data from 600 other DNA markers in this family. More than

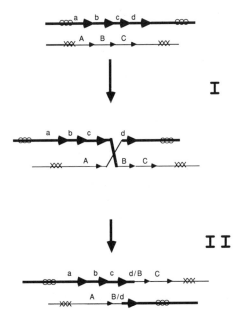

Figure 2. Model of unequal exchange. I. Unequal crossover within repeat units (arrows) takes
place between two alleles at a VNTR locus. II. Following branch migration of the Holliday
structure results in the exchange of flanking markers (XXX and OOO). Note that both
resulting alleles contain different numbers and combinations of repeat units.

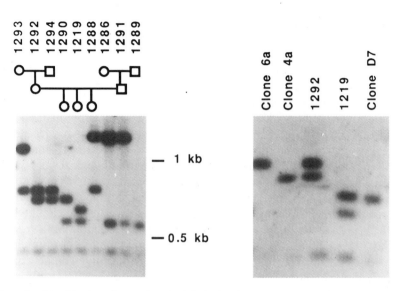

a   **b**

— 1 kb

—0.5 kb

Figure 3a. Identification of a new length-allele by Southern transfer. Individual 1219 does not inherit the proper allele size from the mother, 1292. Lymphocyte DNA (5 ug) was digested with MspI and electrophoresed through a 1.2% agarose gel. The DNA was denatured, transferred to Biotrace RP (Gelman), and probed with ^{32}P-labeled pYNZ22 insert.

Figure 3b. YNZ22 allele sizes present in cosmid clones. Determination of the allele sizes present in the cosmid clones by Southern transfer of MspI-digested cosmids and lymphocyte DNAs probed with pYNZ22 insert. Cosmid 6A and cosmid 4A were isolated from a cosmid library derived from individual 1292. Cosmid D7 was isolated from a library derived from individual 1219.

100 of the markers tested were VNTRs, many of them having more than 10 alleles in the population; all markers except YNZ22 showed Mendelian inheritance within the family.

To investigate the hypothesis that a new allele had been generated by an unequal exchange between homologous regions, we had to be able to distinguish between the two maternal chromosomes. If we were to find polymorphism between the two maternal alleles on both flanks of the tandem repeats, we could determine whether a crossover event had taken place between those flanking markers during generation of the new allele. We also wanted to investigate the structural nature of the size change, and possibly determine the break points. We chose to approach these analyses by cloning all of the YNZ22 alleles of the individuals involved in the generation event, followed by sequence analysis of that DNA.

Total genomic cosmid libraries were constructed using partial digests of DNA from individual 1219 (daughter) and 1292 (mother) with

Sau3AI. Isolation of each allele was confirmed in the cosmids by sizing the cloned alleles on Southern-blots and comparing their sizes to those of the individuals' genomic alleles (Fig. 3b). All YNZ22 clones corresponded in size to that of one of the two alleles in each individual. One cosmid corresponding to each allele was then sequenced; the data are presented in Fig. 4. The two maternal alleles contained respectively four (clone 6A) and three (clone 4A) tandem copies of a 70-bp sequence. The daughter's new allele contained two copies (clone D7) of the 70-bp sequence. All of the alleles sequenced contained perfect 70-bp repeat units with no sequence divergence, so no breakpoints could be determined within the tandem repeat. Three single-base substitutions were identified on each flank between the two maternal alleles. The daughter's new allele contained both flanks consistent with the mother's three repeat allele. The results demonstrate that the new allele was generated by loss of one repeat unit, without a concomitant reciprocal exchange of the flanking DNA sequences. By examining the inheritance of sequence substitutions in the flanking regions of the VNTR locus, we have shown that no apparent exchange of flanking DNA sequences occured between homologous chromosomes when the new allele was generated.

We can, therefore, exclude simple models of crossing-over that require flanking markers to be recombinant. However, our data do not allow us to exclude models that envision two independent crossing-over events within the short interval from the VNTR to the flanking variant bases, or gene-conversion events that allow single exchanges to be resolved into either parental or recombinant flanking markers. Furthermore, our results are also consistent with a variety of pathways for generation of the new allele, such as unequal sister-chromatid exchange, polymerase slippage, and loopout deletion.

The true nature of the mutation described here remains unknown, but does not involve a simple crossover event resulting in exchange of flanking DNA sequences. The study of more events that generate new alleles would help us to distinguish among several other models.

Unequal Crossing Over Between Homologous Chromosomes is not the Major Mechanism Involved in the Generation of New Alleles at VNTR Loci

Our initial results seemed to contradict the model of unequal crossing over, but as only one new allele was characterized we undertook to investigate more events.

We therefore studied locus D1S7, detected by minisatellite probe λMS1; this locus shows the highest mutation rate of any VNTR locus so far described, with 5% of gametes carrying new length-alleles (Jeffreys et al., 1988). Having probed the CEPH reference families with

```
6A    GTCGGAAGAG CGGGGCAGGG AGAGAAAGGT CGAAGAGTGA AGTGCACAGG    50
4A    ---------- ---------- ---------- ---------- ----------
D7    ---------- ---------- ---------- ---------- ----------

                                    _____
6A    AGGGCAAGGC GGTCCTCACC CTGCCTGGGC TGGGGCAGGG CTGTGAGACC   100
4A    ---------- --C-----GA ---------- ---------- ----------
D7    ---------- --C-----GA ---------- ---------- ----------

                                              >_____
6A    CTCCCTTACA GAAGCAATGA GGGCTTGAGG AGGGGGTTAG GGGCCTGGGC   150
4A    ---------- ---------- ---------- ---------- ----------
D7    ---------- ---------- ---------- ---------- ----------

6A    TGGGGCAGGG CTGTGAGACC CTCCCTTACA GAAGCAATGA GGGCTTGAGG   200
4A    ---------- ---------- ---------- ---------- ----------
D7    ---------- ---------- ---------- ---------- ----------

                 >_____
6A    AGGGGGTTAG GGGCCTGGGC TGGGGCAGGG CTGTGAGACC CTCCCTTACA   250
4A    ---------- ---------- ---------- ---------- ----------
D7    ---------- --*******  ********** ********** **********

                                              >_____
6A    GAAGCAATGA GGGCTTGAGG AGGGGGTTAG GGGCCTGGGC TGGGGCAGGG   300
4A    ---------- ---------- ---------- --*******  **********
D7    ********** ********** ********** ********** **********

6A    CTGTGAGACC CTCCCTTACA GAAGCAATGA GGGCTTGAGG AGGGGGTTAG   350
4A    ********** ********** ********** ********** **********
D7    ********** ********** ********** ********** **********

      =>
6A    GGGCAGTAAG TTAACTTGGG AGGCCGATGT GGGGGAACGC TGAAGAATAA   400
4A    **-------- ---------- ---------- ---------- ----------
D7    **-------- ---------- ---------- ---------- ----------

6A    AGACTGTGGG CACAGCAGAC CCCTGGGCA GGTTAGCAAT GCACAGACTC   450
4A    ---------- ---------- ---------- ---------- ----------
D7    ---------- ---------- ---------- ---------- ----------

6A    CACTGC ACAAGATACA TGGGGCGGAG GGGCTCTTTG GAAGTCGGGGGAGG   500
4A    ---------- ---------- ---------- ---------- ----------
D7    ---------- ---------- ---------- ---------- ----------

6A    GGGATCTGAC TCTATCAAAA AGAGAAAAGA TAAAAGAGAT GGGGTCAGAG   550
4A    ---------- ---------- ---------- ---------- ----------
D7    ---------- ---------- ---------- ---------- ----------

6A    AGGCCC TGCATAGTCT GGATTTGGCC TACTAGAGCT CCAGTTTTCTTGCC   600
4A    ---------- ---------- ---------- ---------- ----------
D7    ---------- ---------- ---------- --- ------- ----------

6A    GCTAGCCTTT GTTCCAGGGA TTCTGTTATT CACAAATGCC TTCCTGCTCG   650
4A    ---------- ---------- ---------- ---------- ----------
D7    ---------- ---------- ---------- ---------- ----------

6A    GTAGAAGAAC TCCTATTCAT CCGTCAAAGC CCAGCTCTAG GAGCACTTAT   700
4A    ---------- ---------- ---------- ---------- ----------
D7    ---------- ---------- ---------- ---------- ----------

6A    GAGGAGCCTC TCCACCTCCC CTCAACACTC CGTAACTTCC ACCTCAGGCC   750
4A    ---------- ---------- ---------- ---------- ----------
D7    ---------- ---------- ---------- ---------- ----------

6A    GCAGCACGTC CCTCGGGCGA GATGTCGGGG TCCGCCTTCC CTAGAGACCG   800
4A    ---------- ---------- ---------- ---------- ----------
D7    ---------- ---------- ---------- ---------- ----------

6A    GCAGCTCCGG AGGCCTCCGT TTTCCTCTTT GGGACCTCAC TGCTTCTAAA   850
4A    ---------- ---------- ---------- ---------- ----------
D7    ---------- ---------- ---------- ---------- ----------

6A    GTGAGGATGC TATGGGGAGA CATGGGAGAG GGAGTGGGTA AGATGGATCC   900
4A    ---------- ---------- ---------- ---------- ----------
D7    ---------- ---------- ---------- ---------- ----------

6A    --/    /-- GGATCCGCCG TCAGCCGGGC CCGGGGCTTT CGACATGCCC    40
4A    --/    /-- ---------- ---------- ---------- ----------
D7    --/    /-- ---------- ---------- ---------- ----------

6A    CCCAGGTGAG TCCTCGAGCC GGGGACCGGG AGGGACGGGG GACCCAGGGA    90
4A    --------G- ---------- ---------- ---------- ----------
D7    --------G- ---------- ---------- ---------- ----------

6A    CAGCCCGGTC CTCGATAGCG ATCGGCCAGA CACTCGGGGC CATCTTCTGG   140
4A    ---------- ---------- ---------- --------TG ----------
D7    ---------- ---------- ---------- --------TG ----------

6A    CGCGGGTGCC CCATCGCCGGC TGGCGGCTGG CGTTCAGGGC TCCGGGTGTC   190
4A    ---------- ---------- ---------- ---------- ----------
D7    ---------- ---------- ---------- ---------- ----------

6A    GTCCCTTTCG GACTCCAGGA CCACGGGCC                          219
4A    ---------- ---------- ---------
D7    ---------- ---------- ---------
```

Table 1. Pairwise LODscores between λMS1 and chromosome 1p markers

Locus	Theta	LOD
CMM8	0.040	53.4
YNZ2	0.060	45.8
Rh	0.052	37.4
CMM8.1	0.038	26.3
THI54	0.139	15.4

Recombination events were ascertained by probing the CEPH reference families with the minisatellite probe λMS1. For other chromosome 1 markers, data were obtained in 40 CEPH families and 19 additional Utah pedigrees (O'Connell et al., 1989). Maximum lod scores (LOD) and recombination frequencies (Theta) were computed with the computer program LINKAGE (Lathrop et al., 1984).

λMS1, we performed pairwise linkage analysis (Lathrop et al., 1984) that revealed LOD scores ranging from 15.4–53.4 (Tab. 1) for five chromosome 1p probes, and confirmed the previous data from somatic cell hybrid and in situ hybridization mapping that localized λMS1 to 1p33–35 (Wong et al., 1987; Royle et al., 1988). The CMM8.1 and CMM8 data were obtained by analysis of a site polymorphism and a VNTR, identified by two separate fragments derived from the CMM8 cosmid. Since no recombination between these two polymorphisms was observed, the genotypic information from them was analyzed as a haplotype (designated CMM8). Multipoint linkage analysis (Lathrop et al., 1984) determined that the odds in favor of the linear order THI54-YNZ2-λMS1-CMM8-Rh (Fig. 5) were 8×10^8:1 over the next most likely order. The closest markers flanking λMS1, YNZ2 and CMM8, exhibit recombination frequencies (theta) of 0.068 and 0.039 respectively, with λMS1. Thirty-six new alleles have been observed previously with λMS1 (Jeffreys et al., 1988); the parents from whom 12 of the new alleles were derived were doubly informative for the flanking markers. Eleven of the 12 new alleles were nonrecombinant. Table 2 summarizes the data and shows an assignment of parental origin of the new allele and our estimate of the number of 9-bp repeats gained or lost at this locus in each mutational event.

We perfomed statistical analyses on our data in terms of three possible models. The first model attributes all events generating new alleles to unequal crossing over between homologs within the λMS1 repeat units. Under such a model all cases showing nonrecombinant flanking markers must be derived by a second crossover event, or an

Figure 4. Sequence data for the three alleles of YNZ22. 6A and 4A are the two maternal alleles, and D7 is the new allele. The dashed line represents perfect identity between the alleles. The asterisks represent bases not present in the allele. Base substitutions are indicated with the appropriate base. Arrowheads above the sequence delineate the tandem repeats. The // symbol indicates a 1.5-kb gap in the sequence data.

30

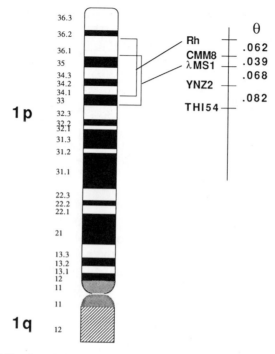

Figure 5. λMS1 Map Location. λMS1 was localized and ordered with respect to other marker loci on Chromosome 1p by linkage analysis. Physical locations of Rh and λMS1 are indicated on the karyogram. Although the genetic map is scaled in Morgans, recombination fractions (θ) are indicated for each interval.

Table 2. New alleles at locus λMs1 (D1S7)

Kindred-individual	Parental origin and allele sizes[a]		New allele size[a]	Change[b]	R/NR[a]
2–07	P	4.32/*12.67	13.61	+105	NR
102–05	P	3.38/*3.86	3.82	−4	NR
102–12	P	*3.38/3.86	3.78	+44	NR
1340–06	M	*3.39/8.90	3.42	+3	NR
1340–07	M	*3.39/8.90	3.36	−3	NR
1344–10	M	*5.00/6.48	4.76	−28	NR
1345–07	M	4.54/*10.83	11.22	+43	NR
1346–09	M	5.41/*13.73	14.53	+89	NR
1362–03	M	8.17/10.32	9.88	ND	R
1362–17	P	5.11/*5.58	5.74	+18	NR
1418–04	P	1.64/*4.10	4.06	−4	NR
13294–09	P	2.64/*10.37	10.77	+44	NR

New alleles at locus λMS1 (D1S7). [a]M, maternal origin of the new allele; P, paternal origin of the new allele. The chromosome carrying the new allele is recombinant (R) or nonrecombinant (NR) for the CMM8-YNZ2 interval. All allele sizes are in kilobases and refer to HinfI digests. The asterisk denotes the progenitor of the de novo allele identified from the YNZ2-CMM8 haplotype. [b]Estimate of change in number of 9-bp repeats associated with the generation of the new allele. ND, the change in number of repeat units was not determined because this chromosome was recombinant for the YNZ2-CMM8 interval.

uneven number of exchanges, in either of the intervals from λMS1 to the flanking markers. The probability of our data under such a model was calculated to be 1.3×10^{-10} by using the binomial distribution. We rejected this model after performing a two-tailed binomial test with a nominal level of $2P = 0.05$ (Pratt and Gibbons, 1981).

The second model presumes that generation of new alleles is independent from crossing over between homologs, and one may ask whether our observed recombination frequency for the YNZ22-CMM8 interval

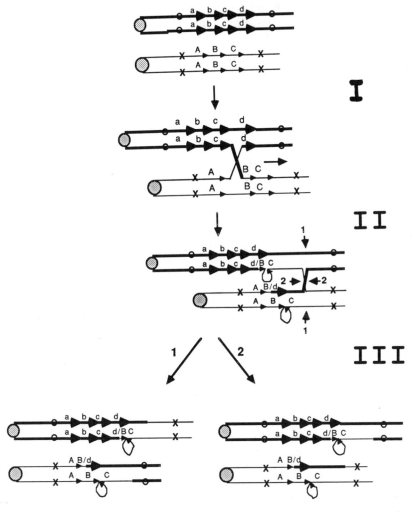

Figure 6. Gene-conversion model. I. An unequal crossover event occurs between strands of homologous chromosomes. II. Branch migration of the resulting Holliday structure occurs. III. Before the Holliday structure migrates beyond the flanking markers it is resolved by cutting strands. Isomerization allows either set of strands to be cut (1 or 2). Depending on whether cuts 1 or 2 are made, either recombinant or parental flanking markers will result.

is consistent with the recombination estimates for the two flanking markers as determined for the general population. The probability of our data under the second model is 0.37; the two-tailed binomial test showed that this type of model is consistent with our data.

The third model is analogous to gene conversion, in which an unequal crossing over event within the repeat units can generate either parental or recombinant flanking markers depending on which strands are used to resolve the resulting Holliday structure (Fig. 6). Because the frequency of recombination between flanking markers associated with gene-conversion is unknown, we decided to let the proportion vary in order to define the limits under which a two-tailed binomial test with a nominal level of $P_L = P_R = 0.025$, can no longer rule out this type of model. We found that if the crossover frequency associated with gene conversion is 38% or lower, we can not reject such a model.

Our approach, using genetic linkage maps to determine flanking markers for locus λMS1, has yielded further evidence that unequal exchange between homologs is not the process most frequently involved in generating new alleles at VNTR loci. Moreover, because our data are consistent with the recombination frequency observed within the YNZ2-CMM8 interval in the general population, the studies have revealed no enhancement of exchange of flanking markers associated with minisatellite mutation. Although we can not rule out the possibility that homologs interact to generate new alleles, we can specify that under the gene conversion model described above, associated crossing over must be biased to account for less than 38% of new alleles generated at λMS1. Our data remain consistent with the previously described models of unequal sister-chromatid exchange at meiosis or mitosis, replication slippage, and insertion/deletion of repeat units as mechanisms for generation of new alleles at VNTR loci.

Evolutionary Analysis of VNTRs

A hypothesis that VNTR loci function as recombination hotspots, could explain the generation of variability as a consequence of a concomitant high rate of unequal exchange. Such a hypothesis would also explain the finding that identical substitutions are usually found within multiple repeat units of a given tandem array (Bell et al., 1982; Ullrich et al., 1982; Goodbourn et al., 1983; Proudfoot et al., 1982; Capon et al., 1983).

Another hypothesis portrays VNTRs as enhancers or regulators of transcription. The basis of this proposal lies in two sets of observations. The first was that all of the original VNTRs were identified adjacent to genes. The second was that the repeats associated with the Ha-*ras* locus were shown to enhance transcription in a chloramphenicol-acetyl-transferase (CAT) assay (Spandidos and Holmes, 1987).

Regardless of which function we attribute to VNTRs, we would expect that if VNTRs are truly functional, their associated hypervariability should be conserved through evolution. If the VNTRs were found to be conserved through evolution at homologous loci, one would have a quick and easy method for finding hypervariable loci in other species, and could use them to construct maps in mammalian species of economic importance. One could monitor the inheritance of specific desired traits, as has been done with many agricultural species such as maize and tomatoes. It has already been shown that hypervariable loci related to the human minisatellite core sequence do exist in other species, yet whether they exist at the homologous loci remains to be seen (Dallas, 1988; Georges et al., 1988; Burke et al., 1989).

Models to explain the sequence relationship that exists among VNTRs have also been suggested, in addition to those that seek to explain VNTR function. Two hypotheses have been put forward to explain the sequence relationship and hypervariable nature of the dispersed loci. The first hypothesis proposes that the core sequence contains an inherent ability to expand tandemly, and that this expansion has occurred at multiple dispersed loci independently.

The second hypothesis is that VNTR loci are related by transposition. Once transposition of the tandem copies of a repeated sequence has occurred, mutation can lead to sequence divergence which may be carried to other repeat units by unequal exchange. This model was proposed by Jeffreys et al. (1985), after they observed a 9-bp direct repeat flanking the tandem repeats at the myoglobin locus. However, when they analyzed the sequence of eight other VNTRs that had been identified by hybridization with the myoglobin repeat sequence, they found that none of the eight possessed flanking direct repeats. They concluded that the loci were probably not related by transposition.

Neither mechanism has been supported or refuted by any strong evidence. We propose that for each mechanism one can make specific predictions about how the locus should have appeared before its tandem expansion. The inherent-activity model would predict that the locus should originally contain exactly one copy of the repeat sequence. The transposition model carries the prediction that the repeat sequence should be completely absent prior to tandem expansion. To test these predictions, one needs to be able to look at a locus before and after the expansion event occurs.

We have attempted to reveal by high-stringency hybridization studies, whether 12 human hypervariable loci are conserved in four nonhuman primate species. All of the VNTR probes identified specific loci in every primate species tested. Ten of the twelve probes detected hypervariable regions, defined as multi-allelic insertion/deletion polymorphism seen

Table 3. Hybridization of 12 human VNTR markers to unrelated nonhuman primate members of 4 species

Probe	Chimpanzee	Gorilla	Orangutan	Gibbon
p68RS2.0	−	−	−	−
pINS310	−	−	−	+
pEJ6.6	+	+	−	−
M27-beta	+	−	−	−
YNZ2	+	+	+	+
YNZ22	−	−	−	−
YNZ23	−	−	+	+
YNZ32	+	+ +	+	+
YNM3	+	+	+	+
YNH24	−	−	+	+
YNH37	+	−	+	−
THH39	−	−	+	+

+ Hypervariable
− Not Hypervariable
+ +Amplified
Each probe was classified as either being hypervariable or not hypervariable. Probe YNZ32 was apparently amplified in gorilla.

among unrelated individuals identified by multiple restriction enzymes, in at least one of the nonhuman primate species tested (Tab. 3). Three marker loci were hypervariable in all four species. On the basis of these results, we estimate that about 83% of the VNTR probes now available will detect a hypervariable locus in at least one of the four species tested. On average about 50% of all VNTR probes will detect hypervariability in any one of the four species. Interestingly, no clear evolutionary pattern exists for the presence or absence of hypervaribility at VNTR loci in the species tested. For example, the locus defined by pEJ6.6 is not hypervariable in gibbon or orangutan, yet it is hypervariable in the two primates that are the closest evolutionarily to human, chimpanzee and gorilla. This would lead us to believe that the hypervariability was generated following the divergence of human, chimpanzee and gorrilla from the other other primates. Other loci cannot be as easily explained. For example, locus YNH24 is hypervariable in human, orangutan, and gibbon but not in gorilla or chimpanzee. This type of result can be explained in many ways; for example, YNH24 might be an ancient hypervariable sequence that predates primate divergence and either it has become quiescent, or one major allele has become fixed in gorilla and chimpanzee independently. It is also possible that the fixation of this locus occurred only once, following divergence of human, chimpanzee, and gorilla from the other primates, and the sequence in human then became active once again. We conclude that because no consistent evolutionary pattern for VNTR hypervariability is evident, it appears that VNTR loci are probably in a continual state of flux, unstably coming and going with time.

Curiously, another locus, YNZ32, identifies an increased amount of homologous DNA, in gorilla as compared with human and other primates. We estimate that 4-fold or greater amplification has occurred at the locus. We do not know whether the amplification that has taken place changed the copy number of the repeat and introduced recognition sites for restriction enzymes, or whether the original locus has dispersed to other loci in the genome. Unless restriction sites for both HaeIII and RsaI were introduced, it appears that the locus has dispersed in gorilla. This question could possibly be resolved by pulsed-field gel analysis, and might shed some light on the processes involved in DNA amplification.

Two probes, pYNZ22 and p68RS2.0, did not reveal hypervariable patterns in any of the nonhuman primate species. This result was consistent with the concept that hypervariability at these loci may have arisen after humans diverged from nonhuman primates. It was also consistent with a particular allele having been fixed in each of the nonhuman primates. We previously sequenced a total of 21 repeat units from six alleles of the YNZ22 locus (Wolff et al., 1988; and unpublished data), and saw no substitutions within those repeat units. In all other VNTR loci that have been sequenced, identical base substitutions are present in multiple copies of the repeat. We might explain our unusual observation with YNZ22 in two ways. First, the ability of the YNZ22 locus to become hypervariable may have arisen fairly recently, allowing humans to generate 14 different allele sizes while all four primate species remained non-polymorphic. This proposed evolutionary change would have been too recent for point mutations to have accumulated within the repeated sequence. Alternatively, this VNTR might be subject to a very active homogenization mechanism such as unequal exchange associated with the locus (Smith, 1976; Dover, 1982); this homogenization mechanism must out-compete the substitution-mutation rate. The explanation that involves a recent event is supported by a secondary observation that in nonhuman primates, no variation in the number of copies of repeats exists at YNZ22 within a given species. These facts lead us to hypothesize that the hypervariable nature of YNZ22 in humans derives from a fairly recent evolutionary event. The YNZ22 locus thus represents an opportunity to test the two proposed hypotheses that attempt to explain the sequence relationship that exists among dispersed VNTR loci.

We began by performing PCR analysis on DNA from one chimpanzee, one gorilla, and one orangutan. To properly size the nonhuman primate alleles we compared them against the YNZ22 alleles of two human individuals (1219 and 1292), whose four homologs had been previously shown to contain one and two, and three and four repeats, respectively by sequence analysis. The size of all nonhuman primate alleles was consistent with one copy of the 70-bp sequence that is

36

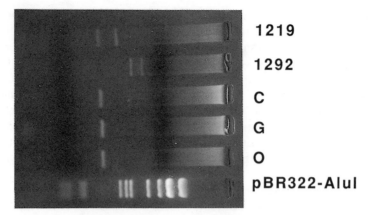

1219

1292

C

G

O

pBR322-AluI

Figure 7. PCR amplification of the YNZ22 locus in three nonhuman primate species. Size markers 1219 and 1292 are human DNA containing one- and two-repeat alleles, and three- and four- repeat alleles respectively. Chimpanzee (C), gorilla (G), and orangutan (O). All three nonhuman primate species contain YNZ22 alleles consistent with the human one-repeat size.

tandemly repeated at the human locus (Fig. 7). These results were consistent with the model by which something inherent in the sequence itself allows expansion to occur independently at dispersed loci.

We next chose to investigate the sequence of the YNZ22 locus in each of the same three nonhuman primates, to see if we could determine why this locus had not expanded into a VNTR in those species. We sequenced the PCR products, and compared them with the sequence of two human alleles (Fig. 8). The only consistent difference found between all of the non-human primates and the human was a deletion of

SEQUENCE OF LOCUS YNZ22 IN PRIMATES

```
HUMAN 1      CACAGGAGGGCAAGGCGGTCCTCACCCTGCCTGGGCTGGGGCAGGGCTGTGAGAC
HUMAN 2      ------------------C-----GA----------------------------
GORILLA      ------------------T-----CA----------------------------
ORANGUTAN    ------------------T-----CA----------------------------
CHIMP        ------------------T-----CA----------------------------


                                                             >
HUMAN 1      CCTCCCTTACAGAAGCAATGAGGGCTTGAGGA*GGGGGTTAGGGGCAGTAAGTT
HUMAN 2      -----------------------------------------------------
GORILLA      --------------------------------T--------------------
ORANGUTAN    --------------------------------T----A---------------
CHIMP        --------------------------------T-------*------------
```

Figure 8. YNZ22 sequence of nonhuman primates. Sequence of the YNZ22 locus in three nonhuman primates compared to that of two human alleles. All three nonhuman primate alleles contain an extra thymine (T) as their only consistent difference compared to human. Three base differences adjacent to the repeat sequence in human are shown in the nonhuman primate sequences to be conserved among them, probably representing the ancestral sequence to that of the human alleles.

one thymine (T) in the human sequence. The deletion of the T does not create a closer match to the VNTR core consensus sequence reported by Nakamura *et al.* (1988a); the deletion in fact disrupts a second consensus core sequence present in all of the nonhuman primates. We propose that unless the loss of the T is responsible for allowing this sequence to duplicate or amplify, the amplification of a locus to two or more copies of the unit sequence is a rate-limiting step in the formation of a VNTR. According to this hypothesis, this rate-limiting step would be the first event in the generation of variability at a VNTR locus; once the locus has a minimum of two copies, it can undergo expansion and contraction to generate variability within the population by one or more of the numerous mechanisms that have been proposed (Wolff *et al.*, 1988, Wolff *et al.*, 1989).

New Directions

Current efforts in the laboratory focus on the new families of polymorphic marker systems. These include the CA/GT repeats of Weber and May (1989) and Litt and Luty (1989) as well as the polymorphic poly A tracts associated with the ALU I repetitive family (Economou *et al.*, 1989). Although the VNTR markers have become strong anchor points on linkage maps, the abundance of these new marker systems should allow higher resolution maps to be constructed. The high resolution maps will allow more access to identification of disease genes for positional cloning.

References

Bell, G. I., Selby, M. J., and Rutter, W. J. (1982) The highly polymorphic region near the human insulin gene is composed of simple tandemly repeating sequences. *Nature* 295: 31–35.

Botstein, D., White, R., Skolnick, M., and Davis, R. W. (1980) Construction of a genetic linkage map in man using restriction fragment length polymorphisms. *Am. J. Hum. Genet.* 32: 314–331.

Burke, T., Davies, N. B., Bruford, M. W., and Hatchwell, B. J. (1989) Parental care and mating behavior of polyandrous dunnocks *Prunella modulais* related to paternity by DNA Fingerprinting. *Nature* 338: 249–251.

Capon, D. J., Chen, E. Y., Levinson, A. D. Seeburg, P. H., and Goeddel, D. V. (1983) Complete nucleotide sequences of the T24 human bladder carcinoma oncogene and its normal homologue. *Nature* 302: 33–37.

Dallas, J. F. (1988) Detection of DNA "fingerprints" of cultivated rice by hybridization with a human minisatellite DNA probe. *Proc. Natl. Acad. Sci. USA* 85: 6831–6835.

Dover, G. (1982) Molecular drive-a cohesive mode of species evolution. *Nature* 299: 111–117.

Economou, E. P., Bergen, A. W., and Antonarakis, S. E. (1989) Novel DNA polymorphic system: Variable poly A tract 3' to Alu I repetitive elements. *Amer. J. Hum. Genet. (Suppl.)* 45: A138.

Georges, M., Lequarre, A. S., Castelli, M., and Vasart, G. (1988) DNA fingerprint in domestic animals using four different minisatellite probes. *Cytogenet. Cell Genet.* 47: 127–131.

38

Goodbourn, S. E. Y., Higgs, D. R. Clegg, J. B., and Weatherall, D. J. (1983) Molecular basis of length polymorphism in the human zeta-globin gene complex. *Proc. Natl. Acad. Sci. USA* 80: 5022–5026.

Jeffreys, A. J., Wilson, V., and Thein, S. L. (1985) Hypervariable 'minisatellite' regions in human DNA. *Nature* 314: 67–73.

Jeffreys, A. J., Royle, N. J., Wilson, V., and Wong, Z. (1988) Spontaneous mutation rates to new length alleles at tandem-repetitive hypervariable loci in human DNA. *Nature* 332: 278–281.

Kan, Y. and Dozy, A. (1978) Antenatal diagnosis of sickle-cell anemia by DNA analysis of amniotic-fluid cells. *Lancet* 2: 910–912.

Lathrop, G. M., Lalouel, J.-M., Julier, C., and Ott, J. (1984) Strategies for multipoint linkage analysis in humans. *Proc. Natl. Acad. Sci. USA* 81: 3443–3446.

Litt, M. and Luty, J. A. (1989) A hypervariable microsatellite revealed by *in vitro* amplification of a dinucleotide repeat within the cardiac muscle actin gene. *Amer. J. Hum. Genet.* 44: 397–401.

Maniatis, T., Hardison, R., Lacy, E., *et al.* (1978) The isolation of structural genes from libraries of eucaryotic DNA. *Cell* 15: 687–701.

Nakamura, Y., Leppert, M., O'Connell, P. Wolff, R., Holm, T., Culver, M., Martin, C., Fujimoto, E., Hoff, M., Kumlin, E., and White, R. (1987) Variable number of tandem repeat (VNTR) markers for human gene mapping. *Science* 237: 1616–1622.

Nakamura, Y., Carlson, M., Krapcho, K., Kanamori, M., and White, R. (1988a) New approach for isolation of VNTR markers. *Am. J. Hum. Genet.* 43: 854–859.

Nakamura, Y., Lathrop, M., O'Connell, P., Leppert, M., Barker, D., Wright, W., Skolnick, M., Kondoleon, S., Litt, M., Lalouel J.-M., and White, R. (1988b) A mapped set of DNA markers for chromosome 17. *Genomics* 2: 302–309.

O'Connell, P., Lathrop, G. M., Nakamura, Y., Leppert, M. L., Ardinger, R. H., Murray, J. C., Lalouel, J. -M., and White, R. (1989) Twenty-eight loci form a continuous linkage map of markers for human chromosome 1. *Genomics* 4: 12–20.

Odelberg, S. J., Plaetke, R., Eldridge, J. R., Ballard, L., O'Connell, P., Nakamura, Y., Leppert, M., Lalouel, J.-M., and White, R. (1989) Characterization of eight VNTR loci by agarose gel electrophoresis. *Genomics* 5: 915–924.

Pratt, J. W. and Gibbons, J. D. (1981) "Concepts of Nonparametric Theory". Springer-Verlag, New York.

Proudfoot, N. J., Gill, A., and Maniatis, T. (1982) The structure of the human zeta-globin gene and a closely linked, nearly identical pseudogene. *Cell* 31: 553–563.

Royle, N. J., Clarkson, R. E., Wong, Z., and Jeffreys, A. J. (1988) Clustering of hypervariable minsatellites in the proterminal region of human autosomes. *Genomics* 3: 352–360.

Smith, G. P. (1976) Evolution of repeated DNA sequences by unequal crossover. *Science* 191: 528–535.

Spandidos, D. A. and Holmes, L. (1987) Transcriptional enhancer activity in the variable tandem repeat DNA sequence downstream of the Ha-*ras*I. *FEBS Lett.* 218: 41–46.

Steinmetz, M., Stephan, D., and Lindahl, K. F. (1986) Gene organization and recombinational hotspots in the Murine Major Histocompatibility Complex. *Cell* 44: 895–904.

Ullrich, A., Dull, T. J., Gray, A., Philips, J. A. III, and Peter, S. (1982) Variation in the sequence and modification state of the human insulin gene flanking regions. *Nucleic Acids Research* 10: 2225–2240.

Vogelstein, B., Fearon, E. R., Kern, S. E., Hamilton, S. R., Preisinger, A. C., Nakamura, Y., and White, R. (1989) Allelotype of colorectal carcinomas. *Science* 224: 207–211.

Weber, J. L. and May, P. E. (1989) Abundant class of human DNA polymorphisms which can be typed using the polymerase chain reaction. *Amer. J. Hum. Genet.* 44: 388–396.

Wolff, R. K., Nakamura, Y., and White, R. (1988) Molecular characterization of a spontaneously generated new allele at a VNTR locus: no exchange of flanking DNA sequence. *Genomics* 3: 347–351.

Wolff, R. K., Plaetke, R., Jeffreys, A. J., and White, R. (1989) Unequal crossingover is not the major mechanism involved in the generation of new alleles at VNTR loci. *Genomics* 5: 382–384.

Wong, Z., Wilson, V., Patel, P., Povey, S., and Jeffreys, A. J. (1987) Characterization of a panel of highly variable minisatellites cloned from human DNA. *Ann. Hum. Genet.* 51: 269–288.

Wyman, A. R. and White, R. (1980) A highly polymorphic locus in human DNA. *Proc. Natl. Acad. Sci. USA* 77: 6754–6758.

DNA Fingerprinting: Approaches and Applications
ed. by T. Burke, G. Dolf, A. J. Jeffreys & R. Wolff

Human VNTR Sequences in Porcine HTF-Islands

B. Brenig* and G. Brem

*Department of Molecular Animal Breeding, Ludwig-Maximilians-University of Munich,
Veterinärstrasse 13, D-8000 München 22, Germany*
*Current address: Institute of Biochemistry, Ludwig-Maximilians-University of Munich,
Karlstrasse 23, D-8000 München 2, Germany*

Summary
Mapping of complex genomes has been influenced substantially by the isolation of locus-specific, but repetitive DNA elements known as VNTRs. Since a high GC-content is characteristic of most of these elements one might expect them to be clustered at least partially in CpG-islands. To address this question we have constructed a porcine liver DNA NotI-linkage library in pUC18 using isolated HTF-islands. HpaII tiny fragments ranging from less than 100 bp to 1 kb in length were randomly selected and analysed by sequencing. As expected from the source of DNA the clones were rich in CpG-, C- and G-content. Several clones exhibiting a repetitive $(GGC)_n$-motif and sequences originally present in human VNTR-markers were shown to detect DNA fingerprints in different species (*Homo sapiens, Sus scrofa domestica, Gallus domesticus, Glycine max, Saccharomyces cerevisiae*). A subset of clones was used for hybridization experiments with porcine DNA and were shown to detect single-copy linkage fragments.

Introduction

In the last decade physical and genetic mapping of genes and chromosomes of mammalian genomes has advanced rapidly. Most of the data come from the dissection of the human genome, which is reflected in the exponentially growing number of isolated and chromosomally assigned genes, polymorphic DNA segments and genetic markers (White *et al.*, 1985; Nakamura *et al.*, 1987). But there is also an increasing interest in the construction of a genetic map of pigs and cattle (Roberts, 1990; Haley *et al.*, 1990).

Considerable progress in gene mapping has been achieved by the use of repetitive DNA elements which are a common property of vertebrate and invertebrate genomes. Polymorphisms arising from tandem arrays of various numbers of such elements can be used both for physical and genetic mapping and identification of individuals (genetic fingerprinting) (Jeffreys *et al.*, 1985a; Jeffreys *et al.*, 1985b; Jeffreys *et al.*, 1985c; Burke *et al.*, 1989; White *et al.*, 1985; Dallas, 1988; Gyllensten *et al.*, 1989). So far, most of the known hypervariable regions consist of short DNA sequence motifs containing a phylogenetically conserved GC-rich "core", or consensus element. Indeed, in some VNTRs the dinucleotide CpG is over-represented.

Although vertebrate DNA is characterized by an under-representation of the dinucleotide CpG, regions of G + C-, and CpG-richness, so-called CpG- or HTF-islands (Bird *et al.*, 1985; Gardiner-Garden and Frommer, 1987; Lindsay and Bird, 1987) are found within the 5′- or promoter regions of housekeeping and some tissue-specific genes (Sargent *et al.*, 1989; Huxley and Fried, 1990). These regions can be separated from the bulk of genomic DNA by digestion with the methylation sensitive restriction endonuclease HpaII, giving rise to fragments of 200–1000 bp in length. Monitoring DNA of known sequences for the presence of HTF-islands is facilitated by calculating the percentage of G + C nucleotides (> 50%) and the ratio of observed/expected (O/E) CpGs (> 0.6) (Gardiner-Garden and Frommer, 1987). Using these criteria HTF-islands have also been identified in 3′-regions and introns of genes. Apart from HTF-islands derived from ribosomal DNA (approx. 20%), it is generally thought that HTF-islands do not contain repetitive DNA. Apparently HTF-islands are rich in rare-cutting restriction enzyme sites (e.g. NotI, EagI, SacII, BssHII, SmaI, NaeI, NarI). Presumably the majority (90–95%) of all NotI-sites in the human genome (approx. $3-4 \times 10^3$) are within HTF-islands (Lindsay and Bird, 1987; Ito and Sakaki, 1989). Since promoters frequently contain GC-clusters, it should be possible to clone linkage fragments located in potential promoter regions of transcribed genes, by isolating HTF-islands and consecutive enriching for fragments containing a specific rare-cutter site, e.g. NotI. These linkage clones may serve as potential genetic markers and will therefore contribute to a genetic map.

Materials and Methods

Isolation and Analysis of Porcine DNA

Unless otherwise specified all standard protocols were carried out as described by Ausubel *et al.* (1987). High-molecular-weight genomic DNA was extracted from deep frozen blood samples (10 ml) by an unpublished protocol of A. J. Jeffreys.

DNA was digested with different restriction enzymes ($2-3$ U/μg DNA) and electrophoresed on 0.6–1% agarose gels. DNA fragments were transferred onto nylon membranes (Hybond-N^{+TM}) for 1 h (50 cm H$_2$O) with 0.4 M NaOH using the VacuGene and hybridized with [α-^{32}P]dCTP labelled pHTF fragments (Feinberg and Vogelstein, 1983) using the random primers DNA labeling system. Nylon membranes were hybridized and washed as described by Church and Gilbert (1984) and exposed on X-Omat XAR-5 films at $-80°$C with intensifying screens.

Preparation and Screening of the HTF-Linkage Library

Figure 1 outlines the experimental strategy for the preparation and construction of the porcine HTF-linkage library. 100 μg porcine liver DNA were restricted to completion with HpaII and electrophoresed on a 1% preparative agarose gel. DNA fragments ≤ 2 kb were recovered and repurified on a 4% NuSieve™ agarose gel. Ca. 10 μg 200–1000 bp fragments were recovered from this gel by electroelution. The 200–1000 bp HTF-fraction was subcloned into the AccI-site of pUC18 (Vieira and Messing, 1982). pUC18 was digested with AccI and dephosphorylated with alkaline phosphatase from calf intestine. Recombinant plasmids were used to transform competent *E. coli* JM83, which were

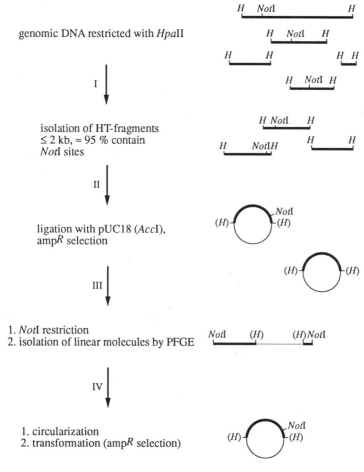

Figure 1. Schematic outline of the protocol used for the construction of the porcine NotI-linkage library.

then grown overnight in 300 ml LB-medium containing ampicillin (50 μg/ml). Approximately 400 μg of recombinant plasmid DNA (pHTF) were prepared. 150 μg were restricted with NotI. Linearized pHTF molecules were separated from circular plasmids by 4% poly-acrylamide-pulsed field gel electrophoresis. Approx. 1 cm thick 4% polyacrylamide gels (acrylamide:N,N′-methylene bisacryl = 30:1) were prepared in $0.5 \times$ TBE (50 mM Tris, 41.5 mM boric acid, 0.5 mM EDTA). Gels were run at 15°C, 100 V, and with pulses of 15 sec for ca. 70 h. Gels were stained with ethidium bromide, and the NotI-linearized bands were excised under UV-light (366 nm). The DNA was electro-eluted in $0.5 \times$ TBE at 120 V for 1 h. 2.5 μg linearized pHTF were circularized in a total volume of 500 μl (5 ng/μl) after purification. Circularized pHTFs (10 ng) were used for transformation of *E. coli* JM83. Ca. 2990 white colonies were obtained corresponding to 3×10^5 recombinants/μg of circularized pHTF. Ligation reactions were per-formed at 16°C as described (Ausubel *et al.*, 1987).

Preparation and transformation of competent *E. coli* JM83 were carried out as described by Hanahan (1987). Transformed *E. coli* JM83 were grown to saturation at 37°C in LB-medium containing 50 μg/ml ampicillin. Plasmid DNA was isolated with Qiagen plasmid kit. A fraction of the pHTF-linkage library was plated to give ≈ 3000 colonies/plate (1.5% LB-agar, 50 μg/ml ampicillin, 20 μg/ml X-gal) and kept at 4°C. Double-stranded plasmid DNA was sequenced as described by Sanger *et al.* (1977) using a T7 DNA polymerase sequencing kit.

DNA from 50 randomly picked colonies was analysed by restriction with XbaI/PstI (separation of insert and vector) and NotI (lineariza-tion). 2 linkage clones (pHTF5, pHTF6) could not be linearized with

Table 1. Parameters for the construction of the NotI-linkage library

Parameter	Units
Starting amount of genomic DNA	100 μg
Amount of size-selected DNA	10 μg
Range of size included	200-1000 bp
Amount of pUC18 (AccI)	50 ng
Amount of HTF-DNA	50 ng
Ligation volume	30 μl
Molar excess of HTF	$\approx 2-3{:}1$
Amount of pHTF from 300 ml culture	400 μg
Starting amount of pHTF	150 μg
Amount of NotI-linearized pHTF	10 μg
Amount of circularized pHTF	2.5 μg
Ligation volume	500 μl
HTF-DNA concentration	5 ng/μl
Total colonies obtained from 10 ng circularized pHTF	2990
Total colonies per μg circularized pHTF	3×10^5
Non-recombinant colonies	$\approx 6\%$

NotI and did not contain an insert. PHTF25 contained an insert of about 420 bp but no NotI restriction site. Therefore we expect a background of approx. 6% of non-recombinants. The sizes of the inserts ranged from 50 bp to ca. 1 kb, with a maximal distribution of clones around 300 bp. Table 1 summarizes the experimental parameters for the construction of the HTF-linkage library.

Results

Structure and Sequence of HTF NotI-Linkage Fragments

The NotI-linkage library was prescreened with a mouse ribosomal DNA probe to eliminate linkage clones containing ribosomal DNA sequences which are known to be highly reiterated (Gardiner-Garden and Frommer, 1987). Out of the remaining clones (ca. 80%) 50 (pHTF1-pHTF50) were randomly selected, recombinant plasmid DNA isolated, and sequenced. Data bank comparison (EMBL Data Bank, MicroGenie) revealed no significant homologies to published sequences. Ca. 30% of all non-ribosomal HTF-clones contained varying numbers (1–4) of the common human VNTR-motif $GGGN_2GTGGGG$ (Nakamura et al., 1987). PHTF32 (264 bp; %G + C: 78.4; O/E CpG: 1.1) e.g. harbours 4 copies of the invariant and 3 copies of the common VNTR-sequence in a tandem array (Tab. 2). Additional directly repeated elements, e.g. $G^{G}/_{C}N_3GGTGGC$, and a $(GGC)_n$-trinucleotide motif similar to that of D14S1, a human chromosome 14 telomeric VNTR-probe, were observed (Fig. 2). The $(GGC)_n$-repeat is a common structural motif in vertebrate genomes and shows extreme inter- and intraspecies-specific size variations (Tautz, 1989).

Table 2. Comparison of VNTR sequence elements in porcine CpG-islands. %G + C and (O/E) CpG-ratio were calculated for pHTF7, pHTF32, and pHTF35. Variations of the VNTR invariant sequence motif (**bold**) within the 11 bp variable VNTR repeat unit (underlined) are shown

	length(bp)	%G + C/CpG-ratio	VNTR sequence
human invariant VNTR			G**NNNN**TGGG
human invariant VNTR			GGG**NN**GTGGGG
pHTF7	225	76.44/1.17	CGGCG**GT**GGGG
			AGAG**TG**TGGGT
			GGGA**GG**TGGGC
pHTF32	264	78.41/1.11	CGGCG**GT**GGGG
			AGAG**TG**TGGGT
			TGTGAGTGGGG
			GGGA**GG**TGGGC
pHTF35	253	77.08/1.07	AGAG**TG**TGGGT
			TGTGAGTGGGG
			GGGA**GG**TGGGC

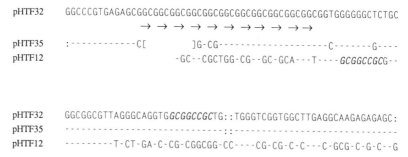

Figure 2. Sequence comparison of pHTF12, pHTF32, pHTF35. Details of the sequences of the fragments used for hybridization are shown. The clones were linearized with PstI and fragments were isolated after digestion. pHTF12 and pHTF35 are aligned to pHTF32. NotI sites are indicated (**bold italic**).: = gap;- = matching bases; [ ] = "spacer".

Detection of DNA Fingerprints in Different Species using pHTF-Linkage Fragments

To analyse cross-reactivity pHTF7, pHTF32, pHTF32, and pHTF35 were selected and used in hybridization experiments with DNA from different species (*Glycine max, Saccharomyces cerevisiae, Gallus domesticus, Homo sapiens*). Figure 3 demonstrates a typical pattern obtained with pHTF32 after hybridization to DNA of two unrelated humans. In order not to detect signals due to unspecific hybridizations of the GC-rich probes, membranes were washed under high stringent conditions (40 mM Na-phosphate, 72°C, 2 × 30 min). Under these conditions pHTF32 detects 19 different bands in DNA restricted with HinfI and approx. 20 resolvable bands after restriction with HaeIII, with fragment sizes ranging from 2–15 kb. The inheritance of the banding patterns was followed in 2 consecutive generations of 3 different families. Individual-specific fingerprints were also obtained using pHTF7, pHTF33, and pHTF35 as probes (data not shown). All of the analysed HTF-linkage clones strongly cross-hybridized with DNA from the above-mentioned species and generated individual DNA fingerprints (Fig. 4). To exclude signals originating from cross-hybridizations with ribosomal sequences, all membranes were rehybridized with a mouse ribosomal probe. The patterns obtained with pMr100 differed significantly from that detected with the linkage fragments (Fig. 3).

Locus-Specificity of Polymorphic Linkage Fragments in Swine

In porcine DNA multiple polymorphisms were detected using different restriction enzymes. At a low molecular size (0.6–1.5 kb) a 5 allelic

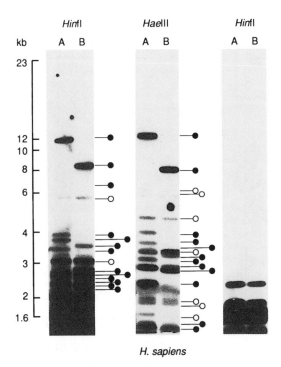

H. sapiens

Figure 3. Interspecies-specificity of HTF NotI-linkage fragments. Autoradiograms of Southern blots from two individuals. Blood DNA samples were prepared from two unrelated individuals (A, B), restricted with HinfI or HaeIII as indicated, blotted onto nylon membranes, and hybridized to ^{32}P-labelled NotI-linkage clone pHTF32. Fragments present in both DNAs are indicated with a line and an open circle, different fragments are marked with a line and a closed circle. The filter with the HinfI digested samples was reprobed with mouse ribosomal DNA (pMr100) as control (left panel). Size standards are indicated in kilobases (kb) on the left.

polymorphism was detected with pHTF7 and pHTF32 after restriction of DNA with HaeIII. A typical hybridization pattern is shown in Fig. 5. Because HaeIII has 4 recognition sites in pHTF32 no polymorphisms were observed above 1.6 kb. Co-segregation of polymorphisms was pursued in 3 consecutive generations of 10 different pig families. DNA fingerprints were obtained after digestion with Sau3AI, RsaI, or AluI, endonucleases that do not cut the linkage fragments (data not shown).

To study locus-specificity of the isolated pHTF-clones at a high molecular size, DNA from different pigs was digested with NotI and separated on pulsed field gels. As expected for linkage probes only two bands at approx. 45 kb and 60 kb were observed corresponding to adjacent NotI fragments (Fig. 6). At this molecular size no polymorphisms were detected.

46

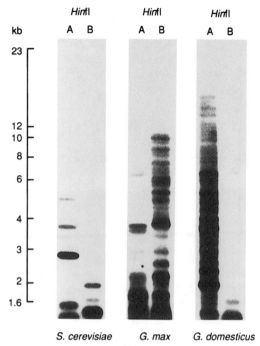

Figure 4. Interspecies-specificity of pHTF NotI-linkage fragments. Autoradiograms of Southern blots from different species (*Saccharomyces cerevisiae*, *Glycine max*, and *Gallus domesticus*). DNA from different strains or species was prepared and two samples (A, B) selected, which gave either the highest or lowest number of fragments respectively. Size standards are indicated in kilobases (kb) on the left.

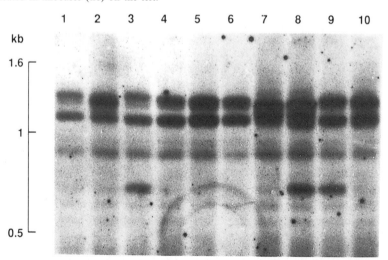

Figure 5. Detection of DNA polymorphisms in pigs. 5 µg DNA isolated from blood samples of different pigs (lanes 1–10) was prepared, digested with HaeIII and electrophoresed on a 0.6% agarose gel. DNA was blotted onto a nylon membrane and hybridized with ^{32}P-labelled linkage clone pHTF32. Size standards are indicated in kilobases (kb) on the left.

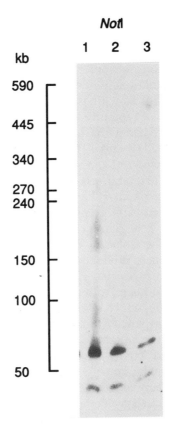

Figure 6. Locus-specificity of pHTF NotI-linkage fragments. Autoradiogram of a Southern blot from three unrelated pigs (lanes 1–3). DNA was isolated from samples of blood, digested with NotI, and electrophoresed on a 1.0% agarose pulsed field gel. Southern blots were hybridized with ^{32}P-labelled linkage clone pHTF35. Size standards are indicated in kilobases (kb) on the left.

Discussion

The construction of linkage and jumping libraries greatly facilitates long-range mapping of complex genomes. We have shown that the use of HTF-islands for the construction of a linkage library is extremely efficient and simple. A total of 50 randomly selected linkage clones have been analysed. A repetitive $(GGC)_n$-motif was found in four clones (pHTF7, pHTF12, pHTF32, pHTF35). Repeats of this kind are present in a number of different mammalian genes and promoter regions, e.g. the bovine gene for epidermal keratin VIb $[(GGC)_3 N_6 (GGC)_2 GGGC GCGGCN_3 GGCN_9 (GGC)_6 N_3 (GGC)_2]$ (Jorcano et $al.$, 1984), the chicken β-4 tubulin gene $[GGCN_2 GGCN_2 (GGC)_2 N_2 GGCN_2 (GGC)_5]$ (Sullivan et $al.$, 1986), the human D14S1 VNTR-probe $[^T/_C pGG_n]$

48

(Nakamura *et al.*, 1987), and the chicken H1 histone gene promoter region [$(GGC)_3CGGCN_2N_2GGCN_{18}GGCN_{14}(GGC)_4$] (Coles *et al.*, 1987). From the data presented here we conclude that approximately 30% of the porcine non-ribosomal HTF-islands preselected for the presence of a NotI restriction site contain varying numbers of VNTR like sequence motifs. Until now it has been assumed that non-ribosomal CpG-islands are devoid of repetitive DNA sequences. However, similar data have been reported from the human α-globin locus. Besides the hypervariable regions (IZHR and 3′HVR) between the ζ-globin genes (ζ and ψζ) and in the 3′-region of the α1-globin gene, there are repetitive DNA elements in the first and second intron of the ζ2-globin gene, which fulfill the criteria of an HTF-island (Goodbourn *et al.*, 1983). Both regions are reported to be responsible for the evolutionary development of the globin gene cluster by gene conversion or unequal crossing-over events. The isolation of a class of porcine CpG-islands harbouring DNA sequences originally found in human VNTR-markers, or repetitive DNA sequences may therefore be an indication for the involvement of CpG-islands in recombinational events.

Compared to the abundance of already available repetitive DNA segments, there is still a paucity of information about the anatomy of animal genomes. A large number of human VNTR or minisatellite probes show no cross-reactivity and consequently cannot be used for linkage analyses in animals. To some extent this is due to the increased evolutionary distance between the different species. Thus, it remains inevitable to isolate species-specific highly polymorphic DNA fragments which can be used for the construction of a precise molecular map.

Acknowledgements
The authors are indebted to Prof. Dr. Ernst-L. Winnacker for providing excellent working conditions in his laboratory. We also want to thank Prof. Dr. Ingrid Grummt for the generous gift of mouse ribosomal DNA (pMr100), Dr. Haralabos Zorbas for DNA from *G. domesticus*, Heidi Feldmann for DNA from *S. cerevisiae*, and Lydia Estermaier for DNA from *G. max*. Simone Jürs is thanked for the preparation of the porcine DNA samples. Appreciation is extended to Sabine Bork for skillful technical assistance. Dr. Haralabos Zorbas, Dr. Avril Arthur, Dr. Brian Salmons, and Dr. Mathias Müller are thanked for critical readings and helpful comments on the manuscript. The nucleotide sequence data reported have been assigned to the EMBL Data Library accession numbers X52499 (pHTF32), X52500 (pHTF7), and X52501 (pHTF35).

References

Ausubel, F. M., Brent, R., Kingston, R. E., Moore, D. D., Smith, J. A., Seidman, J. G., and Struhl, K. (eds.) (1987) Current protocols in molecular biology. Wiley & Sons, New York.

Bird, A. P., Taggert, M., Frommer, M., Miller, O. J., and Mcleod, D., (1985) A fraction of the mouse genome that is derived from islands of nonmethylated, CpG-rich DNA. *Cell* 40: 91–99.

Burke, T., Davies, N. B., Bruford, M. W., and Hatchwell, B. J. (1989) Parental care and mating behaviour of polyandrous dunnocks *Prunella modularis* related to paternity by DNA fingerprinting. *Nature* 338: 249–251.

Church, G. M., and Gilbert, W. (1984) Genomic sequencing. *Proc. Natl. Acad. Sci. USA* 81: 1991–1995.

Coles, L. S., Robins, A. J., Madley, L. K., and Wells, J. R. E. (1987) Characterization of the chicken histone H1 gene complement. *J. Biol. Chem.* 262: 9656–9663.

Dallas, J. F. (1988) Detection of DNA "fingerprints" of cultivated rice by hybridization with a human minisatellite DNA probe. *Proc. Natl. Acad. Sci. USA* 85: 6831–6835.

Feinberg, A. P., and Vogelstein, B. (1983) A technique for radiolabelling DNA restriction endonuclease fragments to high specific activity. *Anal. Biochem.* 132: 6–13.

Gardiner-Garden, M., and Frommer, M. (1987) CpG islands in vertebrate genomes. *J. Mol. Biol.* 196: 261–282.

Goodbourn, S. E. Y., Higgs, D. R., Clegg, J. B., and Weatherall, D. J. (1983) Molecular basis of length polymorphism in the human ζ-globin gene complex. *Proc. Natl. Acad. Sci. USA* 80: 5022–5026.

Gyllensten, U. B., Jakobsson, S., Temrin, H., and Wilson, A. C. (1989) Nucleotide sequence and organization of bird minisatellites. *Nucl. Acids Res.* 17: 2203–2214.

Haley, C. S., Archibald, A., Andersson, L., Bosma, A. A., Davies, W., Fredholm, M., Geldermann, H., Groenen, M., Gustavsson, I., Ollivier, L., Tucker, E. M., and Van de Weghe, A. (1990) The pig gene mapping project–PiGMaP. Proceedings of the 4th World Congress on Genetics Applied to Livestock Production XIII: 67–70.

Hanahan, D. (1987) Techniques for transformation of *E. coli*. In: DNA cloning Vol. I, A practical approach (D. M. Glover, ed.), IRL Press.

Huxley, C., and Fried, M. (1990) The mouse surfeit locus contains a cluster of six genes associated with four CpG-rich islands in 32 kilobases of genomic DNA. *Mol. Cell. Biol.* 10: 605–614.

Ito, T., and Sakaki, Y. (1989) A novel procedure for selective cloning of NotI-linking fragments from mammalian genomes. *Nucl. Acids Res.* 16: 9177–9184.

Jeffreys, A. J., Brookfield, J. F. Y., and Semenoff, R. (1985a) Positive identification of an immigration test-case using human DNA fingerprints. *Nature* 317: 818–819.

Jeffreys, A. J., Wilson, V., and Thein, L. (1985b) Individual-specific "fingerprints" of human DNA. *Nature* 316: 76–79.

Jeffreys, A. J., Wilson, V., and Thein, S. L. (1985c) Hypervariable "minisatellite" regions in human DNA. *Nature* 314: 67–73.

Jorcano, J. L., Rieger, M., Franz, J. K., Schiller, D. L., Moll, R., and Franke, W. W. (1984) Identification of two types of keratin polypeptides within the acidic cytokeratin subfamily I. *J. Mol. Biol.* 179: 257–281.

Lindsay, S., and Bird, A. P. (1987) Use of restriction enzymes to detect potential gene sequences in mammalian DNA. *Nature* 327: 336–338.

Nakamura, Y., Leppert, M., O'Connell, P., Wolff, R., Holm, T., Culver, M., Martin, C., Fujimoto, E., Hoff, M., Kumlian, E., and White, R. (1987) Variable number of tandem repeat (VNTR) markers for human gene mapping. *Science* 235: 1616–1622.

Roberts, L. (1990) An animal genome project? *Science* 248: 550–552.

Sanger, F., Micklen, S., and Coulsen, A. R. (1977) DNA sequencing with chain-terminating inhibitors. *Proc. Natl. Acad. Sci. USA* 74: 5463–5467.

Sargent, C. A., Dunham, I., and Campbell, R. D. (1989) Identification of multiple HTF-island associated genes in the human major histocompatibility complex class III region. *EMBO J.* 8: 2305–2312.

Sullivan, K. F., Havercroft, J. C., Machlin, P. S., and Cleveland, D. W. (1986) Sequence and expression of the chicken β5- and β4-tubulin genes define a pair of divergent β-tubulins with complementary patterns of expression. *Mol. Cell. Biol.* 6: 4409–4418.

Tautz, D. (1989) Hypervariability of simple sequences as a general source for polymorphic DNA markers. *Nucl. Acids Res.* 17: 6463–6471.

Vieria, J., and Messing, J. (1982) The pUC-plasmids, a M13mp7-derived system for insertion mutagenesis and sequencing with synthetic universal primers. *Gene* 19: 259–268.

White, R., Leppert, M., Bishop, D. T., Barker, D., Berkowitz, J., Brown, C., Callahan, P., Holm, T., and Jerominski, L. (1985) Construction of linkage maps with DNA markers for human chromosomes. *Nature* 313: 101–105.

DNA Fingerprinting: Approaches and Applications
ed. by T. Burke, G. Dolf, A. J. Jeffreys & R. Wolff
© 1991 Birkhäuser Verlag Basel/Switzerland

Oligonucleotide Fingerprinting Using Simple Repeat Motifs: A Convenient, Ubiquitously Applicable Method to Detect Hypervariability for Multiple Purposes

J. T. Epplen[a], H. Ammer[a], C. Epplen[a], C. Kammerbauer[a], R. Mitreiter[a], L. Roewer[a], W. Schwaiger[a], V. Steimle[a], H. Zischler[a], E. Albert[b], A. Andreas[b], B. Beyermann[c], W. Meyer[c], J. Buitkamp[d], I. Nanda[e], M. Schmid[e], P. Nürnberg[f], S. D. J. Pena[g], H. Pöche[h], W. Sprecher[i], M. Schartl[j], K. Weising[k] and A. Yassouridis[l]

[a]Max-Planck-Institute for Psychiatry, W-8033 Martinsried, Germany; [b]Kinderpoliklinik University, W-8000 München, Germany; [c]Biology Section Humboldt University, O-1040 Berlin, Germany; [d]Institute for Animal Husbandry and Breeding, Hohenheim University, W-7000 Stuttgart, Germany; [e]Human Genetics Institute University, W-8700 Würzburg, Germany; [f]Medical Genetics Institute Humboldt University (Charité), O-1040 Berlin, Germany; [g]Biochemistry Department, Federal University Minas Gerais, 30161 Belo Horizonte, Brazil; [h]Institute for Legal Medicine Free University, W-1000 Berlin, Germany; [i]Institute for Legal Medicine University, W-3400 Göttingen, Germany; [j]Gene Center, W-8033 Martinsried, FRG; [k]Plant Molecular Biology Group, Institute for Botany University, W-6000 Frankfurt, Germany; [l]Max-Planck-Institute for Psychiatry, W-8000 Munich, Germany

Summary

A panel of simple repetitive oligonucleotide probes has been designed and tested for multilocus DNA fingerprinting in some 200 fungal, plant and animal species as well as man. To date at least one of the probes has been found to be informative in each species. The human genome, however, has been the major target of many fingerprinting studies. Using the probe $(CAC)_5$ or $(GTG)_5$, individualization of all humans is possible except for monozygotic twins. Paternity analyses are now performed on a routine basis by the use of multilocus fingerprints, including also cases of deficiency, i.e. where one of the parents is not available for analysis. In forensic science stain analysis is feasible in all tissue remains containing nucleated cells. Depending on the degree of DNA degradation a variety of oligonucleotides are informative, and they have been proven useful in actual case work. Advantages in comparison to other methods including enzymatic DNA amplification techniques (PCR) are evident. Fingerprint patterns of tumors may be changed due to the gain or loss of chromosomes and/or intrachromosomal deletion and amplification events. Locus-specific probes were isolated from the human $(CAC)_5/(GTG)_5$ fingerprint with a varying degree of informativeness (monomorphic versus truly hypervariable markers). The feasibility of three different approaches for the isolation of hypervariable mono-locus probes was evaluated. Finally, one particular mixed simple $(gt)_n(ga)_m$ repeat locus in the second intron of the HLA-DRB genes has been scrutinized to allow comparison of the extent of exon-encoded (protein-) polymorphisms versus intronic hypervariability of simple repeats: adjacent to a single gene sequence (e.g. HLA-DRB1*0401) many different length alleles were found. Group-specific structures of basic repeats were identified within the evolutionarily related DRB alleles. As a further application it is suggested here that due to the ubiquitous interspersion of their targets, short probes for simple repeat sequences are especially useful tools for ordering genomic cosmid, yeast artificial chromosome and phage banks.

Oligonucleotides Specific for Simple Repeat Motifs are Informative for all Species Tested So Far in the Fungal, Plant and Animal Kingdoms

The existence of repetitive DNA sequences in eukaryotic genomes has been known for more than two decades (Britten and Kohne, 1968). Even after all these years of intense experimentation and scholastic reflection the *raison d'être* of most repetitive elements in the eukaryotic genomes remains elusive. Particularly with respect to the generation and/or maintenance of the class of tandemly organized simple repeat motifs, a coherent or thoroughly convincing concept is lacking. This failure is certainly also due to the fact that these elements exert in almost all cases no sequence-dependent function whatsoever (for a discussion see Epplen, 1988). Bearing this primary shortcoming in mind, simple sequences are evaluated here exclusively as a tool.

In hindsight the design and testing of simple repetitive oligonucleotide probes for DNA profiling could and should have been initiated as early as 1981/82. By then it was already clear that *cloned* simple repetitive $(gata)_n$ and $(gaca)_m$ sequences showed extensive restriction fragment length polymorphisms (RFLP's), e.g. among human individuals (Epplen et al., 1982). Meanwhile Alec Jeffreys and coworkers (1985a, b) have described the "minisatellites" and have established the whole concept of DNA profiling. Finally, in 1986 Ali et al. reported on initial successful efforts to differentiate human genomes using $(GATA)_4$ and $(GACA)_4$ as well as other sequentially related *s*imple *q*uadruplet *r*epeat (*sqr*) oligonucleotides. Additional probes harboring basic repeat motifs of two to six nucleotides were subsequently synthesized and hybridized to DNAs from a variety of sources (Arnemann et al., 1989; Beyermann et al., in press; Buitkamp et al., 1990, 1991; Epplen, 1988; Epplen et al., 1988, 1989; Nanda et al., 1988, 1990a, 1988b, in press; Vogel et al., 1988; Weising et al., 1989, 1990, 1991 (Fig. 1)).

The mere presence and quantity of these simple tandem repeats in the genome can be assessed by a convenient slot-blot hybridization method (Epplen, 1988). This, however, does not in itself prove the suitability of a given probe for DNA fingerprinting. The latter has to be investigated with restriction enzyme digested and electrophoretically separated DNA of the respective species. Even though it is best to start out on the basis of previous experiences using, for example $(GGAT)_4$ for individualization in a novel fish species, the principle of trial and error has to be adopted. Table 1 lists the panel of oligonucleotide probes that have been chemically synthesized for DNA profiling. In common to all the probes are simple repeat motifs of 2–6 bases. The total lengths range from 15 to 24 bases thus allowing the calculation of hybridization and stringent wash temperatures according to the formula T_m (°C) $= [(A + T) \times 2 + (C + G) \times 4] - 5$ (Miyada et al., 1985). This rule applies for a salt concentration of 1M NaCl. In practice, all hybridization and washing

52

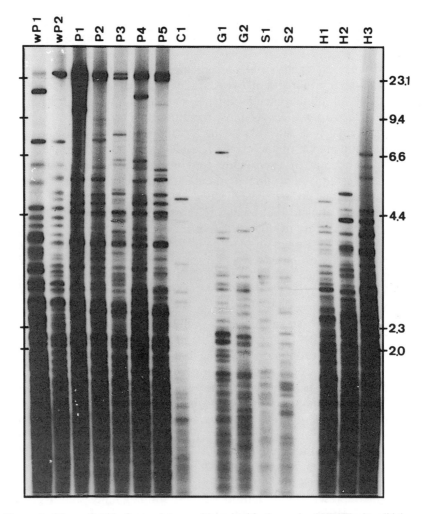

Figure 1. Oligonucleotide fingerprints as obtained with the probe (GGAT)$_4$ in wild boar (wP1–wP2), domestic pigs (P1–P5), camel (C1), goat (G1–G2), sheep (S1–S2) and horse (H1-H3) DNA after HinfI digestion. DNA was prepared from peripheral blood leukocytes, digested and separated by agarose gel electrophoresis (0.7% in Tris-acetate buffer). The gel was dried and hybridized with the ^{32}P-labeled probe under the stringent conditions described previously (Schäfer et al., 1988). Exposure was performed overnight without intensifying screens. Note the inter-individual variability in each species and the species-specific amount and distribution of the ranges of fragment lengths. Molecular weight markers are given in kilobases on the right. Approximately equal amounts of DNA have been applied with the exception of lane P1, which is overloaded. The relationships of the individuals in each species are not known. As can be inferred from our other family data in these species, they appear to include close relatives.

Table 1. Simple repetitive oligonucleotide probes for DNA fingerprinting

Probes	Sequences
(AT)$_8$	atatatatatatatat
(CA)$_8$*	cacacacacacacaca
(CAC)$_5$*°	caccaccaccaccac
(CT)$_8$	ctctctctctctctct
(CT)$_4$(CA)$_5$	ctctctctcacacacaca
(GAA)$_6$*°	gaagaagaagaagaagaa
(GACA)$_4$°	gacagacagacagaca
(GATA)$_2$(GACA)(GATA)$_2$	gatagatagacagatagata
(GATA)$_3$(GACA)$_2$	gatagatagatagacagaca
(GATA)$_4$*°	gatagatagatagata
(GATA)$_{4.5}$, (GATA)$_5$; (GATA)$_{5.75}$*°	18 mer, 20 mer, 23 mer
(GGAT)$_4$°	ggatggatggatggat
(GGGCA)$_3$	gggcagggcagggca
(TCC)$_5$°	tcctcctcctcctcc
(TTAGGG)$_3$	ttagggttagggttaggg

*complementary probe available
°available also in digoxigenated form

steps are conveniently performed in a temperature range from 35°C to 65°C, solely depending on the length and the base composition of the probe.

In order to give a rough overview from our previous experiences, Table 2 lists many of the fungal, plant and animal species that have so far been investigated by oligonucleotide fingerprinting. Depending on the species of interest, different oligonucleotide probes are informative for various purposes. The number of probes, harboring principally different simple repeat sequence motifs, necessary for the study of almost all eukaryotic genomes is 20 or less. It should be mentioned here that no emphasis has been laid on a systematically ordered phylogenetic compilation. In addition to individuality testing in several species, the accumulation of simple repetitive sequences on the heteromorphic sex chromosomes can at the same time be demonstrated by sex-specific hybridization patterns (Nanda et al., 1990a; see also below). Finally, in primates including man (and in several fungi) certain simple sequence motifs are associated with the nucleolus organizer regions (NORs), probably due to their accumulation in the spacer regions between the ribosomal RNA genes (Nanda et al., 1990b; Beyermann et al., unpubl. data).

Oligonucleotide Fingerprinting in Man with Special Reference to Mutation Rates, Tumors and Statistical Evaluations of Paternity Disputes

Following the initial efforts which consecutively used combinations of the (GATA)$_4$ and (GACA)$_4$ and related "mixed" probes, optimal

Table 2. A comprehensive list of eukaryote species fingerprinted by oligonucleotide probes

Species investigated	Informative probes for individualization or identification*, sex chromosome status[+]
Fungi	
Phycomyces blakesleeanus (Zygomycetes)	
Ascochyta rabiei (Ascomycetes)	$(GTG)_5$*, $(GATA)_4$*, $CA)_8$*
Saccharomyces cerevisiae (Ascomycetes)	
Penicillium janthinellum (Ascomycetes)	$(GTG)_5$*, $(CT)_8$*, $(GACA)_4$*
Penicillium citroviridae (Ascomycetes)	$(GTG)_5$*, $(CT)_8$*, $(GACA)_4$*
Penicillium chrysogenum (Ascomycetes)	$(GTG)_5$*, $(CT)_8$*, $(GACA)_4$*
Aspergillus niger (Ascomycetes)	$(GTG)_5$*, $(GACA)_4$*, $(GACA)_4$*
Aspergillus flavus (Ascomycetes)	$(GTG)_5$*, $(GACA)_4$*
Aspergillus oryzae (Ascomycetes)	$(GTG)_5$*, $(GACA)_4$*
Aspergillus batatae (Ascomycetes)	$(GTG)_5$*, $(GACA)_4$*
Aspergillus awamori (Ascomycetes)	$(GTG)_5$*, $(GACA)_4$*
Trichoderma harzianum (Ascomycetes)	$(GTG)_5$*, $(CT)_8$*, $(GACA)_4$*
Candida maltosa (Deuteromycetes)	$(GTG)_5$*, $(GACA)_4$*
Candida utilis (Deuteromycetes)	$(GTG)_5$*, $(GACA)_4$*
Coprinus comatus (Basidiomycetes)	
Plants	
Algae	
Oedogonium spec. (Chlorophyceae)	
Stenogramme interrupta (Rhodophyceae)	
Fucus serratus (Phaeophyceae)	
Liveworts and Mosses	
Marchantia polymorpha (Hepaticae)	
Polytrichum formosum (Musci)	
Ferns	
Equisetum arvense (Equisetatae)	
Polypodum vulgare (Filicatae)	
Osmunda regalis (Filicatae)	
Dodia caudata (Filicatae)	
Gymnosperms	
Taxus baccata (Pinatae)	
Juniperus communis (Pinatae)	
Monocotyledonous Angiosperms	
Echinodorus osiris (Alismatidae)	
Musa acuminata (Liliidae)	$(CA)_8$*, $(GATA)_4$*, $(GTG)_5$*,
Hordeum vulgare (Liliidae)	$(GACA)_4$*, $(GATA)_4$*
Hordeum spontaneum (Liliidae)	$(GACA)_4$*
Asparagus densiflorus (Liliidae)	
Zea mays (Liliidae)	$(GAA)_6$*
Triticum aestivum (Liliidae)	$(GTG)_5$*, $(GATA)_4$*
Secale cereale (Liliidae)	$(GTG)_5$*, $(GATA)_4$*
Chamaedorea cataracterum (Arecidae)	
Dicotyledonous Angiosperms	
Helleborus niger (Magnoliidae)	
Urtica dioica (Hamameliidae)	
Ficus benjamina (Hamameliidae)	
Humulus lupulus (Hamameliidae)	
Aruncus silvester (Rosidae)	
Cicer arietinum (Rosidae)	$(GATA)_4$*, $(CA_8$*, $(GACA)_4$*, $(GTG)_5$*,
and 7 additional species of *Cicer*	$(GGAT)_4$*
Lens culinaris (Rosidae)	$(GATA)_4$*, $(GACA)_4$*, $(CA)_8$*
and additional species	
Simmondsia sinensis (Rosidae)	
Vicia faba (Rosidae)	$(GATA)_4$*, $(GGAT)_4$*

Table 2. (continued)

Species investigated	Informative probes for individualization or identification*, sex chromosome status[+]
Daucus carota sativus (Rosidae)	$(GACA)_4$*, $(GATA)_4$*
Camellia sinensis (Dilleniidae)	
Brassica oleracea (Dilleniidae)	$(GATA)_4$*, $(GTG)_5$*
Brassica napus (Dilleniidae)	$(GATA)_4$*, $(GTG)_5$*
Nicotiana tabacum (Asteridae)	$(GGAT)_4$*, $(TCC)_5$*, $(GTG)_5$*, $(GATA)_4$*
Nicotiana paniculata (Asteridae)	
Nicotiana glutinosa (Asteridae)	
Nicotiana acuminata (Asteridae)	
Lycopersicum esculentum (Asteridae)	$(GGAT)_4$*,$(GTG)_5$*, $(GATA)_4$*
Lycopersicum hirsutum (Asteridae)	$(GGAT)_4$*, $(GAA)_6$*, $(GATA)_4$*
Lycopersicum pimpinellifollum (Asteridae)	$(GGAT)_4$*, $(GAA)_6$*, $(GATA)_4$*
Lycopersicum peruvianum (Asteridae)	$(GGAT)_4$*, $(GAA)_6$*, $(GATA)_4$*
Solanum tuberosum (Asteridae)	$(GATA)_4$*, $(GTG)_5$*, $(GGAT)_4$*, $(TCC)_5$*, $(GACA)_4$*
Helianthus annuus (Asteridae)	
Lactuca sativa (Asteridae)	
Petunia hybrida (Asteridae)	$(GTG)_5$*, $(GATA)_4$*
Silene alba (Carophyllaceae)	
Arabidopsis thaliana (Brassicacaceae)	$(GATA)_4$*
Rumex acetosella (Polygonaceae)	
Beta vulgaris (Chenopodiaceae)	$(GTG)_5$*, $(GATA)_4$*
Beta maritima (Chenopodiaceae)	$(GATA)_4$*, $(GGAT)_4$*
Microseris pygmea (Asteraceae)	$(GACA)_4$*, $(GATA)_4$*
Microseris bigelovii (Asteraceae)	$(GACA)_4$*, $(GATA)_4$*
Animals	
Ciliates	
Stylonychia lemnae	$(GATA)_4$*
Protozoa	
Leishmania brasiliensis	$(CAC)_5$*
Leishmania guyanensis	$(CAC)_5$*
Leishmania mexicana	$(CAC)_5$*
Trypanosoma cruzi	$(CAC)_5$*
Insects	
Poecillimon tessalicus	$(GACA)_4$*, $(GATA)_4$*
Macrosiphum rosae	$(GATA)_4$*
Vertebrates	
Fish	
Latimeria chalumnae	$(CA)_8$*, $(CT)_4(CA)_5$*
Poecilia reticulata	$(GGAT)_4$*, $(GATA)_4$*, $(GACA)_4$[+]
Poecilia sphenops	$(GGAT)_4$*, $(CA)_8$[+]
Poecilia velifera	$(GGAT)_4$*, $(CA)_8$[+]
Poecilia perugiae	$(GGAT)_4$*, $(GACA)_4$
Poecilia formosa	$(GGAT)_4$*, $(GACA)_4$
Poecilia latipinna	$(GGAT)_4$*, $(GACA)_4$*
Xiphophorus maculatus	$(GGAT)_4$*, $(GACA)_4$*
Xiphophorus helleri	$(GATA)_4$*
Xiphophorus montezumae	$(GACA)_4$*
Xiphophorus cortesi	$(GACA)_4$*
Clupea harengus	
Salmo trutta	
Leporinus obtusidens	$(GACA)_4$*
Leporinus elongatus	$(GACA)_4$*
Amphibians	
Bufo melanostictus	$(GACA)_4$*, $(CA)_8$*, $(TCC)_5$*
Necturus maculatus	

56

Table 2. (continued)

Species investigated	Informative probes for individualization or identification*, sex chromosome status[+]
Triturus cristatus carnifex	(GACA)₄*, (CA)₈*
Triturus vulgaris	(GACA)₄*, (CA)₈*
Gastrotheca riobambae	(TCC)₅*
Pyxicephalus adspersus	(GACA)₄*, (GATA)₄*, (CA)₈*, (TCC)₅*
Reptiles	
Elaphe radiata	(GACA)₄[+], (GATA)₄[+]
Elaphe bimulata	(GACA)₄[+], (GATA)₄[+], (GTG)₅*
Elaphe taeniura	(GACA)₄[+], (GATA)₄[+], (GTG)₅*
Python regius	(GATA)₄*
Natrix fasciatus	(GTG)₅*
Boa constrictor imperatus	(GATA)₄*
Birds	
Buteo buteo	(GGAT)₄*
Accipiter nisus	(GGAT)₄*
Accipiter gentilis	(GGAT)₄*
Nyctea scandiaca	(GGAT)₄*
Tinnunculus tinnunculus	(GGAT)₄*
Hierofalco peregrinus	(GACA)₄*
Hierofalco rusticolus	(GACA)₄*
Hierofalco cherrug	(GACA)₄*
Hierofalco biarmicus	(GACA)₄*
Bubo bubo	(GACA)₄*
Milvus milvus	(GGAT)₄*
Aquila chrysaetos	(GACA)₄*, (GGAT)₄*
Gallus domesticus	(GGAT)₄*, (GTG)₅*, (TCC)₅[+]
Columba livia domestica	(GGAT)₄*, (TCC)₅[+]
Phylloscopus collybita	(GACA)₄*, (GGAT)₄*
Periparus ater	(GACA)₄*, (GGAT)₄*
Cyanistes caeruleus	(GACA)₄*, (GGAT)₄*
Erithacus rubecula	(GACA)₄*, (GGAT)₄*
Fringilla coelebs	(GACA)₄*, (GGAT)₄*
Coccothraustes coccothraustes	(GACA)₄*, (GGAT)₄*
Pyrrhula pyrrhula	(GACA)₄*, (GGAT)₄*
Mammals	
Canis lupus familiaris	(CA)₈*, (GTG)₅*
Canis lupus familiaris Sapsari	(GTG)₅*
Equus p. coballus	(CA)₈*, (GGAT)₄*, (GTG)₅*
Sus scrofa domesticus	(AT)₈*, (GGAT)₄*, (TCC)₅*
Ontotragus megaceros	(GTG)₅*
Odocoileus dichotomus (Blastoceros)	(GTG)₅*
Odocoileus bezoarticus (Blastoceros)	(GTG)₅*
Cervus duvauceli	(GTG)₅*
Tragelaphus spekei	(GTG)₅*
Bos taurus	(GTG)₅*, (GGAT)₄*, (CA)₈*
Capra aegagrus hircus	(GACA)₄*, (CA)₈*
Ovis ammon aries	(GACA)₄*, (CA)₈*
Ellobius lutescens	
Cletrionomys glareolus	(GACA)₄*
Apodemus flavicollis	(GACA)₄*
Apodemus sylvaticus	(GACA)₄*
Microtus agrestis	(GGAT)₄*
Rattus norvegicus	(GTG)₅*, (GATA)₄*
Mus musculus	(GGGCA)₃*, (GATA)₄*, (GACA)₄[+]
Cricetulus griseus	(GACA)₄*

Table 2. (continued)

Species investigated	Informative probes for individualization or identification*, sex chromosome status[+]
Nesokia indica	
Nyctalus noctula	$(GATA)_4$
Hylobates syndactylus	$(GTG)_5$*
Macaca mulatta	$(GTG)_5$*, $(GACA)_4$*[&], $(GATA)_4$*
Aotus trivirgatus	$(GACA)$[&]
Pongo pygmaeus	$(GTG)_5$*, $(GACA)_4$[&]
Pan troglodytes	$(GTG)_5$*, $(GACA)_4$[&]
Gorilla gorilla	$(GTG)_5$*, $(GACA)_4$[&]
Homo sapiens	$(GTG)_5$*, $(GACA)_4$[&]

[°] sex differences depending on the population investigated
[&] associated with nucleolus organizers

genetic individualization in the human was subsequently achieved in a single hybridization step using $(CAC)_5$ (Schäfer *et al.*, 1988). Solely depending on the resolving power of the gel system used, all human individuals can be distinguished with the only exception of genetically identical, monozygotic twins. In practice one single case of questionable paternity could not be cleared up because the putative fathers were monozygotic brothers (Pöche *et al.*, in press). The only possibility of solving such a problem would have occurred if one of the men carried a somatic mutation that the other one lacked. Using several informative probes this rare event was not encountered. So far only one case of monozygotic twins with differences in their $(CAC)_5/(GTG)_5$ fingerprints has been encountered (Hundrieser *et al.*, unpubl. data). On the other hand, even questionable paternity cases can be solved beyond any doubt, whereas enzyme polymorphism studies alone would have had insufficient discrimination power or could have yielded false exclusions (Bender *et al.*, 1991).

Nürnberg *et al.* (1989) established the somatic stability of $(CAC)_5/(GTG)_5$ fingerprints in human tissues and determined the mutation rate to be 0.001 per fragment per gamete (from more than 350 meioses). Meanwhile, this rate has been confirmed further by increasing the sample size to over 1500 meioses (unpubl. data). Somatic stability and similar or even reduced mutation rates have been determined for the $(GACA)_4$ fingerprint bands (Roewer *et al.*, 1990). One exception to the rule of somatic stability has been provided by the study of transformed tissues (Lagoda *et al.*, 1989). Indeed, in an extended parallel fingerprint and cytogenetic study on human gliomas versus peripheral blood, $(CAC)_5/(GTG)_5$ proved to be a superior probe for the detection of genomic alterations (Nürnberg *et al.*, in press). Changes of the constitutional patterns were observed in nearly 80% of the gliomas, and subsequent investigation using additional probes detected an additional

15% alterations in the fingerprints. These changed patterns mostly reflect the corresponding chromosomal aberrations seen in a majority of the cells. Concomitantly it was observed that particularly strong fingerprint bands appeared in about 30% of the gliomas when probed with independent simple repeat oligonucleotides. After digestion with several restriction enzymes that recognize unrelated target sequences, consecutive hybridizations with the $(CA)_8$ probe and a cDNA clone containing epidermal growth factor (EGF) receptor gene sequences revealed the comigration of one simple repeat band with one of the amplified EGF receptor gene fragments (Nürnberg et al., in press). Thus gene amplification events can readily be monitored by the set of simple repeat probes due to their ubiquitous distribution all over the human chromosomal complement. During the same investigation it was also noted that the telomere lengths increased during the progression of some glioma tumors as revealed by the telomere repeat probe $(ttaggg)_3$. This is in contrast to the findings of Hastie et al. (1990), who described recently shortened telomeres in colon carcinomas and aged fibroblasts. In conclusion, oligonucleotide fingerprinting is a valuable asset whenever cultured cells of various sources need to be identified (Speth et al., in press).

Since it is not known which bands are allelic to one another in a complex fingerprint pattern, formal genetic models and the corresponding classical statistics cannot be used for the evaluation of *multilocus* fingerprints in paternity analysis. In addition, one or both of the alleles at a locus might well be hidden in the smeared signal below the 2–3 kilobase (kb) range, which is not resolved in routine fingerprint gels. Various new statistical approaches have been worked out for calculating the probability of paternity from multilocus fingerprint profiles (for a discussion see Fimmers et al., 1990; Yassouridis and Epplen, 1991). The primary method relies on the calculation of likelihood ratios based on Bayes' theorem (Evett et al., 1989). This likelihood approach can also be adopted for use in deficiency cases (Yassouridis, in preparation). Despite quite promising results and comparatively widespread recognition – also due to good experiences with the method in practical case work – some restrictions and compromises affect the practicability of the approach of Evett et al. It is not yet known if simpler methods which depend on fewer assumptions will finally prove practicable and if they will meet the necessary standards.

Advantages of the Demonstration of Genetic Individuality by
Oligonucleotide Fingerprinting

It has been repeatedly stated that oligonucleotide fingerprinting is advantageous over related methodologies for a number of reasons: (i) the constant probe quality for up to a million hybridizations from a single batch of chemically synthesized oligonucleotide; (ii) no need for blotting, and therefore no loss of DNA due to non-optimal transfer, by simply fixing the DNA in the dried gel matrix; (iii) the least time-consuming procedure, additionally due to short hybridization times with higher probe concentrations and limited, standardized washing steps; (iv) the possibility of repeated hybridizations of the same gel-fixed DNA – especially in comparison to filter-bound nucleic acids; (v) convenient non-radioactively labeled probes that can also be hybridized directly in the gel (Zischler *et al.*, 1989). Above all, the most significant advantage over other fingerprint probes is the base-specific hybridization under the appropriate conditions: mismatch abrogates hybridization. Therefore, if one follows the established protocols, almost absolute reproducibility is obtained.

Forensic Applications of Genetic Fingerprints

The forensic sciences represent an immediately obvious and broad field of application for individualization techniques, i.e. analyses of stains from spots of blood and sperm, vaginal swabs, sputum, urine, hair roots, decayed tissue remains, parts of bodies, etc. (Pöche *et al.*, 1990; Roewer *et al.*, 1990; Sprecher *et al.*, 1990). Oligonucleotide fingerprinting has been successfully used in actual casework including the identification of burnt bodies (Sprecher *et al.*, in press; Pöche *et al.*, in press). During the successful identification from the DNA of one of the burnt bodies, a remarkable phenomenon was observed that should be brought to the attention of other investigators whilst requiring further experimental clarification: informative $(GTG)_5$ fingerprints of normal intensity could be established, whereas when using the probe $(GACA)_4$ in a rehybridization only extremely weak patterns were observed. This feature is absolutely reproducible using the DNA from the one burnt body. Upon the next hybridization with $(GATA)_4$ the result was again negative, yet $(TCC)_5$ on the other hand yielded the expected intense bands for identification. It thus appears as if the oligonucleotide probes with a high GC content were still hybridizing, whereas this was not true when the GC content was 50% or less. One working hypothesis to explain this probe hybridization behaviour relates to the possible partial denaturation of DNA into single strands under high temperatures depending on the base composition: (i) single-stranded DNA was much more amenable to degradation in the burnt tissue remains, generating at random broken short fragments, (ii) single-stranded DNA would not be cut (as efficiently) and (iii) the subsequent agarose gel electrophoretic separation of unpaired single DNA strands would not result in a distinct pattern but rather in a smeary, background-like signal. These possibilities are considered in ongoing *in vitro* experiments by heating tissues to different temperatures and demonstrating the fingerprint patterns obtained using a variety of simple repetitive oligonucleotides with different GC content. Finally, another source of artifacts has not yet received sufficient attention: the "hidden partial" phenomenon (Nürnberg and Epplen, 1989), where some sites in the genomic DNA need, for completely unknown reasons, an exorbitant amount of enzymatic activity and are not digested.

In this context Roewer et al. (1990) have recently established some general rules for oligonucleotide fingerprinting with respect to the amount and the quality of the DNA samples available, as well as the informativeness of several probes for each purpose. Because of the possibility of repeated rehybridizations, for complicated cases, including those with low amounts of degraded DNA, the results of different oligonucleotides can preferably be combined. Though the virtually absolute sensitivity of the polymerase chain reaction (PCR) cannot be reached by conventional oligonucleotide fingerprinting, the inherent and partly uncontrollable problems of PCR concerning minute contaminations are almost completely avoided. Because of the species-specific accumulation and organisation of these repetitive elements, any large scale contamination of the stain by, for example, blood from a different animal species or fungal overgrowth, can readily be sorted out by hybridizing with the panel of additional simple repeat oligonucleotides identifying the foreign additive. Almost all prokaryotic DNA contaminations would not contain any simple repetitive motifs.

Variation of Simple Repeats in Clonal Vertebrates

Questions concerning genetic identity or relatedness are being asked more and more in the different biological disciplines. In unisexual fish like *Poecilia formosa* (Telostei, Poeciliidae) a single genome is transmitted from mothers to daughters through several generations without genetic recombination (Monaco et al., 1984). For the apomictic breeding of the laboratory lines a single male of an ornamental black molly strain is used which gives the physiological stimulus for gynogenesis, but which does apparently not contribute to the gene pool of *P. formosa*. One of us (MScha) has been keeping two laboratory strains that had been separated for at least seven years. When the fingerprint patterns were compared between several individuals from both strains, a band-sharing rate of 80% or more was observed between the strains (depending on the restriction enzyme/probe combination). Individuals from sibships and their respective mothers show variation limited mostly to very few, truly hypervariable loci (Schartl et al., 1990; Schartl et al., submitted). These results stand in clear contrast to those of a recent study of wild populations by Turner et al. (1990), where the fingerprints of specimens caught at identical and at different locations were described to be comparatively variable.

Heterochromatin and Sex Chromosome Evolution as Studied by Fingerprinting and *in situ* Hybridization Using Non-isotopically Labeled Oligonucleotide Probes

Animal (and plant) sex chromosome evolution probably started on different occasions with a homologous pair of autosomes leading to morphologically differentiated sex chromosomes. In fishes (and reptiles) a range from poorly established to thoroughly heteromorphic gonosomes is sometimes found in closely related species. The guppy fish *Poecilia reticulata* is characterized by the XX/XY sex chromosome mechanism. Yet the Y chromosome can only be identified cytologically on the basis of (i) its heterochromatin staining, (ii) via its Y-chromosomally accumulted (gaca)$_n$ sequences by hybridization *in situ* or (iii) the male-specific high molecular weight smear after electrophoretic separation and hybridization with this same probe (Nanda *et al.*, 1990). All these recent investigations were considerably ameliorated by the fact that non-radioactively labeled oligonucleotide (digoxigenin or biotin) probes could be developed for the *in situ* hybridizations to metaphase chromosomes (Zischler *et al.*, 1989). In addition, the probe (GATA)$_4$ reveals a Y chromosomal polymorphism in the commercially available guppy strains. In contrast, in the closely related genus *Xiphophorus* none of the genetically defined heterogametic situations could be verified consistently with simple repeat probes. Only particular *Xiphophorus* populations produced consistent sex-specific hybridization patterns with (GATA)$_4$. Other poeciliid species (*P. sphenops* var. melanistica and *P. velifera*) harbour different sex-specifically organized repeat motifs in females on the W chromosome (Nanda *et al.*, in press). Thus the accumulation of various diverse simple repeats often accompanies the initial steps of sex chromosome differentiation, leading finally either to male or female heterogamety and the XX/XY or ZZ/ZW sex chromosome mechanism.

Development of Single Copy Probes for Individual Hypervariable Loci from a (CAC)$_5$/(GTG)$_5$ Human Fingerprint

Multilocus fingerprints in humans are composed of many hypervariable bands originating from different loci spread more or less evenly over the genome. Initially, evidence for dispersion originated mainly from *in situ* hybridization data (Zischler et al., 1989; Nanda *et al.*, in press) and from segregation analyses in a limited number of other species (Epplen *et al.*, unpublished). In very few species like, for example, *Ephestia kuehniella* only two or three chromosomal loci harbour long stretches of particular simple repeat motifs (Traut *et al.*, submitted). Whether the allelic signals from one and the same locus in humans appear in the resolvable part of the complex fingerprint pattern or not depends on the multilocus probe used and length variations of the target fragments. As there is much overlap in the size range of alleles at different loci, multilocus profiles do not reveal an individual's genotype at any locus and so maintain genetic confidentiality. The main advantage of locus-specific probes is clearly the defined formal genetic model (and therefore the associated accepted statistical treatment). For other reasons we have tried several approaches to generate such locus-specific markers from the human $(CAC)_5/(GTG)_5$ fingerprint: (i) screening of a complete genomic phage library after partial Sau3AI digestion; (ii) screening of a (partial) phage library containing 3–10 kb fragments after complete MboI digestion; (iii) semi-specific ligation-mediated polymerase chain reaction (PCR), which involves the ligation of adaptors to isolated 3–10 kb fragments followed by PCR using primers specific for the adaptor and the simple $(cac/gtg)_n$ repeat (Zischler *et al.*, 1991; Zischler *et al.*, in preparation). Thereafter the PCR products are conveniently cloned into plasmid vectors.

A major disadvantage of the PCR approach is the comparatively time-consuming evaluation of the informativeness of the isolated probes, which has to be done by hybridization to restriction enzyme digested human DNA from different individuals. In addition, in (i) and (ii) the phages containing simple repeats often showed cloning artifacts in the prokaryotic hosts due to deletions and other complex rearrangements. Approach (iii) also involved prescreening of the clones obtained for the presence of other repetitive elements by separate dot blot hybridization with, for example, human Alu repeats and total human DNA. One of us (HZ) came up with at least four hypervariable locus probes spread over the human genome (heterozygosity rate >90%), several informative probes (heterozygosity >50%) and many clones (25–30) that are less informative or contain additional repeats that confound their exact evaluation even after competition hybridizations, specific oligonucleotide syntheses and hybridizations and other procedures. The apparent evolutionary conservation of individual human hypervariable loci ranges from those found to hybridize to taxa from fish (and chicken) to primates, even under moderately stringent conditions, to no detectable conservation at all. The potential expression of the informative probes is presently being investigated at the mRNA level.

Exonic Polymorphism Versus Intronic Hypervariability: A Mixed Simple Repeat Locus in *HLA-DRB* and Related Genes

Recently, several hypervariable, highly informative single copy loci have been described (see previous section; Nakamura *et al.*, 1987; Jeffreys *et al.*, 1989; Vergnaud, 1989; Armour *et al.*, 1990; Vergnaud *et al.*, 1991; Zischler *et al.*, 1991). In part these have already proven to be useful for many purposes. Yet so far the direct comparison of highly polymorphic protein encoding genes and adjacent simple repeat sequences has been almost completely lacking. The *HLA*-locus (Klein, 1986) in man offers such a possibility: in intron II of the *HLA*-DRB genes, simple repetitive $(gt)_n/(ga)_m$ sequences border exon II (Riess *et al.*, 1990). We therefore started to apply the polymerase chain reaction method to amplify, subclone and determine the sequence of exon II of several *HLA-DRB* loci including in each case the 5'-part of the adjacent intron. In Fig. 2 the experimental strategy is shown; by making use of the sequence information of the *HLA-DRB* sequences several oligonucleotides were constructed to prime specifically either whole groups of related exonic sequences (e.g. *DRw52*) or exclusively the genes coding for one particular serological haplotype (for experimental details see Riess *et al.*, 1990). Two intronic primers anneal 3' to the simple repeat. Being situated essentially at the same target sequence, they vary only in their lengths, and one of them included a convenient restriction enzyme cutting site not present in the original intronic sequence.

From initial analysis on sequencing gels, it was found that a cluster of three to four bands was always amplified from one template DNA due

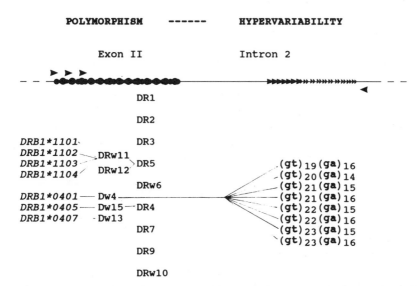

Figure 2. Protein polymorphism versus DNA hypervariability. First a DNA fragment is amplified that contains exon II and the simple repeat locus in the adjacent intron. Amplification primers are symbolized as arrowheads. After cloning, sequence analysis yields from one end protein encoding sequence and from the other the composition of the simple repeat. The serological HLA-DR types can be split into subtypes (DR4-Dw4, Dw15, Dw13 etc.). One exonic sequence (e.g. DRB1∗0401) can be accompanied by many different simple repeats.

to slipping of the Taq polymerase in the simple repeat region. It has been shown previously by electroblot hybridization and sequence determination that these amplified exon–intron region clusters are inherited according to Mendelian laws (Riess et al., 1990). For example, among ten randomly sampled, unrelated individuals typed serologically as *DR4* and subtype *Dw4* (all *HLA-DRB1*0401* exonic sequences), the basic simple repeat structure is completely maintained. Variatons occur exclusively in the numbers of the dinucleotide unit copy numbers: In the $(gt)_n$ stretch, n ranges from 19–23, whereas m in $(ga)_m$ ranges from 14–16. In the *DRw52* group, a more complex basic repeat structure is found: $(gt)_n(ga)_m ca(ga)_3[ca(ga)_3]$. Here, there is abundant variation in the $(gt)_n$ sequence, but significantly less in the first $(ga)_m$ stretch. Table 3 summarizes the group-specific architecture of the simple repeat blocks including data from a baboon (*Papio hamadryas, Paha*) and cattle (*Bos taurus, Bota*). Throughout this table only one intermediately long version is given from several (between 1 and 12) investigated for each *HLA-DRB* gene sequence: invariably the stretch commences with $(gt)_n$, whereby n ranges from 6 in a *DR1* pseudogene intron to 26 in a *DRB1*1401* individual. *This simple dinucleotide organization of the 5′ stretch is interrupted only in two categories*: *DRw53* pseudogenes and *DRB4* introns. On the other hand the following $(ga)_m$ stretches of dinucleotides exhibit numerous deviations, depending on the *HLA-DR* gene investigated. Simple $(ga)_m$ repeats appear in *DRB1*0401*, in the pseudogene associated repeats and in the only baboon sequence. Group-specific simple repeat organization could also be demonstrated across species barriers, for example in a parallel study on *Bota-DR* intronic simple repeats from cattle (H. Ammer, W. Schwaiger, C. Kammerbauer, and J. T. Epplen, submitted). We were particularly surprised by the repeat composition in the *Bota-DR* sequence: the exon exhibits 95% homology to the protein domains of the *DRw52* supergroup and, also, the intronic repeat is virtually indistinguishable from its human counterparts (H. Ammer, W. Schwaiger, C. Kammerbauer, and J. T. Epplen, unpubl. data; genomic *Bota-DR* DNA clone obtained from Groenen et al., 1990; there were several differences in the sequences obtained from the same isolate in independent analyses in two laboratories). Attempts to perform a phylogenetic analysis on the simple repeats for comparison with the results obtained from the *HLA* protein sequences are hampered by the inability of the available alignment programs to perform (meaningful) simple sequence alignments. If the alignments are done "by eye", i.e. introducing gaps to separate the pure $(gt)_n$ and $(ga)_m$ repeats, and thereafter one applies the Felsenstein programs (1989), then more closely related groups are found to cluster together (Epplen, unpublished). Attempts to establish respective algorithms for "meaningful" alignments of simple sequences are presently being evaluated (Zucker et al., in preparation).

Table 3. Intronic simple $(gt)_n(ga)_m$ repeats in DRB and related genes

"DRB" allele/ No. individ./ RFLP	Total repeat length	Repeat composition
HLA-DRB1*0101/2	44 $(gt)_{15}$	aa $(ga)_1$ aa$(ga)_4$
HLA-DRB1*0301/7/3b	64 $(gt)_{15}$	$(ga)_7$ca$(ga)_3$ca$(ga)_5$
HLA-DRB1*0301/1/3a	66 $(gt)_{17}$	$(ga)_6$ca$(ga)_3$ca$(ga)_5$
HLA-DRB1*0302/1/3c	68 $(gt)_{19}$	$(ga)_6$ca$(ga)_3$ca$(ga)_4$
HLA-DRB1*0401/12	74 $(gt)_{22}$	$(ga)_{15}$
HLA-DRB1*0404/1	90 $(gt)_{22}$	$(ga)_{23}$
HLA-DRB1*0405/4	70 $(gt)_{19}$	$(ga)_{16}$
HLA-DRB1*0407/1	76 $(gt)_{21}$	$(ga)_{17}$
HLA-DRB1*0801/4	62 $(gt)_{18}$	$(ga)_7$ca$(ga)_5$
HLA-DRB1*1001/1	48 $(gt)_{16}$	$(ga)_8$
HLA-DRB1*1101/2/5c	74 $(gt)_{20}$	$(ga)_6$ca$(ga)_3$ct$(ga)_6$
HLA-DRB1*1102/1/5c	72 $(gt)_{22}$	$(ga)_3$ca$(ga)_3$ca$(ga)_6$
HLA-DRB1*1103/3/5c	88 $(gt)_{22}$	$(ga)_{11}$ca$(ga)_3$ca$(ga)_6$
HLA-DRB1*1104/1/5c	76 $(gt)_{22}$	$(ga)_5$ca$(ga)_3$ca$(ga)_6$
HLA-DRB1*1104/1/5c	80 $(gt)_{25}$	$(ga)_5$ca$(ga)_3$ca$(ga)_5$
HLA-DRB1*1301/1/6b	76 $(gt)_{17}$	$(ga)_{10}$ca$(ga)_3$cagc$(ga)_5$
HLA-DRB1*1301/3/6a	88 $(gt)_{23}$	$(ga)_{10}$ca$(ga)_3$ca$(ga)_6$
HLA-DRB1*1301/1	92 $(gt)_{25}$	$(ga)_{10}$ca$(ga)_3$ca$(ga)_6$
HLA-DRB1*1401/1	88 $(gt)_{26}$	$(ga)_{11}$ca$(ga)_6$
HLA-DRB1*1501/7	82 $(gt)_{19}$	$(ga)_5$ca$(ga)_4$ca$(ga)_3$ggaa$(ga)_6$
HLA-DRB4*0101/2	70 $(gt)_2$at$(gt)_{11}$	$(ga)_{12}$ca$(ga)_1$ggaa$(ga)_5$
HLA-DRB5*0101/9	74 $(gt)_{21}$	$(ga)_5$ggaa$(ga)_4$ca$(ga)_2$gg$(ga)_1$
HLA-DR1ψ/3	32 $(gt)_6$	$(ga)_{10}$
HLA-DR2ψ/3	48 $(gt)_{13}$	$(ga)_{11}$
HLA-DR4ψ/2	37 $(gt)_2$tt$(gt)_3$	t$(ga)_5$
HLA-DR7ψ/1	35 $(gt)_2$tt$(gt)_3$	t$(ga)_5$
Paha/1	42 $(gt)_{14}$	$(ga)_7$
Bota*/1	86 $(gt)_{20}$	$(ga)_9$ca$(ga)_2$ca$(ga)_2$[ca$(ga)_2]_2$

*Ammer, corrected sequence; but see Groenen et al. (1990)

Finally, apart from these purely evolutionary aspects, our studies also aim at a better understanding of the significance of the invariantly present $(gt)_n(ga)_m$ simple repeat locus, since the latter has been identified in such diverse vertebrate species. Is this ubiquitous presence just a mere coincidence? Does the repeat represent a DNase hypersensitive site? Do particular proteins bind to this DNA region? Are there secondary structures like, for example, Z DNA (Rich et al., 1984) or H DNA (Mirkin et al., 1987) possible in this part of the intron?

In an initial attempt to make use of the information provided by this simple repeat locus, rheumatoid arthritis (RA) patients were also included in the panel of the aforementioned *DRB1*0401* cohort. RA has been found previously to be associated with *HLA-DR4*, especially, for example, with the *DR4Dw4* subtype. In the limited number of probands tested, the structure and the lengths of the simple repeats of RA patients and healthy control individuals did not differ. In principle the same result was found for multiple sclerosis (MS) patients, a disease which is associated with the serologically determined *HLA-DR2* haplotype. The comparison of the *DR2* gene adjacent simple repeats from healthy donors and MS patients revealed neither particular sequence lengths nor configurations (Epplen *et al.*, unpublished).

Future Prospects

It is now well established that loci containing tandem simple repeat sequences are a practically unrestricted source of highly informative markers in eukaryote genomes. Therefore they appear to be *the* tools for a diverse range of applications. In addition to their already recognized utility for individualization, the simple repeat sequences will have an especially major impact in large scale mapping projects, for example in interrelating and ordering cosmids and yeast artificial chromosomes from genomic libraries. One particularly interesting development concerns a veritable hotspot of recombination in the chicken genome, which has initially been identified using the $(GAA)_6$ simple repeat probe: more than 80% of the length alleles in true offspring cannot be matched to either the paternal or to the maternal genotype (Nanda *et al.*, submitted). Meanwhile, from this simple repeat locus an adjacent single copy probe has been isolated by semispecific PCR (see above) as well as genomic phage clones, with the aid of which a detailed, targeted analysis will be performed. But apart from these problem-centred considerations there are also solid grounds to expect, in due course, major breakthroughs in the quest for the actual significance of these enigmatic simple tandem elements. It has been proposed that simple repetitive sequences have weak enhancer-like activity in viral expression systems. Certainly, at least *in vitro*, they are able to occur in special confirmations, i.e. secondary structures, depending on the purine/pyrimidine composition. Yet whether the outcome will be that they represent merely junk and jumble from the evolutionary past or that they bear present day function(s), only the results of diligent experimentation and insightful interpretation in a broad and open conceptual framework will tell.

Acknowledgements
We thank the many collaborating colleagues who contributed information on the informativeness of simple repeat probes in various species not available to us. This work was supported by the VW-Stiftung and the DFG (Ep 7/5-1 and Ep 7/6-1). The simple repetitive oligonucleotides and locus-specific probes are subject to patent applications. Commercial enquiries should be directed to Fresenius AG, W-6370 Oberursel, Germany.

References

Ali, S., Müller, C. R., and Epplen, J. T. (1986) DNA fingerprinting by oligonucleotides specific for simple repeats. *Hum. Genet.* 74: 239–243.

Armour, J. A. L. (1990) Systematic cloning of human minisatellites from ordered array charomid libraries. *Genomics* 8: 501–512.

Arnemann, J., Schmidtke, J., Epplen, J. T., Kuhn, H.-J., and Kaumanns, W. (1989) DNA fingerprinting for paternity and maternity in group O Cayo Santiago-derived Rhesus monkeys at the German Primate Center: results of a pilot study. *Puerto Rico Health Sci. J.* 8: 181–184.

Bender, K., Kasulke, D., Mayerova, A., Hummel, K., Weidinger, S., Epplen, J. T., and Wienker, T. (1991) New mutation versus exclusion at the PI locus: a multifaceted approach in a problematical paternity case. *Hum. Hered.* 814 41: 1–11.

Beyermann, B., Nürnberg, P., Weihe, A., Meixner, M., Epplen, J. T., and Börner, T. (1991) Fingerprinting plant genomes with oligonucleotide probes specific for simple repetitive DNA sequences. *Theor. Appl. Genet.* (in press).

Britten, R. J., and Kohne, D. A. (1968) Repeated sequences in DNA. *Science* 161: 529–540.

Buitkamp, J., Ammer, H., and Geldermann, H. (1991) DNA fingerprinting in domestic animal species. *Electrophoresis* 12: 169–174.

Buitkamp, J., Kühn, C., Zischler, H., Epplen, J. T., and Geldermann, H. (1991) DNA fingerprinting in cattle using oligonucleotide probes. *Anim. Genet.* (in press).

Epplen, J. T. (1988) On simple repeated GATA/GACA sequences: a critical reappraisal. *J. Hered.* 79: 409–417.

Epplen, J. T., Kammerbauer, C., Steimle, V., Zischler, H., Albert, E., Andreas, A., Hala, K., Nanda, I., Schmid, M., Riess, O., and Weising, K. (1989) Methodology and application of oligonucleotide fingerprinting including characterization of individual hypervariable loci. In: Radola, B. J. (ed.) Electrophoresis Forum '89. Bode-Verlag, München, pp. 175–186.

Epplen, J. T., McCarrey, J. R., Sutou, S., and Ohno, S. (1982) Base sequence of a cloned snake W-chromosome DNA fragment and identification of a male-specific putative mRNA in the mouse. *Proc. Natl. Acad. Sci. USA* 79: 3798–3802.

Epplen, J. T., Studer, R., and McLaren, A. (1988) Heterogeneity in the *Sxr (sex-reversal)* locus of the mouse as revealed by synthetic GAT/CA probes. *Genet. Res.* 51: 239–246.

Evett, I. W., Werrett, D. J., and Buckleton, J. S. (1989) Paternity calculations from DNA multilocus profiles. *J. Forensic Sci. Soc.* 29: 249–254.

Felsenstein, J. (1988) Phylogenies from molecular sequences: inference and reliability. *Ann. Rev. Genet.* 22: 521–565.

Fimmers, R., Epplen, J. T., Schneider, P. M., and Baur, M. P. (1990) Likelihood calculations in paternity testing on the basis of DNA-fingerprints. *Adv. Forensic Haemogenet.* 3: 14–16.

Groenen, M. A. M., van der Poel, J. J., R. J. M. Dijkhof, and Giphart, M. J. (1990) The nucleotide sequence of bovine MHC class II *DQB* and *DRB* genes. *Immunogenet.* 31: 37–44.

Hastie, N. D., Demster, M., Dunlop, M. G., Thompson, A. M., Green, D. K., Allshire, R. C. (1990) Telomere reduction in human colorectal carcinoma and with ageing. *Nature* 346: 866–868.

Jeffreys, A. F., Wilson, V., and Thein, S. L. (1985a) Hypervariable "minisatellite" regions in human DNA. *Nature* 316: 67–73.

Jeffreys, A. F., Wilson, V., and Thein, S. L. (1985b) Individual-specific "fingerprints" of human DNA. *Nature* 316: 76–79.

Jeffreys, A. F., Royle, N. J., Wilson, V., and Wong, Z. (1988) Spontaneous mutation rates to new length alleles at tandem repetitive hypervariable loci in human DNA. *Nature* 332: 278–281.

68

Klein, J. (1986) Natural History of the Major Histocompatibility Complex. Wiley, New York.

Lagoda, P. J. L., Seitz, G., Epplen, J. T., and Issinger, O.-G. (1989) Increased detectability of somatic changes in the DNA after probing with "synthetic" and "genome-derived" hypervariable multilocus probes. *Hum. Genet.* 84: 35–40.

McLaren, A., Simpson, E., Epplen, J. T., Studer, R., Koopmann, P., Evans, E. P., and Burgoyne, P. S. (1988) Location of the genes controlling H-Y antigen expression and testis determination on the mouse Y chromosome. *Proc. Natl. Acad. Sci. USA* 85: 6442–6445.

Miyada, C. G., Reyes, A. A., Studencki, A. B., and Wallace, R. B. (1985) Methods of oligonucleotide hybridization. *Proc. Natl. Acad. Sci. USA* 82: 2890–2894.

Monaco, P. J., Rasch, E. M., and Balsano, J. S. (1984) Apomictic reproduction in the Amazon molly, *Poecilia formosa*, and its triploid hybrids. In: Turner, B. J. (ed.) Evolutionary Genetics of Fishes. Plenum Press, New York, pp. 311–328.

Nakamura, Y., Leppert, M., O'Connell, P., Wolff, R., Holm, T., Culver, M., Martin, C., Fujimoto, E., Hoff, M., Kumlin, M., and White, R. (1987) Variable number of tandem repeat (VNTR) markers for human gene mapping. *Science* 235: 1616–1622.

Mirkin, S. M., Lyamichev, V. I., Drushlyak, K. N., Dobrynin, V. N., Filippov, S. A., and Frank-Kamenetskii, M. D. (1987) DNA H form requires a homopurine-homopyrimidine mirror repeat. *Nature* 330: 495–497.

Nanda, I., Neitzel, H., Sperling, K., Studer, R., and Epplen, J. T. (1988) Simple GAT/CA repeats characterize the X chromosomal heterochromatin of *Microtus agrestis*, European field vole (Rodentia, Cricetidae). *Chromosoma* 96: 213–219.

Nanda, I., Deubelbeiss, C., Guttenbach, M., Epplen, J. T., and Schmid, M. (1990b) Heterogeneities in the distribution of $(GACA)_n$ simple repeats in the karyotypes of primates and mouse. *Hum. Genet.* 85: 187–194.

Nanda, I., Feichtinger, W., Schmid, M., Schröder, J. H., Zischler, H., and Epplen, J. T. (1990a) Simple repetitive sequences are associated with differentiation of the sex chromosomes in the guppy fish. *J. Mol. Evol.* 30: 456–462.

Nanda, I., Schartl, M., Feichtinger, W., Epplen, J. T., and Schmid, M. (1990) Early stages of sex chromosome differentiation in fish as analyzed by simple repetitive DNA sequences. *Chromosoma* (in press).

Nanda, I., Schmid, M., and Epplen, J. T. (1991) *In situ* hybridization of nonradioactive oligonucleotide probes to chromosomes. In: Adolph, K. W. (ed.) Advanced Techniques in Chromosome Research. Marcel Decker, New York pp. 117–134.

Nürnberg, P., and Epplen, J. T. (1989) "Hidden Partials" – a cautionary note. *Fingerprint News* 1(4) 11–12.

Nürnberg, P. Roewer, L., Neitzel, H., Sperling, K., Pöpperl, A., Hundrieser, J., Pöche, H., Epplen, C., Zischler, H., and Epplen, J. T. (1989) DNA fingerprinting with the oligonucleotide probe $(CAC)_5/(GTG)_5$: somatic stability and germline mutatations. *Hum. Genet.* 84: 75–78.

Nürnberg, P., Zischler, H., Fuhrmann, E., Thiel, G., Losanova, T., Kinzel, D., Nisch, G., Witkowski, R., and Epplen, J. T. (1991) Co-amplification of simple repetitive DNA fingerprint fragments and the EGF receptor gene in human gliomas. *Genes Chromosomes Cancer* 3 (in press).

Pöche, H., Peters, C., Wrobel, G., Schneider, V., and Epplen, J. T. (1991) Determining consanguinity by oligonucleotide fingerprinting with $(GTG)_5/(CAC)_5$. *Electrophoresis* (in press).

Pöche, H., Wrobel, G., Schneider, V., and Epplen, J. T. (1991) The identification of a charred body by oligonucleotide fingerprinting with the $(GTG)_5$ probe. *DNA Technol. Legal Med.*

Pöche, H., Wrobel, G., Schneider, V., and Epplen, J. T. (1990) DNA fingerprinting with simple repetitive oligonucleotide probes in forensic medicine. *Adv. Forensic Haemogenet.* 3: 122–124.

Pöche, H., Wrobel, G., Schneider, V., and Epplen, J. T. (1990) Oligonucleotid-Fingerprinting mit $(GTG)_5$ und $(GACA)_4$ für die Zuordnung von Leichenteilen. *Archiv Kriminologie* 186: 37–42.

Rich, A., Nordheim, A., and Wang, A. H. Z. (1984) The chemistry and biology of left handed Z DNA. *Ann. Rev. Biochem.* 53: 791–846.

Riess, O., Kammerbauer, C., Roewer, L., Steimle, V., Andreas, A., Albert, E., Nagai, T., and Epplen, J. T. (1990) Hypervariability of intronic simple $(gt)_n(ga)_m$ repeats in *HLA-DRB1* genes. *Immunogenet.* 32: 110–116.

Roewer, L., Nürnberg, P., Fuhrmann, E., Rose, M., Prokop, O., and Epplen, J. T. (1990) Stain analysis using oligonucleotide probes specific for simple repetitive DNA sequences. *Forensic Sci. Internatl.* 47: 59–70.

Roewer, L., Rose, M., Semm, K., Correns, A., Epplen, J. T. (1989) Typisierung gelagerter, hämolysierter Blutproben durch "DNA-Fingerprinting". *Archiv Kriminologie* 184: 103–107.

Schäfer, R., Zischler, H., Birsner, U., Becker, A., and Epplen, J. T. (1988) Optimized oligonucleotide probes for DNA fingerprinting. *Electrophoresis* 9: 369–374.

Schartl, M., Nanda, I., Schlupp, I., Parzefall, J., Schmid, M., and Epplen, J. T. (1990) Genetic variation in the clonal vertebrate *Poecilia formosa* is limited to truly hypervariable loci. *Fingerprint News* 2(4): 22–24.

Speth, C., Epplen, F. T., and Oberbäumer (1991) DNA fingerprinting with oligonucleotides can differentiate cell lines derived from the same tumor. *In Vitro* (in press).

Sprecher, W., Berg, S., and Epplen, J. T. (1990) Identifikation von Blutproben und foetalem Gewebe durch genetisches Fingerprinting. *Archiv Kriminologie* 185: 44–51.

Sprecher, W., Kampmann, H., Epplen, J. T., and Gross, W. (1991) Idenifizierung einer Brandleiche mit Hilfe des DNA-Fingerprinting. *Rechtsmedizin* (in press).

Tautz, D. (1989) Hypervariability of simple sequences as a general source for polymorphic DNA markers. *Nucleic Acids Res.* 17: 6463–6471.

Turner, B. J., Elder, J. F., Laughlin, T. F., and Davis, W. P. (1990) Genetic variation in clonal vertebrates detected by simple-sequence DNA fingerprinting. *Proc. Natl. Acad. Sci. USA* 87: 5653–5657.

Vergnaud, G. (1989) Polymers of random short oligonucleotides detect polymorphic loci in the human genome. *Nucleic Acids Res.* 17: 7623–7630.

Vergnaud, G., Mariat, D., Zoroastro, M., and Lauthier, V. (1991) Synthetic tandem repeats of short oligonucleotides can detect single and multiple polymorphic loci. *Electrophoresis* 12: 134–140.

Vogel, W., Steinbach, P., Djalali, M., Mehnert, K., Ali, S., and Epplen, J. T. (1988) Chromosome 9 of *Ellobius lutescens* is the X chromosome. *Chromosoma* 96: 112–118.

Weising, K., Fiala, B., Ramloch, K., Kahl, G., and Epplen, J. T. (1990) Oligonucleotide fingerprinting in angiosperms. *Fingerprint News* 2: 5–8.

Weising, K., Weigand, F., Driesel, A. J., Kahl, G., Zischler, H., and Epplen, J. T. (1989) Polymorphic simple GATA/GACA repeats in plant genomes. *Nucleic Acids Res.* 17: 10128 [1 page].

Weising, K., Beyermann, B., Ramser, J., and Kahl, G. (1991) Plant DNA fingerprinting with radioactive and digoxigenated oligonucleotide probes complementary to simple repetitive DNA sequences. *Electrophoresis* 12: 159–168.

Weising, K., Ramser, J., Kaemmer, Kahl, G., and Epplen, J. T. (1991) Oligonucleotide fingerprinting in plants and fungi. In: Burke, T., Dolf, G., Jeffreys, A. J., and Wolff, R. (eds) DNA fingerprinting: Approaches and Applications. Birkhäuser, Basel. pp. 312–319. (This volume).

Yassouridis, A., and Epplen, J. T. (1991) On paternity determination from multilocus DNA profiles. *Electrophoresis* 12: 221–225.

Zischler, H., Hinkkanen, A., and Studer, R. (1991) Oligonucleotide fingerprinting with $(CAC)_5$: Non-radioactive in-gel hybridization and isolation of hypervariable loci. *Electrophoresis* 12: 141–145.

Zischler, H., Nanda, I., Schäfer, R., Schmid, M., and Epplen, J. T. (1989) Digoxigenated oligonucleotide probes specific for simple repeats in DNA fingerprinting and hybridization *in situ*. *Hum. Genet.* 82: 227–233.

DNA Fingerprinting: Approaches and Applications
ed. by T. Burke, G. Dolf, A. J. Jeffreys & R. Wolff
© 1991 Birkhäuser Verlag Basel/Switzerland

DNA Fingerprinting of the Human Intestinal Parasite *Giardia intestinalis* with Hypervariable Minisatellite Sequences

P. Upcroft

Queensland Institute of Medical Research, Branston Terrace, Herston, Brisbane, Queensland, Australia 4006

Summary

Individual isolates of the *Giardia duodenalis* group of protozoan intestinal parasites were identified by DNA fingerprinting with hypervariable minisatellite sequences. A morphologically identical parasite is found in some forty different animal species. Although the species name *intestinalis* is reserved for the human isolates, electrophoretic karyotyping suggests that most *duodenalis* isolates fall into the same species grouping. Distinction based upon morphology, restriction endonuclease cleavage of genomic DNA or isoenzyme analysis has not been adequate to identify individual strains. The successful use of hypervariable sequences in the identification of individual human genomes encouraged us to examine the use of these same sequences for the possible identification of parasite isolates. We initially used as a fingerprinting probe the genome of the bacteriophage M13, which has repeated sequences recognising homologous hypervariable sequences in the human genome. The M13 probe recognises a weakly homologous set of hypervariable sequences in *Giardia*. The number of informative bands is comparable to those seen in mammals, since the lower molecular weight bands are also useful. There is considerable divergence in the sequences of individual *Giardia* minisatellites. Some cloned *Giardia* hypervariable sequences are more homologous to M13 than they are to each other. Similar results were observed with the hypervariable repeat sequences 3′ to the human α-globin gene when they were used as a probe to distinguish *Giardia* isolates. The poly(dA-dC).poly(dG-dT) probe which recognises frequent TG tracts in a number of organisms also detects a few variable bands amidst a hybridisation background in the *Giardia* genome. Thus *Giardia* isolates which could not be distinguished by restriction endonuclease cleavage, antibody typing or isoenzyme analysis have been identified by DNA fingerprinting procedures. Detailed analysis of strain movement, resurgence, variation, host range and drug resistance is now possible. Similar families of sequences may be widespread in lower eukaryotes and useful for generating individual specific fingerprints. A procedure for detecting individual parasites is also presented. Since *Giardia* is regarded as the most ancient eukaryote before the occurrence of symbiosis with purple non-sulphur bacteria to generate mitochondria, the identification of hypervariable sequences in the *Giardia* genome should also aid in understanding the mechanism of generation and evolution of these sequences.

Introduction

Giardia intestinalis is the species name given to members of the *Giardia duodenalis* group of flagellated protozoan intestinal parasites which infect man. The parasite can cause severe diarrhoea and malabsorption. Some 280 million people are infected worldwide at any one time (*Bull. WHO*, 1987). All members of the *duodenalis* group are morphologically indistinguishable, although the literature cites over forty different species which were named after the animal host in which they were found (Boreham *et al.*, 1990; Kulda and Nohynkova, 1978). Very few biological and biochemical characteristics have been found to distinguish different isolates or possible variation between them (Capon *et al.*, 1989; Upcroft *et al.*, 1989a and c; Upcroft *et al.*, 1990). No significant or consistent correlation with isoenzyme patterns, surface proteins or immunoreactive reagents with host range, pathogenicity, geographic location, or any other distinguishing characteristic, has been established (for review Upcroft *et al.*, 1989c). This does not imply that such specialisation does not exist, but that these techniques are inadequate to distinguish the possible differences. Broad and overlapping grouping of *Giardia* isolates has been established using isoenzyme analysis; however, these groups do not correspond to any other biological characteristic (Andrews *et al.*, 1989; Cedillo-Rivera *et al.*, 1990).

Isoenzyme typing of parasites

One of the commonly used methods for attempting to identify protozoan parasites is isoenzyme analysis. Changes in electrophoretic mobility of housekeeping enzymes determine different isotypes and the assay of multiple enzymes generates a pattern which is presumed to identify that isolate. This follows a long history of usage in the higher eukaryotes, where the technique was used to disclose a greater extent of polymorphism than was believed to exist (see Ruffié, 1987, for a recent review). The method has been of considerable value but suffers from limitations, particularly in the scope of application and in a number of assumptions regarding its validity, which are often overlooked.

A major issue confronting the use of isoenzyme analysis in categorising isolates is that the assay is for the expression of housekeeping enzymes (which usually have only one or two variants), rather than the genetic constitution of the organism. In most cases where isoenzyme assays have been used, and particularly in parasitology, the copy number of these genes is unknown. Moreover the regulation of these genes and switching capacity between possible alleles is also unknown, i.e., limited phenotypic data were equated with an unknown genotype. One of the most controversial issues in parasitology, that of the existence of distinct pathogenic strains of *Entamoeba histolytica*, is based upon isoenzyme typing and highlights the problem. Thus laboratory isolates can be induced to switch between pathogenic and non-pathogenic isoenzyme types by enviromental factors such as associated bacterial flora (Andrews *et al.*, 1990; Mirelman, 1987; Mirelman *et al.*, 1986a and b; Sargeaunt, 1987). Isoenzyme typing has been invaluable for primary detection of invasive parasites in the field, but the implications of possible plasticity of gene expression and genetic content have not been addressed.

A second example is in the use of isoenzyme analysis to infer that there is no sexual exchange in the protozoan parasites (Tibayrenc *et al.*, 1990): gene expression was regarded as synonymous with genetic content. Previously published data from long term cultures of *Entamoeba, Giardia, Leishmania, Naegleria, Plasmodium, Trichomonas* and *Trypanosoma* were used to examine their population structures. Each of these parasitic protozoa was then regarded as clonal in origin because of its isoenzyme type, even though its genotype was not examined; in one case cited, the original data were used to argue that the populations were in fact a mixture of genotypes (Fenton *et al.*, 1985). The data used for *Giardia* assumed that the organism is diploid, although *Giardia* has two equally sized nuclei and other workers (Adam *et al.*, 1988) have argued from pulsed field gel electrophoretic karyotyping that *Giardia* is polyploid. Our own karyotyping data suggest that it is more likely that *Giardia* has two diploid nuclei, with extensive transposition and gene conversion capacity (Upcroft *et al.*, 1988; Upcroft *et al.*, 1989a and c; Upcroft *et al.*, 1990a; Upcroft, P., unpublished data). Most of the samples had been in culture for so long that it is not surprising that a single type may have predominated over the original population. Furthermore, the efficiency of establishment in culture of most parasites from an isolate is so low that the outgrowing population would be close to clonal even though the original isolate may have contained a mixture of many different genotypes. The sample numbers are so low that interpretation of the original population heterogeneity, or lack thereof, is at best dubious. If each isolate is different, clonal, stable and no sexual exchange takes place, how did each of the isolates arise; through genetic variation? At the other extreme, isoenzyme analysis has been used to support the existence of sexual exchange in *Entamoeba histolytica* (Sergeaunt *et al.*, 1988); gene copy number, switching capacity or allele frequencies were not examined. The critical information needed in each of these examples was the genotype(s), i.e., a measure of the genetic content for relevant markers, of the source material.

Electrophoretic karyotyping

Characterisation of the karyotype has been the classical method for defining species. Each species has a distinct, usually stable set of chromosomes which can be identified by size and number. In rare cases, such as bats, whole genera and families may have the same gross chromosome topography (Baker and Bickham, 1980; Ruffié, 1987). For the protozoan parasites it has been very difficult to establish specific karyotypes because the chromosomes do not condense in the manner

73

that allows chromosome spreading in the higher eukaryotes. The introduction of pulsed field gel electrophoretic systems (Schwartz and Cantor, 1984; Vollrath et al., 1988) has now enabled some measure of chromosome size and number to be determined in the few protozoan parasites that have been studied (Bishop and Akinsehinwa, 1990; Conover and Brunk, 1986; Giannini et al., 1990; Gibson and Borst, 1986; Kemp et al., 1987; Spithill and Samaras, 1987; Upcroft et al., 1989a, b and c; Van der Ploeg et al., 1985). In some protozoan parasites, e.g., the trypanosomes, there is a range of different electrophoretic karyotypes corresponding to the activation of variant surface glycoproteins (VSGs) by diverse mechanisms which enable the parasite to evade the host's immune defence mechanisms (Borst and Greaves, 1987; Gibson and Borst, 1986). These characteristics may prove useful in distinguishing individual isolates, but the VSGs switch at a frequency which is not related to antibody pressure (Lamont et al., 1986) and hence change with time. Similar diversity has been described in the electrophoretic karyotypes of the *Plasmodia* (Kemp et al., 1987; Van der Ploeg et al, 1985) and *Leishmania* (Giannini et al., 1986; Spithill and Samaras, 1987) although the mechanisms and frequencies of change are not as well characterised. Characterisation of some stocks of *Leishmania donovani* has been achieved with genomic DNA heterogeneity and electrophoretic karyotype (Bishop and Akinsehinwa, 1990). We have observed only two major electrophoretic karyotypes in almost 100 worldwide isolates of *Giardia* from man and animals with which he may associate, e.g., sheep, cat, beaver, dog, muskrat (P. Upcroft, unpublished data; Upcroft et al., 1989a, c), although some differences have been observed in unusual animals (DeJonckheere et al., 1990) and abnormally growing isolates (Upcroft et al., 1989a and unpublished data). Recent data suggest that there may be a stable base or subset of chromosomes for *Leishmania* also (Giannini et al., 1990). Thus the classification of protozoans by karyotype may be possible in some cases, e.g., *Giardia* and *Leishmania*, but the complex mechanisms that others have evolved to evade the host immune defences, or maintain their enviromental niche, may preclude such a traditionally simple approach in protozoans such as *Blastocytsis, Plasmodium, Trypanosoma* and *Tetrahymena*, which appear to have an enormous variety of karyotypes (Bishop and Akinsehinwa, 1990; Borst and Greaves, 1987; Conover and Brunk, 1986; Giannini et al., 1990; Kemp et al., 1987; Upcroft et al., 1989a, b and c). The distinction between species, subspecies and isolates on a large scale, i.e., outside the laboratory, offers a considerable challenge.

DNA Fingerprinting

The identification of an unlinked locus in the human genome, which is highly polymorphic in length due to DNA rearrangements (Wyman and

White, 1980), allowed the concept of genetic mapping with restriction fragment length polymorphisms to be formulated (Botstein *et al.*, 1980). A number of such loci were subsequently discovered associated with cloned genes (Higgs *et al.*, 1981; Jeffreys, 1987; Nakamura *et al.*, 1987b). Sequence data from these highly variable regions demonstrated that they were tandemly repeated sequences which varied in number (Bell *et al.*, 1982) (VNTRs; Nakamura *et al.*, 1987b) and were termed hypervariable minisatellites (Jeffreys *et al.* 1985a).

A number of these minisatellites from different loci form a family of related sequences (Goodbourn *et al.*, 1983; Jeffreys *et al.*, 1987). A member of this family of minisatellites from the myoglobin gene was used to select cross-hybridising clones from a recombinant DNA library (Jeffreys *et al.*, 1985). This single-locus polymorphic sequence therefore can detect multiple loci with variable numbers of tandem repeats, and is the basis for the concept of DNA fingerprinting (Jeffreys, 1987; Jeffreys *et al.*, 1985b). Many cloned single-locus hypervariable sequences are not members of this family and have extreme ranges of dG.dC content (Dover, 1989; Nakamura *et al.*, 1987b) and repeat lengths, which in some cases include simple sequence motifs (Hamada and Kakunaga, 1982; Kashi *et al.*, 1990; Schäfer *et al.*, 1988; Tautz *et al.*, 1986). These simple sequence repeats were found at diverse frequencies in a variety of organisms including yeast (Hamada *et al.*, 1982; Tautz and Renz, 1984). The more complex hypervariable sequences have been found embedded in other repeated motifs (Armour *et al.*, 1989; Jeffreys *et al.*, 1990; Nakamura *et al.*, 1987a).

A different source of DNA variability has been found at the species and higher taxonomic levels in the rRNA genes, in particular the variable expansion segments in the non-conserved regions of the 16S rRNA gene. These differences have been sufficient to distinguish the four major *Plasmodium* species that infect man (Waters and Mc-Cutchan, 1989), and to corroborate the thesis that *Giardia* is the most ancient living example of the amitochondrial eukaryotes before they symbiosed with purple non-sulphur bacteria to generate mitochondria (Cavalier-Smith, 1987; Sogin *et al.*, 1989). Our own data on the sequence of the 28S rRNA gene is consistent with the ancient nature of *Giardia* (Healey *et al.*, 1990). Using this phylogenetic approach, and assuming the validity of a reliable molecular clock (see Cavalier-Smith, 1989; Dover, 1987; Hancock and Dover, 1988; Hancock *et al.*, 1988; Holmquist *et al.*, 1988), *Giardia* is closer to the bacteria than to the plants and animals (Cavalier-Smith, 1989).

Even unexpected sources of DNA probes have been found to detect hypervariable sequences in vertebrates, e.g., the M13 genome generated fingerprints in man and cows, due to a repeated motif in gene 3 (Vassart *et al.*, 1987), with the consensus sequence GAGGGTGGXGGXTCT or a truncated version, GGXGGXGGXTCT.

This frequency of occurence and variety of hypervariable sequences suggested a way of addressing the issues of parasite identification by DNA fingerprinting, if equivalent motifs occur in the protozoa and are hypervariable. The appearance of some cross-hybridising motifs in vertebrates, plants, yeast and other organisms (Hamada *et al.*, 1982; Ryskov *et al.*, 1988; Tautz and Renz, 1984) indicated that the basic sequences are probably widespread in biology.

To examine the potential for DNA fingerprinting in *Giardia*, the M13 genome was chosen as a probe primarily because of its availability, apparent simplicity of use and published sequence data (Upcroft *et al.*, 1989c and 1990b). Figures 1 and 2 illustrate the use of ^{32}P-labelled M13 as a fingerprinting probe to distinguish randomly chosen isolates of *Giardia duodenalis*. Different gel and hybridisation conditions and exposure times in each figure demonstrate the range in quality of the derived fingerprint. The fingerprints of *Giardia* BRIS/83/HEPU/106 and BAC1 when cleaved with HinfI in Figs. 1 and 2 have the same profile in each gel. HinfI and RsaI were found to yield the most informative fingerprints (Upcroft *et al.*, 1990) with this probe. The low molecular weight bands are also informative for *Giardia* while they are ignored in the higher eukaryotes (e.g. Jeffreys *et al.*, 1985b). This system is sufficiently sensitive to identify individual isolates which could not be distinguished in any other way (Archibald *et al.*, 1990). We have shown that the M13 fingerprint is stable in cultured parasites for at least two years. The fingerprint of each isolate appears to be unique and of the same complexity as cloned isolates, which is consistent with the general lack of complexity in isoenzyme data. Since every cultured isolate we have examined is different the initial source must be of mixed genotype; if the variety of fingerprints resulted from extensive rearrangement during isolation each fingerprint would be uninterpretable because of its complexity. Each isolate is therefore the result of the final outgrowth of a single type of parasite from the initial population. We are addressing the issue of the extent of mixed populations, whether the outgrowth is a random event or whether there are certain types selected in each isolate. Recently the M13 probe has also been used to detect RFLPs between two strains of the malaria parasite, *Plasmodium falciparum* (Rogstad *et al.*, 1989).

Another probe that we used to fingerprint *Giardia* was the hypervariable region found flanking the 3' terminus of the α-globin gene cluster in man (Higgs *et al.*, 1981; Jarman *et al.*, 1986). We again used random priming for labelling the double stranded plasmid pα3'HVR.64 instead of *in vitro* transcription (Fowler *et al.*, 1988). More bands are seen here (Fig. 3) than with the M13 probe and there are more low molecular weight hybridising bands. Calf thymus and human DNA display a comparable number of bands to *Giardia*, but with quite different profiles. Under the conditions used there is a higher cross-hybridising background with the human sample.

76

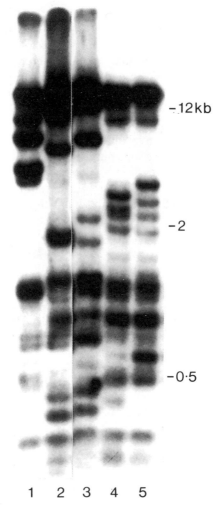

Figure 1. DNA fingerprinting of the *Giardia intestinalis* genome with the M13 genome. *Giardia* genomic DNA was cleaved with Hin fI and electrophoresed in 0.7% SeaKem GTG agarose (Marine Colloids). The DNA was transferred to Hybond nylon membrane (Amersham) by the Southern procedure. The probe was single-stranded M13 DNA labelled with ^{32}P-dCTP by random priming (Upcroft *et al.*, 1990b). Lanes 1, 2, 3, 4 and 5 contained DNA from *Giardia* isolates BAC1, WB1B, BRIS/83/HEPU/106 and two Brisbane human isolates, respectively.

As a third type of fingerprinting probe, we used simple repeat sequences that have been found in the higher eukaryotes. These repeats are found spread throughout genomes at extreme ranges of frequency (Hamada *et al.*, 1982; Tautz and Renz, 1984), and classes have been shown to be subject to extensive turnover, probably by slipped-strand mispairing (Tautz *et al.*, 1986). Although their frequency is usually too

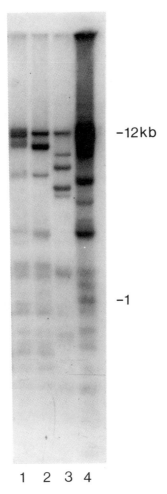

-12kb

-1

1 2 3 4

Figure 2. DNA fingerprinting of *Giardia intestinalis* with the M13 genome. Conditions were similar to Fig. 1 except that the lower molecular weight bands were separated further, the probe was not as intensively labelled and the exposure was for a shorter period. Lanes 1, 2, 3 and 4 contained DNA from *Giardia* isolates BRIS/83/HEPU/106-2ID$_{10}$, BRIS/83/HEPU/106, BAC1 and Ad2, respectively.

high to be of use in mammals, polyTG, for example is highly polymorphic in a variety of vertebrates (Kashi *et al.*, 1990). These simple sequence repeats are being used extensively for mapping libraries of cosmids and YACs because of the lowered complexity of the sample and maintenance of the hypervariability between locations. Because of the small size of the *Giardia* genome, simple repeat sequences may also be of use; poly(dA-dC).poly(dG-dT), for example, yields a fingerprint pattern with a high background (data not shown). Such sequences may prove to be useful as a supplementary system to probes such as M13 and the α-globin 3′ repeat, and in some situations may be as informative. Highly repeated telomeric sequences in the *Leishmania* genome vary in copy number between isolates and may be another example of this class of DNA (Ellis and Crampton, 1989).

The *Giardia* genome therefore contains representative hypervariable sequences of the same types found in the higher eukaryotes. The mechanisms for the formation and variability of these sequences are still not understood. This is most likely due to the limited number of hypervariable sequences that have been examined in any detail, although mechanisms such as slipped-strand mispairing and intra-allelic unequal crossing-over still appear to be contenders (Jeffreys *et al.*, 1988a and 1990). Since *Giardia* is regarded as an extremely ancient eukaryote, the fundamental nature of the generation of hypervariable sequences must also have arisen very early in the eukaryotic lineage. It may lie in the DNA polymerase enzyme itself since simple repeats can be synthesized by purified *Escherichia coli* polI (Kornberg, 1980) and slipped-strand mispairing has been shown to cause replication mini-insertions in the IS2 insertion sequence (Ghosal and Saedler, 1978). A detailed comparison of the variety and structure of the *Giardia* hypervariable sequences should provide some insight into their generation and evolution.

Detection of genetic heterogeneity and identification of individual parasites in uncultured populations

A number of issues regarding the ability to detect different taxonomic or heirarchical levels and the overlap of these categories with biological traits of the organisms now arise. For example, if protozoa do not have a sexual phase, how does one define a species biologically, taking into account the difficulties addressed above? If phenotypic switching occurs, how is an isolate defined and how can epidemiological parameters be monitored? If culturing conditions impose selection upon the population, how can the original genetic heterogeneity be determined? Does the act of culturing or animal passage cause a common phenotypic switch, or select the strain most amenable to that enviroment? Do parasites in culture represent the population that causes the disease?

The identification of individual parasites prior to culturing would answer the basic question as to the extent of genotypic difference in a single isolate, and by extrapolation, larger populations. An extension of this would be monitoring the spread of particular populations (or strains) in man or animals, such as reinfection after drug treatment, potential zoonoses, drug resistant strains, and virulent or pathogenic strains in communities. Since the fingerprint is stable, criticisms of phenotypic switching, contamination of isoenzyme types in cloned lines

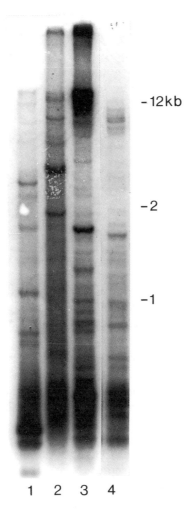

$-12kb$

-2

-1

1   2   3   4

Figure 3. DNA fingerprinting of *Giardia intestinalis* with the hypervariable sequence which is 3′ to the α–globin cluster in man. Target DNA was cleaved, electrophoresed and transferred as in Fig. 1. The probe DNA, pα3′HVR.64 (Fowler *et al.*, 1988), was labelled with ^{32}P-dCTP by random priming (Upcroft *et al.*, 1990b). Lanes 1, 2, 3 and 4 contain DNA from calf thymus, human lymphocytes, *Giardia* isolates Ad2 and BRIS/83/HEPU/106 respectively.

when examining phenotypic switching and the investigation of possible sexual exchange can be monitored at the genetic level. I outline below my approach using fingerprinting to address these issues.

Since each fingerprint appears to be unique to each cloned isolate, and a number of bands have so far been sufficiently hypervariable to be different in every isolate examined, each of these single locus hypervariable sequences should be adequate to identify an isolate, in a manner comparable to that described for human single locus probes (see Jeffreys *et al.*, 1990 and references therein). Since the *Giardia* genome is approximately 150 times smaller than the human genome, the argument used to defend this approach should be stronger for the parasite genome (Jeffreys *et al.*, 1985a and b; 1988a). Cloning such hypervariable sequences and their flanking regions would yield unique primer sequence data so that oligonucleotide primers could be constructed to amplify the intervening hypervariable region by the polymerase chain reaction (Saiki *et al.*, 1988). A single amplified band would be detected in a cloned parasite isolate or in a population that had been founded from a single parasite. If the population was mixed, there would be a different band for each genotype present. Since each different hypervariable sequence will be amplified, all possible representatives should be detectable, depending on the number of amplification cycles. Confirmatory data could be obtained by amplification of more than one locus. Although amplification of some hypervariable sequences has shown technical difficulties, suitable conditions have been described for a few human sequences (Horn *et al.*, 1989; Jeffreys *et al.*, 1988b and 1990). Furthermore, the M13 repeat sequence does not appear to suffer from these difficulties (Bellamy *et al.*, 1990), although amplification of longer sequences may. We have now constructed plasmid libraries from gel enriched regions containing hypervariable bands from *Giardia* stock BRIS/83/HEPU/106. The first two clones which hybridised to the M13 repeat contain flanking sequences which should be appropriate for constructing PCR primers. Since the informative hypervariable bands in *Giardia* are not particularly long when compared with those in the human genome, they should be more suitable both for cloning in recombination deficient hosts (Kurnit, 1989) and for PCR amplification. Variation and mutation in these hypervariable sequences (see Jeffreys *et al.*, 1990 and the references therein) can also be analysed in outgrowing populations. The combination of isolate-specific fingerprints and the identification of individual parasites and populations should allow ready characterization of strain variation, movement and resurgence. Correlation with traits such as zoonotic potential, drug resistance, host range, virulence and pathogenesis is now possible.

Acknowledgements
I would like to thank the National Health and Medical Research Council of Australia for support, my colleagues Jacqui Upcroft and Peter Boreham, and Roger Mitchell and Andrew Healey for excellent technical assistance.

References

Adam, R. D., Nash, T. E., and Wellems, T. E. (1988) The *Giardia lamblia* trophozite contains sets of closely related chromosomes. *Nucl. Acids Res.* 16: 4555–4567.
Andrews, B. J., Mentzoni, L., and Bjorvatn, B. (1990) Zymodeme conversion of isolates of *Entamoeba histolytica*. *Trans. R. Soc. Trop. Med. Hyg.* 84: 63–65.

Andrews, R. H., Adams, M., Boreham, P. F. L., Mayrhofer, G., and Meloni, B. P. (1989) *Giardia intestinalis*: electrophoretic evidence for a species complex. *Int. J. Parasitol.* 19: 183–190.

Archibald, S. C., Mitchell, R. W., Upcroft, J. A., Boreham, P. F .L., and Upcroft, P. (1991) Variation between human and animal isolates of *Giardia* as demonstrated by DNA fingerprinting. *Int. J. Parasitol.* 21: 123–124.

Armour, J. A. L., Wong, Z., Wilson, V., Royle, N. J., and Jeffreys, A. J. (1989) Sequences flanking the repeat arrays of human minisatellites: association wiht tandem and dispersed repeat elements. *Nucl. Acids Res.* 17: 4925–4934.

Baker, R. J. and Bickham, J. W. (1980) Karyotypic evolution in bats: Evidence of extensive and conservative chromosomal evolution in closely related taxa. *Syst. Zool.* 29: 239–253.

Bell, G. I., Selby, M. J., and Rutter, W. J. (1982) The highly polymorphic region near the human insulin gene is composed of simple tandemly repeating sequences. *Nature* 295: 31–35.

Bellamy, R., Inglehearn, C., Lester, D., Hardcastle, A., and Bhattacharya, S. (1990) Better fingerprinting with PCR. *Trends Genet.* 6: 32.

Bishop, R. P., and Akinsehinwa, F. (1990) Characterization of *Leishmania donovani* stocks by genomic DNA heterogeneity and molecular karyotype. *Trans. R. Soc. Trop. Med. Hyg.* 83: 629–634.

Boreham, P. F. L., Upcroft, J. A., and Upcroft, P. (1990) Changing approaches to the study of *Giardia* epidemiology: 1681–2000. *Int. J. Parasitol.* 20: 479–487.

Borst, P., and Greaves, D. R. (1987) Programmed gene rearrangements altering gene expression. *Science* 235: 658–667.

Botstein, D., White, R. L., Skolnick, M., and Davis, R. W. (1980) Construction of a genetic linkage map in man using restriction fragment length polymorphisms. *Am. J. Hum. Genet.* 32: 314–331.

Bulletin WHO (1987) Public health significance of intestinal parasitic infections. 65: 575–588.

Cavalier-Smith, T. (1987) The origin of eukaryote and archaebacterial cells. In: Endocytobiology III, Annal. New York Acad. Sci. 503: 17–54.

Cavalier-Smith, T. (1989) Archaebacteria and archezoa. *Nature* 339: 100–101.

Capon, A. G., Upcroft, J. A., Boreham, P. F. L., Cottis, L. E. and Bundesen, P. G. (1989) Similarities of *Giardia* antigens derived from human and animal sources. *Int. J. Parasitol.* 19: 91–98.

Cedillo-Rivera, R., Enciso-Moreno, J. A., Martinez-Palomo, A., and Ortega-Pierres, G. (1989) *Giardia lamblia*: isoenzyme analysis of 19 axenic strains isolated from symptomatic and asymptomatic patients in Mexico. *Trans. R. Soc. Trop. Med. Hyg.* 83: 644–646.

Conover, R. K. and Brunk, C. F. (1986) Macronuclear DNA molecules of *Tetrahymena thermophila*. *Mol. Cell Biol.* 6: 900–905.

DeJonckheere, J. F., Majewska, A. C., and Kasprzak, W. (1990) *Giardia* isolates from primates and rodents display the same molecular polymorphism as human isolates. *Mol. Biochem. Parasitol.* 39: 23–29.

Dover, G. A. (1987) DNA turnover and the molecular clock. *J. Mol. Evol.* 26: 47–58.

Dover, G. A. (1989) Victims or perpetrators of DNA turnover? *Nature* 342: 347–348.

Ellis, J., and Crampton, J. (1989) A simple, highly repetitive sequence in the *Leishmania* genome. In: Hart, D. T. (ed.) Leishmaniasis. The Current Status and New Strategies for Control, Plenum Press, New York, pp. 589–596.

Fenton, B., Walker, A., and Walliker, D. (1985) Protein variation in clones of *Plasmodium falciparum* detected by two dimensional electrophoresis. *Mol. Biochem. Parasitol.* 16: 173–183.

Fowler, S. J., Gill, P., Werret, D. J., and Higgs, D. R. (1988) Individual specific DNA fingerprints from a hypervariable region probe: alpha-globin 3'HVR. *Hum. Genet.* 79: 142–146.

Ghosal, D., and Saedler, H. (1978) DNA sequence of the mini-insertion IS2–6 and its relation to the sequence of IS2. *Nature* 275: 611–617.

Giannini, S. H., Schittini, M., Keithly, J. S., Warburton, P. W., Cantor, C. R., and Van der Ploeg, L. H. T. (1986) Karyotype analysis of *Leishmania* species and its use in classification and clinical diagnosis. *Science* 232: 762–765.

Giannini, S. H., Curry, S. S., Tesh, R. B., and Van der Ploeg, L. H. T. (1990) Size-conserved chromosomes and stability of molecular karyotype in cloned stocks of *Leishmania major*. *Mol. Biochem. Parasitol.* 39: 9–22.

Gibson, W. C., and Borst, P. (1986) Size-fractionation of the small chromosomes of *Trypanozoon* and *Nannomonas* trypanosomes by pulsed field gradient gel electrophoresis. *Molec. Biochem. Parasitol.* 18: 127–140.

Goodbourn, S. E. Y., Higgs, D. R., Clegg, J. B., and Weatherall, D. J. (1983) Molecular basis of length polymorphism in the human zeta-globin gene complex. *Proc. Natl. Acad. Sci. USA* 80: 5022–5026.

Hamada, H., and Kakunaga, T. (1982) Potential Z-DNA forming sequences are highly dispersed in the human genome. *Nature* 298: 396–398.

Hamada, H., Petrino, M. G., and Kakunaga, T. (1982) A novel repeated element with Z-DNA-forming potential is widely found in evolutionarily diverse eukaryotic genomes. *Proc. Natl. Acad. Sci. USA* 79: 6465–6469.

Hancock, J. M., and Dover, G. A. (1988) Molecular coevolution among cryptically simple expansion segments of eukaryotic 26/28S rRNAs. *Mol. Biol. Evol.* 5: 377–391.

Hancock, J. M., Tautz, D., and Dover, G. A. (1988) Evolution of the secondary structures and compensatory mutations of the ribosomal RNAs of *Drosophila melanogaster*. *Mol. Biol. Evol.* 5: 393–414.

Healey, A., Mitchell, R., Upcroft, J. A., Boreham, P. F. L., and Upcroft, P. (1990) Complete nucleotide sequence of the ribosomal RNA tandem repeat unit from *Giardia intestinalis*. *Nucl. Acids Res.* 18: 4006.

Higgs, D. R., Goodbourn, S. E. Y., Wainscoat, J. S., Clegg, J. B., and Weatherall, D. J. (1981) Highly variable regions of DNA flank the human α globin genes. *Nucl. Acids Res* 9: 4213–4224.

Holmquist, R., Miyamoto, M. M., and Goodman, M. (1988) Higher primate phylogeny - Why can't we decide? *Mol. Biol. Evol.* 5: 201–216.

Horn, G. T., Richards, B., and Klinger, K. W. (1989) Amplification of a highly polymorphic VNTR segment by the polymerase chain reaction. *Nucl. Acids Res.* 17: 2140.

Jarman, A. P., Nicholls, R. D., Weatherall, D. J., Clegg, J. B., and Higgs, D. R. (1986) Molecular characterisation of a hypervariable region downstream of the human α-globin gene cluster. *EMBO J* 5: 1857–1863.

Jeffreys, A. J. (1987) Highly variable minisatellites and DNA fingerprints. *Biochem. Soc. Trans.* 15: 309–317.

Jeffreys, A. J., Neumann, R., and Wilson, V. (1990) Repeat unit sequence variation in minisatellites: a novel source of DNA polymorphism for studying variation and mutation by single molecule analysis. *Cell* 60: 473–485.

Jeffreys, A. J., Royle, N. J., Wilson, V., and Wong, Z. (1988a) Spontaneous mutation rates to new length alleles at tandem-repetitive hypervariable loci in human DNA. *Nature* 332: 278–281.

Jeffreys, A. J., Wilson, V., Neumann, R., and Keyte, J. (1988b) Amplification of human minisatellites by the polymerase chain reaction: towards DNA fingerprinting of single cells. *Nucl. Acids Res.* 16: 10953–10971.

Jeffreys, A. J., Wilson, V., and Thein, S. L. (1985a) Hypervariable 'minisatellite' regions in human DNA. *Nature* 314: 67–73.

Jeffreys, A. J., Wilson, V., and Thein, S. L. (1985b) Individual-specific 'fingerprints' of human DNA. *Nature* 316: 76–79.

Kashi, Y., Tikochinsky, Y., Genislav, E., Iraqi, F., Nave, A., Beckman, J. S., Gruenbaum, Y., and Soller, M. (1990) Large restriction fragments containing poly-TG are highly polymorphic in a variety of vertebrates. *Nucl. Acids Res.* 18: 1129–1132.

Kemp, D. J., Thompson, J. K., Walliker, D., and Corcoran, L. M. (1987) Molecular karyotype of *Plasmodium falciparum*: conserved linkage groups and expendable histidine-rich protein genes. Proc. Natl. Acad. Sci. USA 84: 7672–7676.

Kornberg, A. (1980) DNA Replication. W. H. Freeman, San Francisco, pp. 143–150.

Kulda, J., and Nohynkova, E. (1978) Flagellates of the human intestine and the intestines of other species. In: Krier, J. P. (ed.), Parasitic Protozoa, vol. 2, Academic Press, New York, pp. 1–138.

Kurnit, D. M. (1989) *Escherichia coli recA* deletion strains that are highly competent for transformation and for in vivo phage packaging. *Gene* 82: 313–315.

Lamont, G. S., Tucker, R. S., and Cross, G. A. M. (1986) Analysis of antigen switching rates in *Trypanosoma brucei*. *Parasitol.* 92: 355–367.

Mirelman, D. (1987) Effect of culture conditions and bacterial associates on the zymodemes of *Entamoeba histolytica*. *Parasitol. Today* 3: 40–43.

Mirelman, D., Bracha, R. Chayen, A., Aust-Kettis, A., and Diamond, L. S. (1986a) *Entamoeba histolytica*: Effect of growth conditions and bacterial associates on isoenzyme patterns and virulence. *Exp. Parasitol.* 62: 142–148.

Mirelman, D., Bracha, R., Wexler, A., and Chayen, A. (1986b) Changes in isoenzyme patterns of a cloned culture of nonpathogenic *Entamoeba histolytica* during axenization. *Infect. Immum.* 54: 827–832.

Nakamura, Y., Julier, C., Wolff, R., Holm, T., O' Connell, P., Leppert, M., and White, R. (1987a) Characterization of a human 'midisatellite' sequence. *Nucl. Acids Res.* 15: 2537–2547.

Nakamura, Y., Leppert, M., O'Connell, P., Wolff, R., Holm, T., Culver, M., Martin, C., Fujimoto, E., Hoff, M., Kumlin, E., and White, R. (1987b) Variable number of tandem repeat (VNTR) markers for human gene mapping. *Science* 235: 1616–1622.

Rogstad, S. H., Herwaldt, B. L., Schlesinger, P. H., and Krogstad, D. J. (1989) The M13 repeat probe detects RFLPs between two strains of the protozoan malaria parasite *Plasmodium falciparum*. *Nucl. Acids Res.* 17: 3610.

Ruffié, J. (1987) The Population Alternative. Penguin Books, Middlesex.

Ryskov, A. P., Jincharadze, A. G., Prosnyak, M. I., Ivanov, P. L., and Limborska, S. A. (1988) M13 phage DNA as a universal marker for DNA fingerprinting of animals, plants and microorganisms. *FEBS Lett.* 233: 388–392.

Saiki, R. K., Gelfand, D. H., Stoffel, S., Scharf, S. J., Higuchi, R., Horn, G. T., Mullis, K. B., and Erlich, H. A. (1988) Primer-directed enzymatic amplification of DNA with a thermostable DNA polymerase. *Science* 239: 487–491.

Sargeaunt, P. G. (1987) The reliability of *Entamoeba histolytica* zymodemes in clinical diagnosis. *Parasitol. Today* 3: 37–40.

Sargeaunt, P. G., Jackson, T. F. H. G., Wiffen, S. R., and Bhojnani, R. (1988) Biological evidence of genetic exchange in *Entamoeba histolytica*. *Trans. R. Soc. Trop. Med. Hyg.* 82: 862–867.

Schäfer, R., Zischler, H., and Epplen, J. T. (1988) $(CAC)_5$, a very informative oligonucleotide probe for DNA fingerprinting. *Nucl. Acids Res.* 16: 5196.

Schwartz, D. C., and Cantor, C. R. (1984) Separation of yeast chromosome-sized DNAs by pulsed field gradient gel electrophoresis. *Cell* 37: 65–75.

Sogin, M. L., Gunderson, J. H., Elwood, H. J., Alonso, R. A., and Peattie, D. A. (1989) Phylogenetic meaning of the kingdom concept: an unusual ribosomal RNA from *Giardia lamblia*. *Science* 243: 75–77.

Spithill, T. W., and Samaras, N. (1987) Genomic organisation, chromosomal location and transcription of dispersed and repeated tubulin genes in *Leishmania major*. *Molec. Biochem. Parasitol.* 24: 23–37.

Tautz, D., and Renz, M. (1984) Simple sequences are ubiquitous repetitive components of eukaryotic genomes. *Nucl. Acids Res.* 12: 4127–4138.

Tautz, D., Trick, M., and Dover, G. A. (1986) Cryptic simplicity in DNA is a major source of genetic variation. *Nature* 322: 652–656.

Tibayrenc, M., Kjellberg, F., and Ayala, F. J. (1990) A clonal theory of parasitic protozoa: The population structures of *Entamoeba*, *Giardia*, *Leishmania*, *Naegleria*, *Plasmodium*, *Trichomonas*, and *Trypanosoma* and their medical and taxonomical consequences. *Proc. Natl. Acad. Sci. USA* 87: 2414–2418.

Upcroft, J. A., Boreham, P. F. L., and Upcroft, P. (1989a) Geographic variation in *Giardia* karyotypes. *Int. J. Parasitol.* 19: 519–527.

Upcroft, J. A., Dunn, L. A., Dommett, L. S., Healey, A., Upcroft, P., and Boreham, P. F. L. (1989b) Chromosomes of *Blastocystis hominis*. *Int. J. Parasitol.* 19: 879–883.

Upcroft, J. A., Healey, A., Mitchell, R., Boreham, P. F. L., and Upcroft, P. (1990a) Antigen expression from the ribosomal repeat of *Giardia intestinalis*. *Nucl. Acids Res.* 18: 7077–7081.

Upcroft, P., Boreham, P. F. L., and Upcroft, J. A. (1988) The genome of *Giardia intestinalis*. In: Advances in *Giardia* Research, Wallis, P. M. and Hammond, B. R. (eds.), University of Calgary Press, Calgary, pp. 147–152.

Upcroft, P., Mitchell, R., and Boreham, P. F. L. (1990b) DNA fingerprinting of the intestinal parasite *Giardia duodenalis* with the M13 phage genome. *Int. J. Parasitol.* 20: 319–323.

84

Upcroft, P., Upcroft, J. A., and Boreham, P. F. L. (1989c) Genetic variation in *Giardia intestinalis* in comparison with other unicellular organisms. In: Ko, R. C. (ed.), Immumonogical and Molecular Basis of Pathogenesis in Parasitic Diseases, University of Hong Kong, Hong Kong, pp. 155–168.

Van der Ploeg, L. H. T., Smits, M., Ponnadurai, T., Vermeulen, A., Meuwissen, J. H. E. Th., and Langsley, G. (1985) Chromosome-sized DNA molecules of *Plasmodium falciparum*. *Science* 229: 658–661.

Vassart, G., Georges, M., Monsieur, R., Brocas, H., Lequarre, A. S., and Cristophe, D. (1987) A sequence in M13 phage detects hypervariable minisatellites in human and animal DNA. *Science* 235: 683–684.

Vollrath, D., Davis, R. W., Connelly, C., and Hieter, P. (1988) Physical mapping of large DNA by chromosome fragmentation. *Proc. Natl. Acad. Sci. USA* 85: 6027–6031.

Waters, A. P., and McCutchan, T. F. (1989) Rapid, sensitive diagnosis of malaria based on ribosomal RNA. *Lancet* i: 1343–1346.

Wyman, A. R., and White, R. (1980) A highly polymorphic locus in human DNA. *Proc. Natl. Acad. Sci. USA* 77: 6754–6758.

DNA Fingerprinting: Approaches and Applications
ed. by T. Burke, G. Dolf, A. J. Jeffreys & R. Wolff
© 1991 Birkhäuser Verlag Basel/Switzerland

Human Variable Number of Tandem Repeat Probes As a Source of Polymorphic Markers in Experimental Animals

B. de Gouyon[1], C. Julier[1], P. Avner[2], M. Georges[3], and M. Lathrop[1]

[1]*Centre d'Etude du Polymorphism Humain, 27 rue Juliette Dodu, Paris 75010, France;*
[2]*Institut Pasteur, Paris, France;* [3]*Genmark Inc, Salt Lake City, Utah, USA*

Summary
Human VNTR (Variable Number of Tandem Repeat) markers are examined as a source of polymorphism for linkage studies in inbred strains of mice and rats. High frequencies of cross-hybridization are found under fingerprinting conditions that detect many distinct minisatellite loci in these species. Linkage studies suggest that minisatellite markers are widely distributed in the mouse genome, in contrast to humans where they are clustered, particularly in telomeric regions. Human VNTR probes can be used to screen in mouse genomic libraries to isolate mouse specific VNTR sequences. Some of these sequences reveal fingerprint patterns under stringent hybridization conditions.

Introduction

Linkage studies of animal models provide one way to identify candidate genes, or candidate chromosome regions for investigations of human genetic diseases. Models of multifactorial traits, or of rare monogenetic diseases for which informative human families are difficult to collect may be particularly useful. For example, mice from NOD (non-obese diabetic) strain have a high frequency of diabetes closely resembling insulin-dependent diabetes (IDDM) in humans, and crosses involving the NOD inbred strain show that genes outside of the MHC region are involved in disease susceptibility (Wicker *et al.*, 1987). If these genes are localized in the mouse, the homologous regions can be studied in human families by applying a high density of polymorphic markers, to identify or exclude the possibility of similar genes acting in human IDDM (Hyer *et al.*, in press).

As in human studies, a prerequisite for successful linkage investigation in animals is the availability of a sufficient number of polymorphic markers throughout the genome. In the mouse, although many restriction fragment length polymorphisms have been defined, these are often diallelic and usually they are not polymorphic between all inbred strains (Elliot, 1989; Roderick and Guidi, 1989). Studies of RFLPs with many restriction enzymes may be needed to cover a reasonable portion of the mouse genome. Fewer polymorphisms have been described in the rat, where similar problems are encountered.

Recently, both minisatellite (Jeffreys et al., 1985; Jeffreys et al., 1987) and microsatellite (CA) tandem-repeat polymorphisms (Weber and May, 1989; Litt and Luty, 1989) have been applied to overcome these limitations. Both types of markers have been shown to have a wide-distribution in the mouse genome, and to efficiently detect polymorphisms between many different laboratory strains (Julier et al., 1990; Love et al., 1990).

In this paper, we review some recent work on the use of human VNTR probes to detect multiple minisatellite polymorphisms as fingerprint patterns in mouse and rat crosses. Possible applications to linkage studies of experimental animal models are discussed.

Cross-hybridization of Human VNTR Probes to Mouse and Rat DNA

In human, single locus minisatellite sequences, or VNTRs have been isolated by screening genomic libraries of human DNA with oligomeric sequences derived from many different sources (Wong et al., 1986; Nakamura et al., 1987b; Nakamura et al., 1988). Many human VNTRs cross-hybridize and reveal complex minisatellite polymorphisms, or genetic fingerprints in other species (Jeffreys et al., 1987; Georges et al., 1988). Although fingerprint patterns present some problems of interpretation, potentially, they can provide an abundance of information for linkage studies, in appropriately designed crosses (Georges et al., 1990). Rapid screening of the genome is possible, since these probes may detect several unlinked loci in a single hybridization experiment (Julier et al., 1990).

To investigate the potential of human VNTRs and other probes for studies in inbred mouse lines, we initially examined 28 human probes, and one probe containing mouse DNA homologous to the Drosophila per gene under conditions designed to detect cross-hybridization to minisatellite sequences. DNA was digested with TaqI, HaeIII or HinfI. With the latter two restriction enzymes, Southern blots were prepared under electrophoresis conditions that allowed retention of fragments > 1.5 kb, conditions which are known to favour the resolution of minisatellite bands. TaqI was used to detect cross-hybridization to fragments that are frequently out by the other restriction enzymes, and to allow detection of loci other than VNTRs. Electrophoresis conditions allowed retention of Taq 1 fragments > 0.5 kb. Further technical details are provided in Julier et al. (1990). Under low stringency hybridization conditions >80% of these probes hybridized to mouse DNA, and > 48% detected complex fingerprint patterns. Eleven probes were selected and characterized on 26 recombinant inbred mouse lines (DNA obtained from the Jackson laboratory), six of those probes were also hybridized to rat DNA, extracted from tissues under standard conditions.

Figure 1 shows the results of the hybridization for one human VNTR (EFD134.7) characterized in 5 inbred rat strains. DNA samples for the

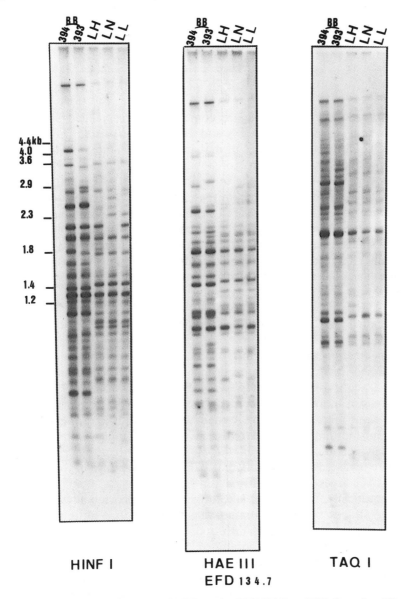

Figure 1. Hybridization of a human VNTR probe (EFD134.7) to DNA from five different inbred strains of rat digested by three restriction enzymes. The designation of the rat strains is explained in the text.

Table 1. Polymorphic differences detected between five different inbred strains of rat by 6 human VNTR probes (YNH37.3, YNG24, RMU3, CMM101, EFD134.7, YNZ2), and three restriction enzymes (TaqI, HaeIII, HinfI). The designation of the rat strains is given in the text; + indicates the strain with the band present; − indicates the strain with the band absent

−/+	LL	LH	LN	393	394
LL		17	31	133	124
LH	15		37	137	132
LN	24	33		139	129
393	75	82	78		47
394	72	80	73	49	

experiment shown in this figure come from three related strains for studies of hypertension (LH, LN and LL), and two related strains of BB rats (393 and 394) for studies of diabetes. In this figure polymorphisms are observed as the presence of a band in one inbred line, and the absence in one or more of the other inbred lines. As evident from the figure, non-related strains exhibit a greater number of polymorphic differences than do closely related strains. However, with the study of 6 different probes and three restriction enzymes a substantial number of polymorphisms could be detected between all the strains. Table 1 shows a summary of these results. Similar data have been provided for inbred mouse lines in Julier et al. (1990).

Segregation and Linkage of Fingerprint Bands

The segregation of minisatellite bands detected as described above can be studied in domestic or experimental animal crosses to establish linkage relationships (Julier et al., 1990; Georges et al., 1990). Large pedigrees of individuals from a common progenitor (or representatives of the same inbred strain) are needed because bands that migrate to the same location may represent distinct loci in families obtained from outbred individuals. For domestic animal, half-sib pedigrees in which one sire is mated to many different dams have been useful for mapping studies (Georges et al., 1990).

Theoretically, fully-inbred laboratory strains provide advantages for linkage studies because of homozygosity at all loci, although several practical questions must be addressed. Offspring from different parents of the same strain can be combined, but it is advisable to study the parental population in order to confirm complete inbreeding and to eliminate hypermutable loci. Whenever possible, this should include DNA from the true parents, rather than other representatives of the same inbred line, as these individuals may be separated by many generations. Despite this caution, we have found that it is usually possible to detect and eliminate significant new mutations, even if the true parents cannot be studies. Another potential difficulty is the existence of comigrating bands that are indistinguishable in a hybridization experiment but represent alleles at different loci. Usually, these will appear as deviations from Mendelian segregation in offspring populations.

Backcross, F_2 or recombinant inbred (RI) lines can all provide information on linkage relationships, although the interpretation of the information in each type of cross is somewhat different (see next section). Initially, we chose to characterize an RI panel of 26 animals formed by a cross of C57BL/6 with DBA/2J (B × D), for which an unlimited supply of DNA could be obtained. The disadvantage of RI crosses is that the detection of linkage is possible only at short distances because of the limited number of animals available for study, and the many opportunities for recombination before fixation of alleles. In this panel, linkage can be judged to be significant at observed recombination rates of < 4%.

Figure 2 shows an example of the pattern of segregation observed in the B × D panel with human VNTR probes. As discussed in Julier *et al.* (1990), when we characterized 11 probes with 3 restriction enzymes, the segregation pattern of 346 polymorphic bands could be identified in the panel. Although we observed several instances of new mutations (in which a band that was not present in either parent appeared in one of the RI lines), these did not create practical difficulties in interpretation. Linkage analysis showed that the minisatellite bands could be classified into 166 systems that exhibited no recombination. Each system represents a single locus, or group of tightly linked loci. After further accounting for significant linkage (recombination < 4%), the bands were classified into 101 linkage groups.

A database of 142 independent markers with known chromosomal localizations that were polymorphic in the RI panel was available for study. These data permitted us to assign 38 of the minisatellite groups by linkage to other markers. Figure 3 shows the resulting minisatellite linkage maps for the polymorphisms between the two parental strains. Presumably, the unassigned minisatellite linkage groups are in regions of the mouse chromosomes shown in Fig. 3 that have insufficient coverage for the detection of linkage.

Strategies for Linkage Studies

Linkage studies of disease or other traits can be conducted with minisatellite, fingerprint type polymorphisms if appropriate pedigrees can be obtained. For example, Georges *et al.* (1990) have reported linkage between a gene responsible for muscular hypertrophy in cattle, and a

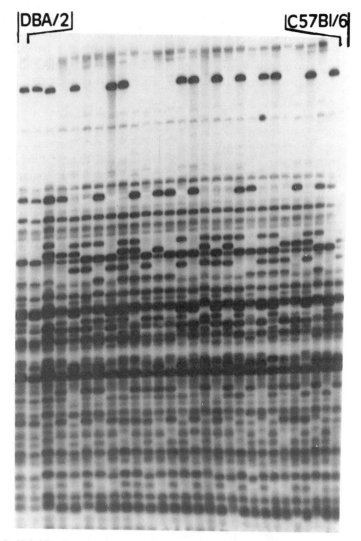

|DBA/2| |C57Bl/6|

Figure 2. Hybridization of a human VNTR probe (pYNZ2) to DNA from 26 recombinant inbred, and parent strains as described in the text. The DNA was digested with HinfI. Reprinted with permission from Julier *et al.* (1990).

fingerprint band detected with a human probe EFD134.7. Muscular hypertrophy is a recessive trait, and in this instance, the pedigree consisted of 67 half-sibs obtained from a bull that was an obligate heterozygote carrier. A chromosomal assignment will be possible from this linkage result, only if a cloned DNA fragment containing the sequence corresponding to the band can be isolated for further study.

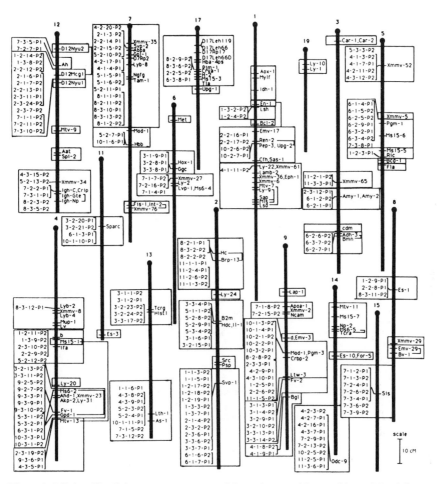

Figure 3. Minisatellite linkage maps constructed from data on 26 recombinant inbred lines. Reprinted with permission from Julier *et al.* (1990).

In pedigrees from inbred lines, the fingerprint bands can be mapped as described above, and the information can be used to obtain a direct assignment of a trait locus. Minisatellite linkage maps of the form shown in Fig. 3 are strain specific, since allelic systems cannot be directly identified in different inbred lines. However, once a map has been constructed for a particular strain, it can be used for linkage studies in other crosses with the same strain. In particular, the C57BL/6 minisatellite map in Fig. 3 has been used in our laboratory to investigate linkage to diabetes in a (C57BL/6 × NOD) × NOD backcross panel.

Since the presence of a minisatellite band is a dominant trait (unless densitometry is attempted to distinguish the presence of one or two copies of the band), only a subset of polymorphisms may be useful for

92

A. Backcross panel

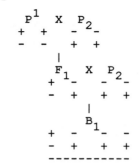

B. F₂ panel

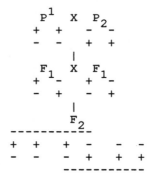

Figure 4. Examples of the segregation of two different minisatellite bands in backcross (Fig. 4A) or F_2 panels (Fig. 4B). P_1 and P_2 indicate the two parental lines; other generations are denoted F_1, B_1, and F_2. The presence of a band is indicated as + and the absence as −. Dashed lines (above or below genotypes) indicates genotype classes that cannot be distinguished.

linkage studies of a particular cross. For example, the diabetes susceptibility genes in NOD appear to be recessive in nature, and backcross panels are largely used for their study. As shown in Fig. 4a, linkage information for diabetes is contributed only by bands present in the donor (non-NOD) and absent in NOD. In F_2 crosses, the situation is different as illustrated in Fig. 4b.

In the mouse, interspecific crosses involving *M. Spretus* are frequently established for the construction of linkage maps, because of the large number of genetic differences that can be detected between this subspecies and laboratory mice (Avner *et al.*, 1988). Similarly, *M. Spretus* is an ideal candidate for development of a minisatellite map. Crosses with *M. Spretus* are potentially a very powerful means for high-resolution localization of the genes responsible for disease in experimental models, and the minisatellite *M. Spretus* map will be useful in such studies. A backcross between *M. Spretus* and NOD has been obtained by one of us (PA) for study of diabetes, and panels of first and second backcross animals are presently being characterized with these probes.

Other Topics

Since human VNTR frequently detect minisatellite polymorphisms in other species, they can be used to screen in genomic libraries for isolation of locus-specific VNTR markers. To test this strategy in the mouse, a library of HaeIII fragments in the size range of 3–6kb was constructed (by MG) and screened with some human VNTR probes. Several independent clones, selected by hybridization to the same human VNTR probe, exhibited the same complex fingerprint pattern when hybridized to mouse DNA from different strains under stringent hybridization conditions (Fig. 5). Independent segregation of several polymorphisms was observed in the RI panel.

Further studies are being undertaken to characterize the nature of the polymorphisms detected in this way, and to determine if other clones that detect multiple loci can be obtained by screening with different human VNTRs. The availability of such probes would increase the reproducibility of fingerprint studies in many laboratories, and therefore, aid in the interpretation of linkage results.

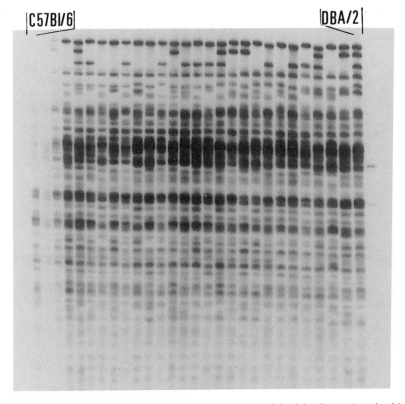

Figure 5. Hybridization of a mouse probe (isolated as explained in the text) under high stringency conditions to DNA from 26 recombinant inbred strains, digested by HinfI.

Conclusions

Our studies have demonstrated that human VNTR loci can provide an abundant source of polymorphism for linkage studies in inbred laboratory strains of mice and rat. Strain-specific minisatellite maps can be constructed, and applied to the localization of genes important for susceptibility in models of human genetic disease. In contrast to the human genome, in which VNTR loci are often clustered in telomeric regions, minisatellite polymorphisms seem to widely disperse in the mouse genome, as shown in Fig. 3.

Despite potential difficulties of interpretation, we have found that complex minisatellite patterns provide highly reproducible linkage results. Corresponding bands can be identified in different experiments if DNA from the parental strains are always included for hybridization. Frequent new mutations that could have a significant effect on results are limited, and readily identified. In some instances, it may be possible to screen genomic libraries with human VNTRs to identify new probes that detect multiple polymorphisms.

References

Avner, P., Amar, L., Dandolo, L., and Guenet, J.-L. (1988) *Trends in Genetics* 4: 18–23.

Elliot, R. W. (1989). In: Genetic Variants and Strains of the Laboratory Mouse, 2nd Edition, M. Lyon and A. Searle, Oxford University Press, 537–558.

Georges, M., Lathrop, M., Hilbert, P., Marcotte, A., Schwers, A., Swillens, S., Vassart, G., and Hanset, R. (1990) *Genomics* 6: 461–476.

Georges, M., Lequarre, A.-S., Castelli, M., Hanset, R., and Vassart, G. (1988) Cytogenet, *Cell Genet* 47: 127–131.

Hyer, R. N., Julier, C., Buckley, J. D., Trucco, M., Rotter, J., Spielman, R., Barnett, A., Bain, S., Boitard, C., Deschamps, I., Todd, J. A., Bell, J. I., and Lathrop, G. M., *Am. J. Hum. Genet.*, In press.

Jeffreys, A., Wilson, V., Kelly, R., Taylor, B., and Bullfield, G. (1987) *Nuc. Aci. Res.* 15: 2832–2837.

Jeffreys, A., Wilson, V., and Thein, S. (1985) *Nature* 67–73.

Julier, C., de Gouyon, B., Georges, M., Guenet L., Nakamura, Y., Avner, P., and Lathrop, G. M. (1990) Proc. Natl. Acad. Sci. USA 87: 4585–4589.

Litt, M., and Luty, J. A. (1989) *Am. J. Hum. Genet.* 44: 397–401.

Love, J. M., Knight, A. M., McAleer, M. A., and Todd, J. A. (1990) *Nuc. Aci. Res.* 14: 4605–4616.

Nakamura, Y., Leppert, M., O'Connell, P., Wolff, R., Holm, T., Culver, M., Martin, C., Fujimoto, E., Hoff, M., Kumlin, E., and White R. (1987b) *Science* 235: 1616–1622.

Nakamura, Y., Carlson, M., Krapcho, K., Kanamori, M., and White R. (1988) *Am. J. Hum. Genet.* 43: 854–859.

Roderick T. H., and Guidi J. N. (1989). In: Genetic Variants and Strains of the Laboratory Mouse, 2nd edition, M. Lyon and A. Searle, Oxford University Press, 663–772.

Weber, J. L., and May, P. E. (1989) *Am. J. Hum. Genet.* 44: 388–396.

Wicker, L. S., Miller, B. J., Coker, L. Z., McNally, S. E., Scott, S., Mullen, Y., and Cooper, M. C. (1987). *J. Exp. Med.* 165: 1639–1647.

Wong, Z., Wilson, V., Jeffreys, A. J., and Thein, S. L. (1986) *Nuc. Aci. Res.* 14: 4605–4616.

DNA Fingerprinting: Approaches and Applications
ed. by T. Burke, G. Dolf, A. J. Jeffreys & R. Wolff
© 1991 Birkhäuser Verlag Basel/Switzerland

DNA Fingerprinting: The Utilization of Minisatellite Probes to Detect a Somatic Mutation in the Proteus Syndrome

C. E. Schwartz[a], A. M. Brown[a], V. M. Der Kaloustian[b], J. J. McGill[c], and R. A. Saul[a]

[a]*Greenwood Genetic Center, One Gregor Mendel Circle, Greenwood, S.C. 29646, U.S.A.;* [b]*Montreal Children's Hospital, Montreal, Canada;* [c]*Royal Children's Hospital, Brisbane, Australia*

Summary
Syndromes with localized or segmental abnormalities have been proposed to be the result of a somatic mutation leading to the presence of somatic mosaicism in the tissue. The Proteus syndrome, with its hemihypertrophy, macrodactyly and exostoses, has features which would indicate that the phenotype results from such events. The success of utilizing DNA fingerprint probes to detect somatic mutations in cancer raised the possibility that a similar approach might be successful in an investigation of two patients with the Proteus syndrome. Single band differences were detected with the probe 33.6 in a pair of monozygotic twins discordant for Proteus and in a comparison of tissue from normal and affected areas in another patient. These findings would appear to confirm the hypothesis that the Proteus syndrome results from a somatic mutation. Furthermore, the results indicate that DNA fingerprinting may offer a valuable technique for identifying probes for investigations of similar syndromes.

Introduction

Wiedemann *et al.* in 1983 reported on 4 boys with unusual growth disturbances manifested by partial gigantism of the hands and/or feet, pigmented nevi, hemihypertrophy, subcutaneous hamartomatous tumors and macrocephaly. They coined the term "Proteus syndrome", after the Greek god Proteus who could assume various shapes (polymorphous) to escape threatening situations. Reports as early as Graetz in 1928 have been subsequently identified with the Proteus syndrome, and numerous reviews (Clark *et al.*, 1987; Hotamisligil, 1990; Viljoen *et al.*, 1987) have catalogued the "protean" manifestations of this condition.

The Proteus syndrome is one of the many hamartomatous syndromes that are characterized by the abnormal proliferation of mature cells (often with one cellular element predominating) that normally occur in certain tissues or organs. This abnormal proliferation, by virtue of its overgrowth, can affect the normal morphologic and functional development of the "host" tissue.

The most celebrated patient identified with the Proteus syndrome was Joseph Merrick, known as the Elephant Man in London in the late 1800's. Cohen (1987) noted that Mr. Merrick's features (macrocephaly, hyperostosis of the skull, asymmetric long bone hypertrophy, macro-dactyly, thickened skin and subcutaneous tissue with plantar hyper-plasia) are consistent with the Proteus syndrome and the lack of documented cafe-au-lait spots are inconsistent with neurofibromatosis, the previously suspected diagnosis. Undoubtedly, Mr. Merrick repre-sents one of the most extreme examples of the Proteus syndrome.

Recent reports by Happle (1987) and Hall (1988) propose somatic mosaicism secondary to postzygotic mutation as an explanation for conditions with segmental or localized abnormalities. A somatic muta-tion in which 100% of the cells were affected might likely lead to death of the organism. Fetal survival would be possible if the mutation was present in a mosaic state and occurred during development (Happle 1987). Multiple factors will then determine how that mutation affects the phenotype such as the time the mutation occurs during develop-ment, the cell or tissue type containing the mutation and the viability of the mutant line. Features in these conditions with segmented or local-ized abnormalities would include random distribution of overgrowth, patient to patient variability for tissues affected, wide range of pheno-type and cessation of changes after puberty. The Proteus syndrome fits these features and would therefore appear to be consistent with the hypothesis that the disorder results from a somatic mutation.

Until recently it was difficult to pursue analysis of conditions resulting from a spontaneous somatic mutation. However, recent developments in molecular biology have presented researchers with a means by which to access the role of somatic mutations in the development of these syndromes. Jeffreys et al. (1985) have described hypervariable "mini-satellite" regions of repetitive DNA embedded in unique-sequence DNA that are dispersed throughout the human genome. These loci exhibit multiple allelic variation in the number of tandem repeats and assort independently. Probes containing the 'core' sequences for these minisatellite regions are able to detect more than 20 bands (Jeffreys et al., 1985a) and as such, generate a "DNA fingerprint" of an individual. These fingerprint probes, by detecting many dispersed hypervariable loci simultaneously, are well suited to search for a marker linked to a disease locus (Jeffreys et al., 1986) and also somatic mutations. Indeed, this approach has proved successful in the identification of somatic changes in cancer (Thein et al., 1987; Fey et al., 1988; Armour et al., 1989) and the detection of differences in fetal and trophoblast samples (Butler et al., 1988).

With this background, it would seem appropriate to utilize DNA fingerprinting as a means to detect a somatic mutation involved in giving rise to the Proteus syndrome. Our approach was to study two different cases, each of which provided a separate means of conducting the search. The first involved utilization of monozygotic twins discordant for the Proteus syndrome; the second involved studying samples of affected and unaffected tissue from a single patient. Preliminary results indicate that the fingerprinting technique may offer a sensitive means for identifying markers for the investigation of this syndrome.

Methods

Case Report 1

D.T., the first of twins born at 29 weeks, weighed 3240 grams at birth, whereas his unaffected twin weighed 1350 grams. Abnormalities at birth included large right pinna, large flaccid abdomen with prominent venous patterning, cryptorchidism and equinovarus deformities. He developed progressive lipodystrophy and lipomatosis. Initial motor development was delayed but mental development appears to be normal. The brothers were clearly different at 5 and 10/12 years of age as seen in Fig. 1. D.T.'s height was less than the 2nd percentile and his weight was at the 25th percentile. Craniofacial features included triangular facies with decreased subcutaneous tissue and asymmetric palate and tongue with verrucous projections from the side of the tongue. Massive fatty deposits of the anterior and posterior trunk and buttocks, lipodystrophy of the extremities and wide turned-in feet were also present. Chromosome analysis was 46, XY. His twin has normal mental and physical development. HLA haplotypes and red cell antigens were identical for the two boys.

Case Report 2

G.V. was born at term and weighed 3093 grams with a birth length of 53 cm. No abnormalities at birth were noted. Within the first year of life, a mild degree of asymmetry and pigmented skin lesions in the neck were noted. By the time he began walking at 15 months of age, he had definite genu valgum which persisted despite corrective procedures. Asymmetry has become more pronounced with time. Skeletal changes have included abnormalities of the feet with a bony protuberance on the dorsum of the left foot as well as a valgus deformity and enlargement of both halluces. Craniofacial features were normal except for an unusually large maxillary incisor and a conjunctival dermoid on the right eye. A radiologic survey revealed dysplastic thoracic and lumbar vetebrae, genu valgum with remodeling defect in the distal femora and gigantism and dysplasia of the first tarsal, metatarsal and first toe bilaterally. Computerized tomography scan of the abdomen and pelvis revealed lipomatosis.

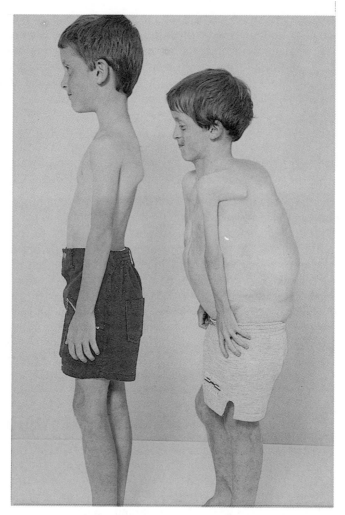

Figure 1. Affected twin (D.T.) on right with Proteus syndrome and normal twin brother (M.T.) on left.

DNA Isolation

Genomic DNA was isolated from peripheral blood obtained from D.T., his twin and his parents using standard procedures previously described (Schwartz *et al.*, 1986). For patient G.V., genomic DNA was isolated from two skin fibroblast cell lines established from biopsies taken from an unaffected and an affected area (Miller *et al.*, 1988).

DNA Digestion

Genomic DNA was digested with five-fold excess of enzyme using specifications supplied by the supplier except that 0.1 M spermidine-

HCl, in 0.1 M Tris-HCl, pH 7.0 (4 μl/100 μl reaction volume) was added to the incubation. Completeness of digestion was assessed prior to use for Southern transfer as outlined previously (Schwartz et al., 1986). The digested samples were electrophoresed in 1% agarose in Tris-acetate buffer (0.04 M Tris-acetate, 0.001 M EDTA) for approximately 48 hours with two changes of the running buffer. The separated DNA fragments were transferred to nylon membrane (Nylon 66, MSI) according to Schwartz et al. (1986). The membranes were prehybridized at 37°C for at least 2 hours in 50% formamide, 3XSSC, 5% dextran sulfate, 0.2% SDS and 50 μg/ml heparin.

DNA Probes and Hybridization Procedure

The minisatellite probes used were 33.15, containing the core minisatellite sequence from λ33.15 in a 600 bp Pst-AhaIII fragment subcloned into M13mp19 previously digested with PstI and SmaI and 33.6, containing the core minisatellite sequence from λ33.6 in a 720 bp HaeIII fragment subcloned into the SmaI site of M13mp8 (Jeffreys et al. 1985a).

Probes 33.15 and 33.6 were labelled either of two ways. The first utilized single-stranded DNA which was then ^{32}P labelled by primer extension as described by Jeffreys et al. 1985. Just as efficient was the utilization of the insert purified from the RF form of the probes after digestion with the appropriate enzymes. Both probes were labelled with ^{32}P-dCTP (New England Nuclear, 3000 Ci/mmole) to a specific activities of 3×10^8 cpm/μg. The minisatellite probes were hybridized to prehybridized filters at 37°C in 50% formamide, 3X SSC, 5% dextran sulfate, 0.2% SDS and 200 μg/ml heparin. The filters were exposed to Kodak XAR-5 film backed by Lightning Plus intensifying screen (DuPont) at -70°C.

Results

Molecular analysis of patient D.T. and his twin brother, M.T., was done using both minisatellite probes and their DNA fingerprints were compared. Minisatellite probe 33.15 revealed no differences between the two boys when their DNA was digested with four enzymes, AluI, HaeIII, HinfI and MboI of which ony one, HinfI, is shown in Fig. 2A. These results are consistent with the fact that D.T. and M.T. are monozygotic twins. However, when the second minisatellite probe, 33.6 was used, some differences were observed with AluI and HaeIII (Fig. 2B). The differences amounted to the presence of a band in D.T. which was apparently absent in M.T. A similar size band is present in the paternal lane, but it likely represents another locus.

100

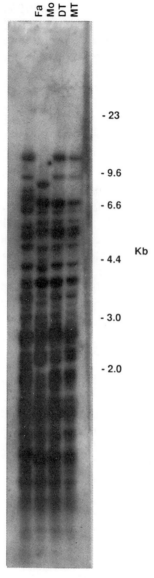

Figure 2. A. Genomic DNA of D.T., brother M.T. and parents digested with HinfI and probed with 33.15.

A similar analysis was conducted using patient G.V. In this case, the analysis involved DNA isolated from fibroblasts obtained from affected and unaffected regions of his body. Both minisatellite probes were uninformative with enzymes HinfI, AluI and HaeIII. At this point, other enzymes not usually employed in DNA fingerprint investigations were used. As before probe 33.15 was uninformative. However, probe

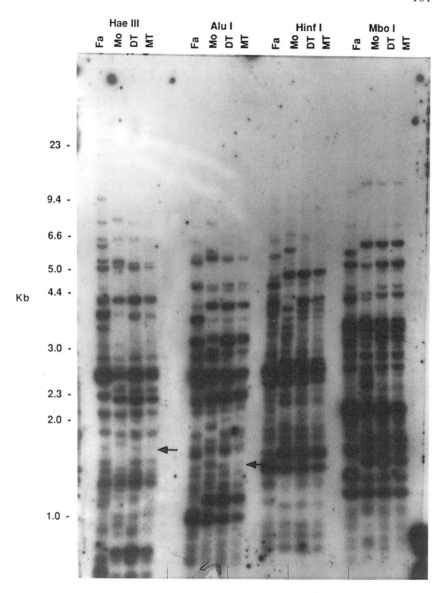

Figure 2. B. Genomic DNA of the same people digested with various enzymes and probed with 33.6.

33.6 revealed the absence of a band in the EcoRI DNA fingerprint from the affected area (Fig. 3). Furthermore, the pattern observed with PvuII revealed the absence of one band and the appearance of a novel band in the DNA fingerprint generated from the affected tissue. These results are not likely due to methylation variation as the two enzymes are insensitive to the presence of methyl groups.

102

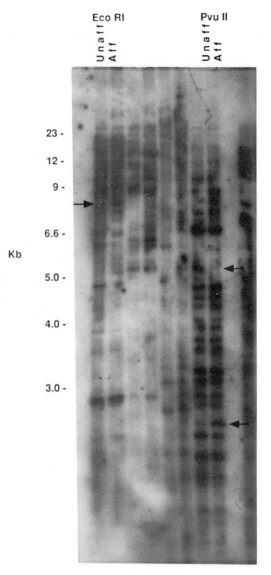

Figure 3. Genomic DNA of patient G.V. digested with various enzymes and probed with 33.6. Unaff = DNA from normal tissue and aff = DNA from affected tissue.

Discussion

The Proteus syndrome, with its unusual overgrowth (hemi-hypertrophy, exostoses, macrodactyly, rugated soles) and tissue dysplasias (hemangiomas, lipomas, nevi), is likely to represent a syndrome which arises by virtue of a somatic mutation, a hypothesis further supported by the monozygotic twins discordant for the Proteus syndrome reported herein. A "prezygotic" mutation affecting either one of the germ cells contributing to the zygote should theoretically affect both twins. Somatic changes which cause cellular proliferation and tissue overgrowth are thought to be caused by either loss of chromosomal regions by deletion, mitotic non-disjunction or mitotic recombination or by point mutations. Minisatellite probes which are able to recognize many hypervariable loci which are thought to have high mutation rates (Jeffreys *et al.*, 1985b; Jeffreys *et al.*, 1988) would appear to provide a means of obtaining a marker for these sites of somatic mutation.

The analysis of D.T. and his twin brother indicate that there probably exists a difference in their DNA fingerprints, generated by probe 33.6, of a single band even though they are monozygotic twins. The difference is present with two enzymes, HaeIII and AluI (Fig. 2B) and the band sizes are comparable for both enzymes. A difficulty does exist in that the band present in D.T. and absent in his normal brother appears to migrate at the place as a band in the paternal lane (HaeIII, Fig. 2B). Most likely the band in D.T.'s fingerprint does not represent the same locus present in the father but rather represents a novel band which arose by a length change of a pre-existing minisatellite locus due to either unequal crossover or sister-chromatid exchange. As the observations are made on DNA obtained from blood, the results might indeed reflect an early developmental somatic mutation event at a minisatellite. This is consistent with the presentation of D.T. who exhibited the Proteus syndrome at birth and who suffers from extensive involvement of various tissues.

The finding of a potential marker for the somatic mutation involved in the Proteus syndrome is supported by the results obtained in a second patient. In this case, the analysis with the same minisatellite probe 33.6 revealed the absence of one band in the DNA fingerprint obtained from the affected tissue (Fig. 3) using two different enzymes which are insensitive to methylation. The disappearance of one band and the appearance of a novel band in the PvuII fingerprint presents strong evidence that probe 33.6 is detecting a somatic mutation event (Thein *et al.*, 1987; Fey *et al.*, 1988; Armour *et al.*, 1989; Kelly *et al.*, 1989).

The approach described in this article should be applicable to other syndromes thought to be the result of somatic mutations, such as Klippel-Trenaunay-Weber syndrome, Sturge-Weber, Ollier disease, and Maffuci syndrome (Hall, 1988). Analysis using monozygotic twins discordant for the syndrome of interest is a good starting point but it should be noted that the approach is not overly efficient. Calculations estimate that for every 500 fingerprint changes observed in a comparison of twin fingerprints, only one will represent a disease related event (R. A. Wells, pers. commun.). This is based on the rather high mutation rate observed for some minisatellite loci (Jeffreys *et al.*, 1985b; Jeffreys *et al.*, 1988). This is borne out by the fact that a separate analysis of twins discordant for the Klippel-Trenaunay-Weber syndrome did not reveal any differences (Schwartz, unpublished data). Therefore, any study should include analysis of affected and unaffected tissue obtained from individual patients. This approach has proved successful in an analysis of cancer patients (Thein *et al.*, 1987; Fey *et al.*, 1988; Armour *et al.*, 1989) and as described in this report was informative for the Proteus syndrome. However, a note of caution needs to be applied to this approach. Due to the nature of many of these overgrowth syndromes, utmost care must be taken to insure that tissue from an unaffected area is truly unaffected.

Acknowledgements
We wish to thank Dr. Alec Jeffreys for providing probes 33.6 and 33.15. This work was supported in part by grants from the Self Foundation, the Abney Foundation, the Duke Endowment and the South Carolina Department of Mental Retardation.

References

Armour, J. A. L., Patel, I., Thein, S. L., Fey, M. F., and Jeffreys, A. J. (1989) Analysis of somatic mutations at human minisatellite loci in tumors and cell lines. *Genomics* 4: 328–334.

Butler, W. J., Schwartz, C. E., Sauer, S. M., Wilson, J. T., and McDonough, P. G. (1988) Discordance in deoxyribonucleic acid analysis of fetus and trophoblast. *Am. J. Obstet. Gynecol.* 158: 642–645.

Clark, R. D., Donnai, D., Rogers, J., Cooper, J., and Baraitser, M. (1987) Proteus syndrome: An expanded phenotype. *Am. J. Med Genet.* 27: 99–117.

Cohen, R. D. (1988) Understanding Proteus syndrome, unmasking the elephant man and stemming the elephant fever. *Neurofibromatosis* 1: 260–280.

Fey, M. F., Wells, R. A., Wainscoat, J. S., and Thein, S. L. (1988) Assessment of clonality in gastrointestinal cancer by DNA fingerprinting. *J. Clin. Invest.* 82: 1532–1537.

Graetz, H. (1928) Über einen Fall von sogenannter totaler halbseitiger Körperhypertrophie. *Z Kinderheikd* 45: 381–403.

Hotamisligil, G. S. (1990) Proteus syndrome and hamartoses with overgrowth. *Dysmorph. Clin. Genet.* 4: 87–102.

Hall, J. G. (1988) Somatic mosaicism: Observations related to clinical genetics. *Am. J. Hum. Genet.* 43: 355–363.

Happle, R. (1986) The McCune-Albright syndrome: A lethal gene surviving by mosaicism. *Clin Genet* 29: 321–324.

Jeffreys, A. J., Wilson, V., and Thein, S. L. (1985a) Hypervariable 'minisatellite' regions in human DNA. *Nature* 314: 67–73.

Jeffreys, A. J., Wilson, V., and Thein, S. L. (1985b) Individual-specific "fingerprints" of human DNA. *Nature* 316: 76–79.

Jeffreys, A. J., Wilson, V., Thein, S. L., Weatherall, D. J., and Ponder, B. A. J. (1986) DNA "fingerprints" and segregation analysis of multiple markers in human pedigrees. *Am. J. Hum. Genet.* 39: 11–24.

Jeffreys, A. J., Royle, N. J., Wilson, V., and Wong, Z. (1988) Spontaneous mutation rates to new length alleles at tandem-repetitive hypervariable loci in human DNA. *Nature* 332: 278–281.

Kelly, R., Bulfield, G., Collick, A., Gibbs, M., and Jeffreys, A. J. (1989) Characterization of a highly unstable mouse minisatellite locus: evidence for somatic mutation during early development. *Genomics* 5: 844–856.

Miller, S. A., Dykes, D. D., and Pulesky, H. F. (1988) A simple salting out procedure for extracting DNA from human nucleated cells. *Nucl. Acid. Res.* 16: 1215.

Schwartz, C. E., McNally, E., Leinwand, L., and Skolnick, M. H. (1986) Linkage of a myosin heavy chain locus to an anonymous single copy locus (D17S1) at 17p13. *Cytogenet. Cell Genet.* 43: 117–120.

Thein, S. L., Jeffreys, A. J., Gooi, H. C., Cotter, F., Flint, J., O'Conner, J. T. J., Weatherall, D. J., and Wainscoat, J. S. (1987) Detection of somatic changes in human cancer DNA by DNA fingerprint analysis. *Brit. J. Cancer* 55: 353–356.

Vilgsen, D. L., Nelson, M. M., de Jong, G., and Beighton, P. (1987) Proteus syndrome in Southern Africa: Natural history and clinical manifestations in six individuals. *Am. J. Med. Genet.* 27: 87–97.

Widemann, H. R., Burgion, G. R., Aldenhoff, P., Kunze, J., Kaufmann, H. J., and Schrig, E. (1983) The Proteus syndrome. *Eur. J. Pediatr.* 140: 5–12.

DNA Fingerprinting: Approaches and Applications
ed. by T. Burke, G. Dolf, A. J. Jeffreys & R. Wolff
© 1991 Birkhäuser Verlag Basel/Switzerland

Genetic Variability of a Satellite Sequence in the Dipteran *Musca domestica*

A. Blanchetot

Department of Biochemistry, University of Saskatchewan, Saskatoon, Sask. S7N 0W0, Canada

Summary
A *Musca domestica* satellite sequence that hybridizes to multiple sites in the genome has been characterised. The cloned sequence, with a structural organisation similar to other satellite DNA, contains at least 50 repeats of a 24 bp unit. This insect-specific satellite hybridizes to sites occurring at different locations and represents only a minor fraction of the *M. domestica* genome. Polymorphic fragments from DNA isolated from single houseflies have been obtained, generating individual specific DNA fingerprints. In addition, pedigree analysis confirms that the DNA fingerprint fragments are inherited in a Mendelian fashion and originate from multiple, apparently dispersed, loci.

Introduction

Despite the interactive and sometimes crucial role played by certain insects relative to human, animal and plant species, almost nothing is known about these organisms at the molecular level. In a broad sense, my research interest is orientated towards the development of molecular tools to attempt genomic mapping of insects having medical or economic importance. The construction of a physical map is usually achieved through the use of genetic markers such as random restriction fragment length polymorphisms (RFLPs), sequences isolated from cDNA and genomic libraries and sequences that are evolutionarily conserved between organisms. The search for genetic markers revealing polymorphism has led to the discovery of tandemly repeated sequences, originally termed minisatellites (Jeffreys *et al.*, 1985). Such sequences, with a structural organisation similar to that of satellite DNA, provide very informative markers that can be exploited in genetic studies of previously unknown organisms. Minisatellite structures have been detected in a wide variety of organisms and intensively studied in the human genome (Jeffreys *et al.*, 1987). Details obtained by sequence analysis show that the extensive genetic variability in minisatellite structures is directly related to variation in copy number of tandem repeats. To my knowledge, most of the work in this area has been focused on higher animal species and investigations on minisatellite sequences related to the further development of the molecular genetics of insects have not yet been reported. In this paper, I will present recent results on the characterisation and the properties of a satellite sequence isolated from the dipteran *Musca domestica* (housefly).

Characterisation of a Specific DNA Satellite Sequence in *M. Domestica*

Only a few minisatellite probes had been discovered at the time when this project was initiated. Assuming an evolutionary conservation of minisatellite sequences, it is sensible to test the probing potential of minisatellite sequences on distantly related organisms. In this context, minisatellite probes, derived from the intron sequence of the myoglobin gene and the M13 sequence, which are effective in vertebrates, were used for hybridization on insect genomes. In my hands none of these probes were found adequate to detect minisatellite structures in insects.

The present study arose from the characterisation of recombinant λsc isolated from a *M. domestica* genomic library having sequence homology with a sodium channel gene. The deduced restriction map of λsc reveals that a 2.3 kb Sst-EcoRI fragment hybridizes to a dispersed family of repeated sequences. For example, genomic Southern blotting of DNA extracted from different housefly strains digested with restriction enzymes such as HindIII and EcoRI, reveals complex profiles of bands. By comparison, similar experiments, using other restriction enzymes such as HinfI and AluI, resolve the complex hybridizing pattern into much simpler profiles and show clear evidence of polymorphic variations among the *M. domestica* strains (see Fig. 1). Interestingly, the 2.3 kb fragment reveals also by Southern hybridization sequence homology in distantly related insect species including honeybees and mosquitoes but not in *Drosophila* (Blanchetot, 1989). Early experiments by DNA reassociation have shown that the *Musca domestica* genome is organised with short repetitive sequences (0.3 kb) that are adjacent to 1–2 kb of unique copy sequences (Hough-Evans *et al.*, 1980). This type of structural organization is representative of those observed in the genomes of most higher animals (Crain *et al.*, 1976).

The plasmid recombinant pHMd.24, containing the 2.3 kb fragment, was characterised by restriction mapping and sequence analysis. The 2.3 kb insert, refractory to digestion by most restriction enzymes, contains unique sequences adjacent to a repeated sequence harboring at least 50 arrays of a 24 bp unit. Each unit is characterised by an internal AluI cleavage site. The different repetitive units show no deletions or insertions and only 9 nucleotide differences occur within the first 10 units when compared to the consensus sequence (Blanchetot, 1991). The structural organisation of pHMd.24, containing short tandemly repeated sequences, is similar to that of satellite

108

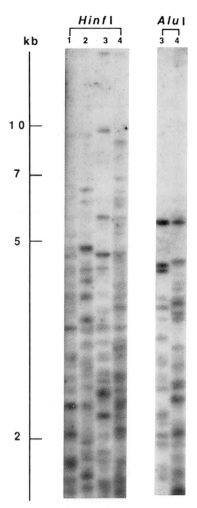

Figure 1. Southern blot analysis of *Musca domestica* strains. 5 µg of genomic DNA from *M. domestica* were digested with HinfI and AluI and probed with pHMd24. Hybridization conditions are as described in Jeffreys *et al.* (1985). The number above the lane indicates the name of the housefly strain. 1: Learn, 2: NAIDM, 3: Hirokawa, 4: WHO.

DNA found in other insect genomes (Lohe *et al.*, 1987). A search of the EMBL database has failed to reveal homology with other repeated sequences.

It might be worth noting that the amplified arrays of 24 bp sequences can potentially fold into a series of repeated stem loop structures flanked by direct repeats. No evidence has been obtained to support the existence of such configurations *in vivo*, and it is unknown whether these structures might have a biological function. An interesting but speculative possibility is that amplified stem loop structures represent a biological buffer to release topological constraints in DNA (Htun and Dahlberg, 1989).

Variability of the *M. Domestica* Satellite in Individual Houseflies

A consensus sequence deduced by comparison of the different 24 bp units was used to construct a synthetic satellite formed by ligation and then amplification, by the polymerase chain reaction, of this 24 bp array. This new probe was hybridized to blots of HaeIII-digested genomic DNA from unrelated housefly individuals. The resulting autoradiographs, shown in Fig. 2, reveal that the probe hybridized to a large number of fragments. The different housefly patterns reveal exten-

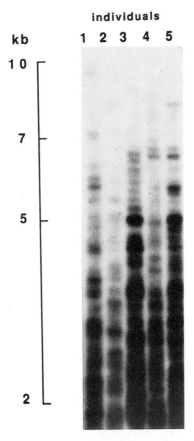

Figure 2. DNA fingerprint profiles from unrelated houseflies. DNA extracted from single houseflies was digested with HaeIII and hybridized with the 24 bp satellite sequence. Experimental conditions related to DNA extraction from individual flies, agarose gel electrophoresis separation and hybridization procedures are described elsewhere (Blanchetot, 1991).

sive variations in band positions and intensities. Each housefly has a unique profile that clearly resembles human and animal DNA fingerprints (Jeffreys *et al.*, 1985; Jeffreys *et al.*, 1987). The nature of the sequence organisation detected by the 24 satellite probe is not yet known. Preliminary results indicate that the clones, isolated after screening a *M. domestica* library with the synthetic 24 satellite probe, are organised as tandemly repeated sequences. Therefore, the genetic polymorphism revealed by the satellite probe, at each locus, appears to be due to variation in the number of copies of the repeated unit at each locus. Quantitative data deduced from the analysis of various individual housefly profiles shows that the DNA fingerprints are derived from an average of 24 variable hybridizing fragments spanning from 2 kb to 10 kb. Thus, the repeated sequences detected by this probe only represent 0.006% of the *M. domestica* genome (Hough-Evans *et al.*, 1980). In addition, the amount of variability among individuals, estimated by measuring the relative position of each band, shows that about 33% of fragments are shared between unrelated houseflies.

Segregation Analysis of the *M. domestica* Satellite Sequence

The segregation of fragments detected by hybridization with the 24 bp satellite sequence has also been determined by pedigree analysis. Fig. 3 represents a HaeIII DNA profile analysis of 6 offspring obtained by mating a male and a virgin female housefly. Fragments directly transmitted from the mother to the offspring are readily identified. Even though the DNA from the father was lost during the extraction process, and thus unavailable for the analysis, paternally-derived fragments can be identified by their presence in some of the offspring while absent in the mother. The pairwise comparison of maternal and paternal fragments from this small family shows no clear evidence of allelism or linkage. In addition, no specific fragment has been found to segregate with sex.

The segregation of bands in the offspring was compared against the expected binomial distribution (Tab. 1). Fragments that were transmitted to all the offspring and considered to be homozygous were ignored. In a similar fashion, single fragments present only in the mother but not transmitted to any sib were also excluded. In this pedigree, 14 distinct maternal and 10 paternal fragments were transmitted to an average of 50% and 43.3% of the offspring. The number of fragments received by each sib approximates the expected binomial distribution and is also consistent with a 1:1 segregation. To my knowledge, this is the first report on the development of a DNA fingerprint in an insect genome that is revealed by an endogenous satellite sequence.

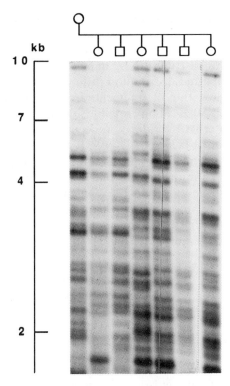

Figure 3. Segregation analysis of fragments in a housefly pedigree. Southern blot of HaeIII digested DNA of a female and six offspring were probed with the 24 bp synthetic satellite sequence. Square and round symbols represent males and females, respectively.

Table 1. Segregation of parental DNA fragments in a housefly family detected by the 24 bp satellite sequence. The expected distribution of fragments is given for a transmission frequency of 50%

Number of offspring receiving the fragment (s)	Maternal fragments		Paternal fragments	
	Observed	Expected	Observed	Expected
0	0	0.2	0	0.2
1	2	1.3	4	0.9
2	4	3.2	1	2.3
3	5	4.3	2	3.1
4	3	3.2	3	2.3
5	1	1.3	0	0.9
6	0	0.2	0	0.2
Transmission frequency (SD)	50.0% (3.5%)		43.3% (4.2%)	

112

In addition to advantageous genetic attributes such as high fertility, *M. domestica* has only three chromosomes. Thus, it is possible that an excess of linkage exists among loci detected by the satellite probe giving rise to an observed distribution of unlinked loci that deviates significantly from the expected distribution ratio of 50%. Such data would be obtained by cosegregation of fragments from a large housefly sibship.

It appears that the 24 bp satellite represents an insect-specific sequence since no cross-hybridization occurs with vertebrate and other invertebrate species. It has recently been found that the design of a new synthetic satellite formed by amplification of 24 bp arrays, having a structural organisation identical to that of *M. domestica* 24 bp satellite but different in nucleotide sequence, acted as a locus-specific probe revealing extreme polymorphism in human and domestic animals. It is tempting to propose that repeats of short sequences flanking an inverted repeat might represent additional minisatellite structures in mammalian genomes.

Acknowledgements
This work was supported by the Centre National de la Recherche Scientifique (France) and the Medical Research Council of Canada.

References

Blanchetot, A. (1989) Detection of highly polymorphic regions in insect genomes. *Nucleic Acids Res.* 17: 3313.

Blanchetot, A. (1991) A *Musca domestica* satellite sequence detects individual polymorphic regions in insects genome *Nucleic Acids Res.* 19: 929–932.

Crain, W. R., Davidson, E. H., and Britten, R. J. (1976) Contrasting patterns of DNA sequence arrangement in *Apis mellifera* (Honeybee) and *Musca domestica* (Housefly). *Chromosoma* 59: 1–12.

Hough-Evans, B. R., Jacobs-Lorena, M., Cummings, M. R., Britten, R. J., and Davidson, G. H. (1980) Complexity of RNA in eggs of *Drosophila melanogaster* and *Musca domestica*. *Genetics* 95: 81–94.

Htun, H., and Dahlberg J. E. (1989) Topology and formation of triple-stranded H-DNA. *Science* 243: 1571–1576.

Jeffreys, A. J., Wilson, V., and Thein, S. L. (1985) Hypervariable "minisatellite" regions in human DNA. *Nature* 314: 67–73.

Jeffreys, A. F., Wilson, V., Wong, Z., Royle, N., Patel, I., Kelly, R., and Clarkson R. (1987) Highly variable minisatellites and DNA fingerprints. *Biochem. Soc. Trans.* 15: 309–317.

Jeffreys, A. J., and Morton, D. B. (1987) DNA fingerprints of dogs and cats. *Anim. Genet.* 18: 1–15.

Lohe, A. R., and Brutlag, D. L. (1987) Adjacent satellite DNA segments in *Drosophila*. Structure of junctions. *J. Mol. Biol.* 194: 171–179.

DNA Fingerprinting: Approaches and Applications
ed. by T. Burke, G. Dolf, A. J. Jeffreys & R. Wolff
© 1991 Birkhäuser Verlag Basel/Switzerland

Analysis of Population Genetic Structure by DNA Fingerprinting

M. Lynch

Department of Biology, University of Oregon, Eugene, Oregon 97403, USA

Summary
DNA fingerprint similarity is now being used widely to make inferences about the genetic structure of natural and domesticated populations, often with little regard to the limitations of such data. This paper provides an overview of the statistical theory of DNA fingerprint analysis with special focus on applications to natural populations for which little if anything is known about the detailed genetics of the DNA profiles. Approaches to estimating individual and population homozygosity, effective population size, population subdivision, and relatedness are reviewed, and issues concerning the biases and sampling properties of the statistics are discussed.

Introduction

Tests of a number of ideas in evolutionary biology, particularly in areas of social organization and kin selection, require accurate estimates of relatedness between individuals, of levels of individual and population homozygosity, and of the degree of population subdivision. Similar types of measurements are needed in programs of genetic conservation, as are estimates of pedigree structure and of breed differentiation. Traditionally, isozymes and blood proteins have been exploited for these purposes, their advantage being a clear Mendelian interpretation of banding patterns on gels. However, protein markers also have significant disadvantages, most notably the need for different protocols for each locus, relatively low levels of detectable polymorphisms, and the weak statistical power associated with loci exhibiting low degrees of variation.

Thus, it comes as no surprise that hypervariable DNA-fingerprinting loci (Jeffreys *et al.*, 1985b, c; Jeffreys *et al.*, 1990) have been embraced widely as a sort of mother lode of genetic information for studies on the structure of natural and domesticated populations. Because such loci usually exist as dispersed families with a common core sequence, multiple restriction fragment length polymorphisms can be visualized simultaneously on the same gel. This has substantial economic advantages. Moreover, the high levels of allelic diversity at DNA-fingerprinting loci imply a maximum amount of information obtained per unit effort. However, one pays a price for these apparent advantages. Although multilocus profiles have a Mendelian basis, an exact genetic interpretation is usually beyond reach. Without laborious breeding experiments, specific bands cannot be associated with particular loci, and fragments with low molecular weights will usually go undetected. Consequently, locus specific gene and genotype frequencies, the basis for all conventional population genetic analyses, are usually unknown. Alternative analytical procedures are required for DNA-fingerprinting studies.

This paper provides an overview of the statistical issues associated with DNA-fingerprint analysis, some of which are covered in more technical detail in Lynch (1988, 1990), Brookfield (1989), and Cohen (1990). Special attention will be given to the types of inferences that can be made in the absence of information or assumptions on allele-frequency distributions, since this will usually be a necessity for the practitioner. There are many aspects of sample preparation, gel running and reading that can lead to problems before the analysis of data even begins (Lander, 1989), but these are ignored below in order to focus on the essential mathematical issues.

The DNA-Fingerprint Phenotype

The fundamental units of data in a DNA-fingerprinting study are the numbers of bands exhibited in individual lanes n_x, where x denotes an individual, and the number of shared bands for pairs of individuals n_{xy}. Although most applications of DNA-fingerprinting have focused on aspects of band sharing, the average number of bands can also provide useful information about population structure. If L is the average number of loci sampled, then the expected number of bands for individual x is

$$E(n_x) = L(2 - H_x), \qquad (1)$$

where H_x is the homozygosity of individual x at an average fingerprinting locus. This formula also applies to the average number of bands for random members of a population when the subscript x is dropped and H is taken to be the mean homozygosity in the population. Since homozygosity is related to the level of inbreeding, a simple investigation of the average number of bands in fingerprint profiles may be a useful means of assessing variation in inbreeding within and between populations.

It seems reasonable to assume that the length variants at VNTR loci are effectively neutral, in which case Malécot's (1948) recursion equation

$$H_t = (1 - \mu)^2 \left[\frac{1}{2N} - \left(1 - \frac{1}{2N}\right) H_{t-1} \right] \qquad (2)$$

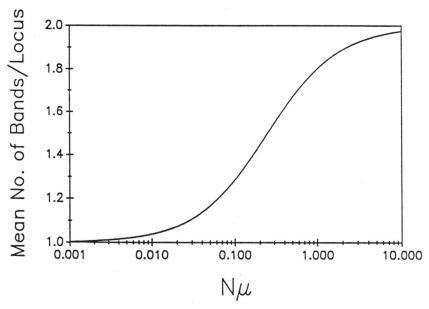

Figure 1. Relationship between the mean number of bands per locus for individuals from a population in drift-mutation equilibrium and the composite parameter $N\mu$. N is the effective population size and μ is the gametic mutation rate.

where N is the effective population size and μ is the mutation rate (to a new discernible length variant), can be used to project the dynamics of homozygosity under the joint influence of mutation and random genetic drift. Under drift-mutation equilibrium, $\hat{H} \simeq 1/(1 + 4N\mu)$, and the expected number of bands per individual is

$$E(\hat{n}) = L\left(\frac{8N\mu + 1}{4N\mu + 1}\right). \tag{3}$$

Figure 1 shows that the average number of bands is a most sensitive indicator of the composite parameter $N\mu$ when the latter is in the range of 0.1 to 1.0.

Rearrangement of Equation (3) yields an estimator for the effective size of an equilibrium population in terms of the average number of bands per individual,

$$N = \frac{\bar{n} - L}{4\mu(2L - \bar{n})}. \tag{4}$$

Application of this formula requires an estimate of the mutation rate, which can in principle be obtained by observing the incidence of nonparental bands in progeny (Jeffreys *et al.*, 1985c, 1988; Gyllensten *et al.*, 1989; Georges *et al.*, 1990). Direct estimation of the number of loci sampled is more difficult in the absence of breeding experiments, but to

116

a first approximation $L \simeq \bar{n}(4 - \bar{S})/4(2 - \bar{S})$, where $\bar{S}$ is the average fraction of shared bands for pairs of nonrelatives (Lynch, 1990). Noting that this expression tends to slightly underestimate L and substituting into Equation (4),

$$N = \frac{4 - 3\bar{S}}{8\mu\bar{S}}$$ (5)

provides an upwardly biased estimate for the effective population size.

It is also possible to use Equation (2) to project the change in the mean number of bands per individual as a population becomes progressively inbred. Letting H_0 be the homozygosity at time zero, $\lambda = 1 - (1/2N)$, and $\phi = \lambda(1 - \mu)^2$,

$$H_t = \frac{(1 - \mu)^2}{2N} + H_0\phi^t + \frac{1}{2N - 1}\sum_{i=2}^{t}\phi^i.$$ (6)

Figure 2 shows the response of the mean number of bands to prolonged periods of small effective size as a function of the conventional inbreeding coefficient $f_t = 1 - \lambda^t$. Note that the expected number of bands declines with increasing f in a roughly linear manner. However, $\bar{n}$ is not

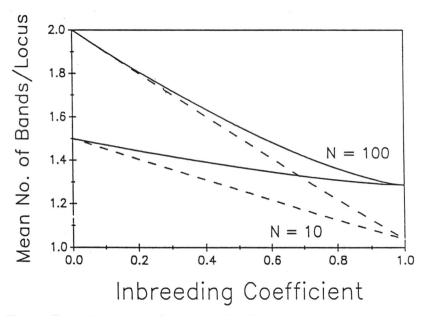

Figure 2. The transient response of the mean number of bands per locus to prolonged periods at effective population sizes of 10 (dashed lines) and 100 (solid lines). The upper and lower sets of lines refer to situations in which the base population is 100% and 50% heterozygous respectively. As the inbreeding coefficient approaches one, a new equilibrium mean number of bands per locus is established, as defined by Equation (3). The inbreeding coefficient can be viewed as the expected homozygosity at more stable loci that have a negligible chance of mutating over periods of time $< t/N$ generations.

just a function of f. Larger populations have higher values of $\bar{n}$ than smaller populations with the same inbreeding coefficient due to the greater opportunities for the replenishment of alleles by mutation in the former. Note also that a substantial amount of heterozygosity can exist at VNTR loci in an equilibrium population, due to high mutation rates, while more stable loci are essentially 100% inbred (and homozygous, as denoted by f).

The Similarity Index

The similarity index for individuals x and y is the number of common bands in their fingerprint profiles divided by the average number of bands exhibited by both individuals,

$$S_{xy} = \frac{2n_{xy}}{n_x + n_y}. \tag{7}$$

For randomly mating population in genetic equilibrium within and between loci, the expected similarity for random pairs of individuals is

$$E(S) = \frac{\sum_{k,i} p_{ki}^2(2 - p_{ki})}{L} \tag{8}$$

where p_{ki} is the frequency of the ith allele at the kth locus (Jeffreys et $al.$, 1985a; Lynch, 1988). This shows that the average similarity does not have a conventional interpretation from the standpoint of population genetics. Since the term $(2 - p_{ki})$ is always greater than one, the mean similarity always overestimates the population homozygosity, $\Sigma_{k,i}\, p_{ki}^2/L$, with the inflation being approximately two-fold when all alleles are rare. Usually, all alleles are not rare, in which case the bias is not so great, but in any event its magnitude cannot be determined in the absence of information on allele frequencies.

On the other hand, over a broad spectrum of gene frequency distributions, the mean similarity does approximate the average identity-in-state-of pairs of individuals (Lynch, 1990). For any locus, identity-in-state of pairs of individuals is either, 0, 0.5, or 1.0, depending on whether the genotypes share 0, 1, or 2 of the same genes. For random members of a panmictic population,

$$E(I) = E(S) - \frac{\sum_{k,i} p_{ki}^3(1 - p_{ki})}{L} \tag{9}$$

Thus, as in the case of population homozygosity, the similarity index is always an upwardly biased estimator of I. However, the bias is a function of the cubed gene frequencies, attaining a maximum of

$E(S) - E(I) = 0.125$ when $p = 0.5$ for all alleles, which is substantially less than $E(S) - E(H) = 0.25$ which arises under the same conditions.

Recalling that the mean similarity somewhat overestimates the population homozygosity and substituting for the expected value of the latter under drift-mutation equilibrium,

$$N = \frac{1 - \bar{S}}{4\mu S} \qquad (10)$$

provides a downwardly biased estimate of the effective population size. An advantage of this expression over Equation (5) is that its derivation does not require an assumption regarding the number of loci sampled. Together, the two formulae should provide an order-of-magnitude approximation of the effective population size, provided the assumption of drift-mutation equilibrium is met.

Sampling Variance of the Basic Statistics

DNA-fingerprint profiles usually involve fairly large numbers (20 to 40) of individually segregating fragments. Thus, by the central limit theorem, the composite statistics n_x and S_{xy} should be roughly normal in distribution, in which case standard statistical procedures can be used to construct confidence limits for the population means of these quantities.

For example, assuming samples have been taken randomly, the sampling variance of the number of bands per individual is simply

$$\text{Var}(n) = \frac{k(\overline{n^2} - \bar{n}^2)}{k - 1}, \qquad (11)$$

where k is the sample size. The sampling variance of the average number of bands is this quantity divided by k. The standard errors of n and $\bar{n}$, which are the square roots of the sampling variances, can be used to construct confidence limits and for other applications associated with hypothesis testing. For instance, if the assumption of drift-mutation equilibrium is valid, populations that differ significantly in $\bar{n}$, must also differ in $N\mu$, most likely in N. The confidence limits for n and $\bar{n}$ are obtained by multiplying the square roots of the sampling variances by the appropriate t values (obtained from any basic statistics text).

Depending on what individuals are used to estimate similarities, computation of the sampling variance of $\bar{S}$ is somewhat more involved. If individuals are used in multiple comparisons (as when all possible combinations of lanes on a gel are compared), the data will not be independent. Similarity measures involving a common member tend to be positively correlated–individuals that happen to exhibit bands that are relatively common in the population will tend to have high similarities with most other individuals, and *vice versa* for those that happen to carry rare alleles. The standard formula for a sampling variance assumes independence of data, and in this case, its application would yield downwardly biased estimates.

To cope with this problem, the investigator has two options. On the one hand, each individual similarity estimate could be based on a unique, nonoverlapping pair of individuals (i.e., 1 and 2, 3 and 4, etc.), and the standard variance formula used. In that case, the total number of similarity estimates can be no greater than half the number of individuals sampled, a sacrifice that most investigators will not want to make. Alternatively similarities can be estimated for arbitrary pairs of individuals, and the sampling variance computed by the formula of Lynch (1990),

$$\mathrm{Var}(\bar{S}) = \frac{k\,\mathrm{Var}(S_{xy}) + 2k'\,\mathrm{Cov}(S_{xy}, S_{xz})}{k^2} \qquad (12)$$

where k is the total number of similarity measures used to estimate $\bar{S}$ and k' is the number of pairs of those measures that share an individual. For example, if all possible comparisons between four individuals have been made $k = 6$ and $k' = 12$. The standard error of $\bar{S}$ is estimated by the square root of this quantity.

In Equation (12), $\mathrm{Var}(S_{xy})$ is the unbiased estimate of the variance of independent similarity measures. It can be estimated with

$$\mathrm{Var}(S_{xy}) = \frac{\sum (S_{wx} - S_{yz})^2}{2k^*}. \qquad (13)$$

where k^* is the number of pairwise comparisons that do not share members. This is a more general formula than that given in Lynch (1990) in that it uses all applicable pairs of data. The sampling covariance of overlapping similarities can also be estimated directly from the data,

$$\mathrm{Cov}(S_{xy}, S_{xz}) = \frac{k^*(\overline{S_{xy} S_{xz}} - \bar{S}^2)}{k^* - 1}, \qquad (14)$$

where k^* is now the number of pairs of comparisons involving shared members. The mean cross-product can be computed most efficiently by focusing on adjacent triplets on gels (i.e., lanes 1, 2 and 3 yield $S_{12} S_{23}$, lanes 4, 5 and 6 yield $S_{45} S_{56}$, etc.) The estimate of $\bar{S}$ to be used in this formula should be based on the same measures as the mean cross-product.

It should be noted that the Equation (12) only estimates the sampling variance of $\bar{S}$ associated with the loci that happened to be included in the survey. It does not account for the error arising from the sampling of a finite number of loci. If, however, one is willing to assume that the sampled loci have gene-frequency distributions that are representative of other such loci throughout the genome, then

$$\mathrm{Var}'(S_{xy}) = \frac{2\bar{S}(1-\bar{S})(2-\bar{S})}{\bar{n}(4-\bar{S})} \tag{15}$$

accounts for this additional source of sampling error (Lynch, 1990). Equation (15) should be used in place of Equation (13) when the mean similarities of different populations are being compared and it is uncertain whether the same loci have been sampled. It should also be used when one is using the set of sampled loci to make inferences about genome wide properties. Since the covariance between similarity measures is proportional to the sampling variance, Equation (14) can be corrected for locus sampling by multiplying by $\mathrm{Var}'(S_{xy})/\mathrm{Var}(S_{xy})$. Although Equation (15) is only a first-order approximation, it tends to slightly overestimate the actual sampling variance and therefore yields conservative estimates for the standard error (Lynch, 1990).

Finally, it should be noted that problems of gel running and reading will ordinarily lead to some sampling variance in the estimates of S_{xy}, particularly when individuals from distant lanes and/or different gels are compared. Such variance has little to do with the genetic properties of the population. Thus, if one has an interest in the true genetic variation of n and/or S, the variance due to technical problems needs to be subtracted from the estimates described above. The latter variance can be estimated by scoring replicate pairs of individuals and computing the variance among replicates.

Population Subdivision

Measures of gene diversity (Nei, 1987) extracted from single-locus analyses are used frequently to estimate measures of population subdivision analogous to Wright's (1951) F-statistics. Strictly speaking, the usual formulae cannot be applied to DNA-fingerprinting data since explicit estimates of gene frequencies are not usually available. Nevertheless, it is possible to test the hypothesis of population subdivision through the use of the similarity index. A question of interest is whether there is significantly less similarity between samples from two populations than expected on the basis of the within-sample similarity. This can be resolved by computing an index of between-population similarity corrected by the within-population similarity,

$$\bar{S}_{ij} = 1 + \bar{S}'_{ij} - \frac{\bar{S}_i + \bar{S}_j}{2}, \tag{16}$$

where $\bar{S}_i$ is the average similarity of individuals within population i, and $\bar{S}'_{ij}$ is the average similarity between random pairs of individuals across populations i and j (Lynch, 1990). When $\bar{S}'_{ij}$ equals the mean similarity in the two populations, $\bar{S}_{ij} = 1$ indicating that the populations are homogeneous. Lynch (1990) presents procedures for computing the sampling variance of $\bar{S}_{ij}$, which are necessary for testing the hypothesis of population subdivision. Problems of nonindependence, mentioned above, need to be accounted for in these computations.

As noted above, the similarity index yields upwardly biased estimates of population homozygosity. Hence, $1 - \bar{S}$ is a downwardly biased estimate of the heterozygosity (or gene diversity). Wright's (1951) index of population subdivision, F_{ST}, is defined to be the fraction of total gene diversity that is attributable to population differentiation. Since the similarity index will bias the estimates of both the within- and between-population homozygosity in the same direction, it seems likely that these biases will nearly cancel out when ratios of the components are employed. Thus, as a first-order approximation,

$$F'_{ST} \simeq \frac{1 - S_b}{2 - S_w - S_b},\qquad(17)$$

where S_b is the average valus of $\bar{S}_{ij}$ over all pairs of populations i, j, and S_w is the average value of $\bar{S}_i$ over all i. F'_{ST} takes on a maximum value of one when populations are fixed for different alleles and a minimum value of zero when there is no subdivision. A standard error for F'_{ST} can be obtained by use of a Taylor expansion approximation that takes into account the sampling variance-covariance structure of S_b and S_w (Lynch, 1990).

Two simple examples show that Equation (17) may work quite well under a broad range of conditions. Suppose that the true mean within-population homozygosity is 0.6 whereas the estimate is $S_w = 0.7$, and the true between-population homozygosity is 0.2 whereas the estimate $S_b = 0.4$. In that case, both the true value of F_{ST} and F'_{ST} are equal to 2/3. If, on the other hand, the between-population homozygosity is 0.8 and $S_b = 0.9$, $F'_{ST} = 1/4$ whereas the actual subdivision is $F_{ST} = 1/3$. In practice, differences of this magnitude are usually well within the bounds of sampling error.

F_{ST} is a measure of inbreeding due to population subdivision. Inbreeding may also result from consanguineous matings within populations. This is usually quantified by Wright's (1951) F_{IS}, which measures the fractional loss of heterozygosity due to local inbreeding on a scale of zero to one. For inbred populations, Equations (1) and (8) generalize to

$$E(n_F) = (1 - F_{IS})E(n_0) + LF_{IS},\qquad(18)$$

$$E(S_F) = E(S_0) + F_{IS}\sum_{k,i} p_{k,i}^2(p_{k,i} - 1)/L.\qquad(19)$$

These formula show that there is no simple way to quantify the degree of local inbreeding from direct observations of the average number of bands per individual or of average similarity. If, however, a sample of the population can be mated randomly, then F_{IS} can be estimated from the observed mean number of bands before ($\bar{n}_F$) and after ($\bar{n}_0$) mating,

$$F_{IS} \simeq \frac{\bar{n}_F - \bar{n}_0}{L - \bar{n}_0}, \tag{20}$$

provided an estimate of the number of loci is available (see above). In the next section, a simple method for testing for consanguineous mating, which does not require an artificial breeding program, is introduced.

For the same reason that the similarity index can be used to obtain a nearly unbiased estimate of F_{ST}, it should also be possible to closely approximate the genetic distance between populations. An analog of Nei's (1972) estimator is

$$D'_{ij} \simeq -\ln\left(\frac{\bar{S}'_{ij}}{\sqrt{\bar{S}_i \bar{S}_j}}\right). \tag{21}$$

An expression for the sampling variance of D'_{ij}, which requires the use of some formulae in Lynch (1990), can be found in Nei (1987).

Under the assumption of drift-mutation-migration equilibrium, relatively simple expressions exist for expected values of F_{ST} and D in terms of migration and mutation rates (Nei, 1987), so Equations (17) and (21) may be of some use in estimating these parameters. For example, for populations that have been completely isolated for t generations, the expected value of D is $2\mu t$.

Estimation of Relatedness

The fact that most individuals in outbred populations have unique DNA-fingerprint profiles has encouraged the belief that the enormous power for identifying individuals would extend to identifying specific kinds of relationships. For a few types of applications, such as identification and/or exclusion of parentage, DNA fingerprinting has, in fact, been highly profitable (Jeffreys et al., 1985a; Wetton et al., 1987; Brookfield, 1989; Burke et al., 1989; Morton et al., 1990). However, successful extensions to more distant relationships are still notably absent.

It stands to reason that the DNA-fingerprint similarity between a pair of individuals should increase with their degree of relatedness. Lynch (1988) showed formally that this relationship is linear in a randomly mating population,

$$E(S) = \bar{\theta} + r(1 - \bar{\theta}), \tag{22}$$

where the relatedness r is the proportion of genes identical by descent between two individuals ($r = 0.5$ for parent-offspring, 0.25 for grandparent-grandchild and half-sibs, etc.), and $\bar{\theta}$ is the fraction of bands shared by nonrelatives. This formula states quite simply that the expected similarity of a pair of individuals is the sum of the probability that genes in the two individuals are identical by descent and the probability that they are not identical by descent but nevertheless identical in state. Equation (22) shows that a regression of S or r should have an intercept equal to $\bar{\theta}$ and a slope equal to $(1 - \bar{\theta})$. Thus, provided the population is panmictic, the regression of S on r, which can be exploited in future studies, does not actually require the availability of pairs of individuals of various known degrees of relatedness. It merely requires an accurate estimate of the mean similarity between nonrelatives, a quantity that should ordinarily be obtainable.

In principle, S would provide a nearly unbiased estimate of r if the investigator were able to exploit a set of loci for which $\bar{\theta}$ is nearly zero. When most loci contain a large number of alleles, the genes in two individuals are unlikely to be identical in state unless they are also identical by descent. Unfortunately, for most existing studies $\bar{\theta}$ is on the order of 0.2 or much greater (Table 1), in which case the similarity index substantially exceeds the relatedness. Moreover, the magnitude of the bias increases with the more distant relationships—the very ones that it had been hoped that DNA fingerprinting would elucidate.

At first sight, one might expect that the bias could be eliminated by inserting the population estimate of $\bar{\theta}$ into Equation (22) and solving for r as a function of S. This should indeed work on average. There are, however, a couple of additional problems if one desires to estimate the relationships between specific pairs of individuals. First, the bias between S_{xy} and r_{xy} is not simply a function of the population average $\bar{\theta}$ but of the fraction of bands that the specific individuals x and y share

Table 1. Average fraction of shared bands for pairs of nonrelatives

Natural populations:		
House sparrows	0.1–0.3	Burke and Bruford 1987
Pied flycatchers	0.2	Wetton et al. 1987
Dunnocks	0.2	Burke et al. 1989
Purple martins	0.2	Morton et al. 1990
Channel Island foxes	0.7–1.0	Gilbert et al. 1990
Humans	0.2	Jeffreys et al. 1985c
Domesticated species:		
Chickens	0.4–1.0	Kuhnlein et al. 1990
Dogs	0.5	Jeffreys and Morton 1986
Cats	0.5	
Cattle	0.3–0.4	Georges et al. 1988
Horses	0.3–0.7	
Pigs	0.5–0.7	

124

with nonrelatives,

$$r_{xy} = \frac{E(S_{XY} - \theta_{xy})}{1 - \theta_{xy}} \tag{23}$$

(Lynch, 1988). θ_{xy} may be greater or less than $\bar{\theta}$ depending upon whether x and/or y carry relatively common or rare alleles. θ_{xy} will often be an unobservable quantity, and even when estimable, will necessarily be less accurate than $\bar{\theta}$. Second, the range of expected similarity values is $\bar{\theta}$ to 1. Consequently, as the average similarity between nonrelatives increases, the regression of S on r becomes shallower, i.e., S becomes a less sensitive indicator of r. This is a serious issue since the variance among similarity values for specific kinds of relationships is quite high, even for parents and offspring if $\bar{\theta}$ is moderately high (Lynch, 1988).

For studies in which there is a moderate amount of similarity between nonrelatives (as is the case in almost all existing studies), there is little question that the distribution of similarity measures from adjacent types of relatives (e.g., parent-offspring vs. grandparent-grandchild) will be broadly overlapping. Assuming that the distribution of similarity is approximately normal, one can use Equation (22) in conjunction with

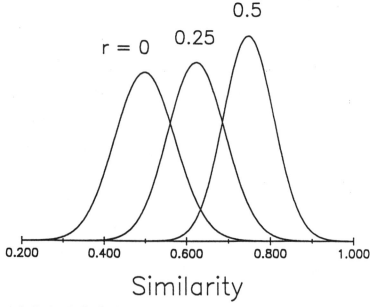

Figure 3. Expected distributions of similarity for individuals with $r = 0$ (nonrelatives), $r = 0.125$ (half-sibs or grandparent-grandchild), and $r = 0.5$ (full-sibs) obtained by use of Equations (15) and (22). It is assumed that the mean similarity of nonrelatives is $\bar{S}_0 = 0.5$ and the mean number of bands is $\bar{n} = 40$.

Equation (15) to derive the expected distributions of S for different degrees of relatedness. Thus, knowing only $\bar{\theta}$ and $\bar{n}$ in advance, it is possible to get a good impression of the possible power of DNA-fingerprinting to resolve issues of relatedness (Fig. 3). For investigators involved in long-term research programs with specific populations, it may be useful to go a step further and directly establish base-line data on the empirical distributions of S for known types of relatives. Such distributions could be used in future investigations to test the hypothesis that a pair of unknown individuals has a similarity value consistent with a specific type of relationship.

Finally, it may be noted that a modification of Equation (24) can be used to generate an estimate of the degree of relatedness between mates:

$$r_m = \frac{\bar{S}_m - \bar{\theta}}{1 - \bar{\theta}}, \tag{24}$$

where $\bar{S}_m$ is the mean number of shared bands for mates. A simple test for consanguineous mating is to evaluate whether $\bar{S}_m$ and $\bar{\theta}$ are significantly different.

Discussion

The main point of this paper has been to show how information extracted from DNA-fingerprint analyses is related to conventional population genetic parameters. Most of the connections that have been pointed out make no assumptions about the gene-frequency distributions for the loci sampled. In principle, much more precise statements can be made if the distribution of genotypes is known since the exact distribution of band number and band-sharing can be expressed in terms of formulae of Chakraborty (1981). However, this possibility will not be realized as long as investigators rely on multilocus probes.

The data from multilocus DNA-fingerprinting studies are somewhat phenomenological in nature, and consequently they do not usually yield unbiased estimators of most of the measures of population structure that routinely appear in population genetic theory. Nevertheless, as noted above, they do provide reasonable first approximations in many cases. It is even possible that the magnitude of the bias from DNA-fingerprinting surveys is relatively small compared to the sampling error that results from more conventional surveys involving isozymes and single-locus RFLPs. That is a useful area for future theoretical and empirical investigation.

Acknowledgements
This work has been supported by NSF grant BSR 86-00487 and PHS grant R01 GM36827-01. Helpful comments were provided by R. Chakraborty, A. Jeffreys, and other conference participants.

126

References

Brookfield, J. F. Y. (1989) Analysis of DNA fingerprinting data in cases of disputed paternity. *IMA J. Math. Appl. Med. Biol.* 6: 111–131.

Burke, T., and Bruford, M. W. (1987) DNA fingerprinting in birds. *Nature* 327: 149–152.

Burke, T., Davies, N. B., Bruford, M. W., and Hatchwell, B. J. (1989) Parental care and mating behavior of polyandrous dunnocks *Prunella modularis* related to paternity by DNA fingerprinting. *Nature* 338: 249–251.

Chakraborty, R. (1981) The distribution of the number of heterozygous loci in an individual in natural populations. *Genetics* 98: 461–466.

Cohen, J. E. (1990) DNA fingerprinting for forensic identification: potential effects on data interpretation of subpopulation heterogeneity and band number variability. *Am. J. Hum. Genet.* 46: 358–368.

Georges, M., Lathrop, M., Hilbert, P., Marcotte, A., Schwers, Swillens, S., Vassart, G., and Hanset, R. (1990) On the use of DNA fingerprints for linkage studies in cattle. *Genomics* 6: 461–474.

Georges, M., Lequarré, A.-S., Castelli, M., Hanset, R., and Vassart, G. (1988) DNA fingerprinting in domestic animals using four different minisatellite probes. *Cytogenet. Cell. Genet.* 47: 127–131.

Gilbert, D. A., Lehman, N., O'Brien, S. J., and Wayne, R. K. (1990) Genetic fingerprinting reflects population differentiation in the California Channel Island fox. *Nature* 344: 764–767.

Gyllensten, U. B., Jakobsson, S., Temrin, H., and Wilson, A. C. (1989) Nucleotide sequence and genomic organization of bird minisatellites. *Nucleic Acids Res.* 17: 2203–2214.

Jeffreys, A. J., Brookfield, J. F. Y., and Semeonoff, R. (1985a) Positive identification of an immigration test-case using human DNA fingerprints. *Nature* 317: 818–819.

Jeffreys, A. J., and Morton, D. B. (1987) DNA fingerprints of dogs and cats. *Animal Genetics* 18: 1–15.

Jeffreys, A. J., Neumann, R., and Wilson, V. (1990) Repeat unit sequence variation in minisatellites: a novel source of DNA polymorphism for studying variation and mutation by single molecule analysis. *Cell* 60: 473–485.

Jeffreys, A. J., Royle, N. J., Wilson, V., and Wong, Z. (1988) Spontaneous mutation rates to new length alleles at tandem-repetitive hypervariable loci in human DNA. *Nature* 332: 278–280.

Jeffreys, A. J., Wilson, V., and Thein, S. L. (1985b) Hypervariable 'minisatellite' regions in human DNA. *Nature* 314: 67–73.

Jeffreys, A. J., Wilson, V., and Thein, S. L. (1985c) Individual-specific 'fingerprints' of human DNA. *Nature* 316: 76–79.

Kuhnlein, U., Zadworny, D., Dawe, Y., Fairfull, R. W., and Gavora, J. S. (1990) Assessment of inbreeding by DNA fingerprinting: development of a calibration curve using defined strains of chickens. *Genetics* 125: 161–165.

Lander, E. S. (1989) DNA fingerprinting on trial. *Nature* 339: 501–505.

Lynch, M. (1988) Estimation of relatedness by DNA fingerprinting. *Mol. Biol. Evol.* 5: 584–599.

Lynch, M. (1990) The similarity index and DNA fingerprinting. *Mol. Biol. Evol.* 7: 478–484.

Lynch, M., and Crease, T. J. (1990) The analysis of population survey data on DNA sequence variation. *Mol. Biol. Evol.* 7: 377–394.

Malécot, G. (1948) Les Mathématiques de l'Hérédité. Masson et Cie. Paris.

Morton, E. S., Forman, L., and Braun, M. (1990) Extrapair fertilizations and the evolution of colonial breeding in purple martins. *Auk* 107: 275–283.

Nei, M. (1972) Genetic distance between populations. *Amer. Natur.* 106: 283–292.

Nei, M. (1987) Molecular evolutionary genetics. Univ. Columbia Press. New York.

Wetton, J. H., Carter, R. E., Parkin, D. T., and Walters, D. (1987) Demographic study of a wild house sparrow population by DNA fingerprinting. *Nature* 327: 147–152.

Wright, S. (1951) The genetical structure of populations. *Ann. Eugen.* 15: 323–354.

DNA Fingerprinting: Approaches and Applications
ed. by T. Burke, G. Dolf, A. J. Jeffreys & R. Wolff
© 1991 Birkhäuser Verlag Basel/Switzerland

Population Genetics of Hypervariable Loci: Analysis of PCR Based VNTR Polymorphism Within a Population

R. Chakraborty[a], M. Fornage[a,b], R. Gueguen[b], and E. Boerwinkle[a]

[a]*Genetics Centers, University of Texas Graduate School of Biomedical Sciences, P.O. Box 20334, Houston, Texas 77225, USA;* [b]*Center for Preventive Medicine, Nancy, France*

Summary
Using a polymerase chain reaction (PCR) based method, genotypes at two hypervariable loci (3' to the Apo-B-structural gene and at the ApoC-II gene) were determined by size classification of alleles. Genotype data at the Apo-B locus (Apo-B VNTR) were obtained on 240 French Caucasians; the sample size for the ApoC-II VNTR was 162. For 160 individuals two-locus genotype data were available. Applications of some recently developed statistical methods to these data indicate that both of these loci are at Hardy-Weinberg equilibrium (HWE) and there is no indication of allelic associations between these two unlinked loci. In addition, the observed numbers of alleles (12 for the Apo-B and 11 for the ApoC-II VNTR loci) are also consistent with their respective expectations based on the observed heterozygosities (76.9% for the Apo-B and 85.9% for the ApoC-II loci) suggesting genetic homogeneity of this population-based sample. The multimodal distribution of allele sizes observed for both loci indicate that the production of new alleles at such VNTR loci may be caused by more than one molecular mechanism. The utility of such highly polymorphic loci for human genetic research and forensic applications are discussed in the context of these findings.

Introduction

A large number of DNA segments in the human genome contain a variable number of short tandemly repeated sequences. To varying degree, the core repeat unit is conserved from one repeat to another. The core sequence may also vary among such loci from 2 bases to several kilobases. The copy number of such core sequences reveals genetic variation several orders larger than that detected by classical serologic and biochemical genetic markers. Since the first demonstration of such a highly polymorphic, locus-specific sequence in the human genome by Wyman and White (1980), numerous hypervariable regions have been described that flank structural loci in the human genome (e.g., Bell *et al.*, 1982; Capon *et al.*, 1983; Goodbourn *et al.*, 1983; Jeffreys *et al.*, 1985; Boerwinkle *et al.*, 1989; Ludwig *et al.*, 1989; reviewed also by Jeffreys and Wolff, this volume). It is now well-recognized that the human genome contains a large number of these polymorphic segments, numbering possibly in the thousands. Several acronyms of such polymorphic systems are proposed. Jeffreys *et al.*

(1985) suggested locus-specific 'minisatellites'; Nakamura *et al.* (1987) coined the term 'VNTR' (Variable Number of Tandem Repeats), while several others (e.g., Balazs *et al.*, 1989; Ludwig *et al.*, 1989) used the terminology 'HVR' (Hypervariable Region) to describe this type of genetic variation. When the core unit is small (di-, tri, or tetra-nucleotide), Edwards *et al.* (1991) called them 'STR'-loci (Short Tandemly Repeated loci).

Since the allelic designation at these VNTR loci can be conveniently defined by the number of repeat units of the core sequence and each locus conforms to simple codominant Mendelian mode of inheritance, such loci are extremely useful for human genetic research. Recent tabulation of genetic markers used for human gene mapping indicates that collectively nearly 50 per cent of all genetic markers belong to this category (Kidd *et al.*, 1989). The increased efficiency of VNTR loci, compared to classical biochemical and RFLP markers, arises from the fact that the number of different alleles found at any VNTR locus is generally much larger. In addition, these VNTR loci have a high heterozygosity (sometimes approaching levels as high as 95–99 percent). As a consequence, detection of recombination between a VNTR locus and a disease gene or other genetic markers is simple because both parents can be heterozygous and provide four distinct alleles at such loci far more commonly than other classical markers.

The presence of large numbers of alleles at VNTR loci also makes them useful in the context of paternity testing and forensic medicine. A growing body of literature suggests that their utility is not only limited to academic circles and biomedical research; criminal justice and social welfare agencies can also benefit immensely from the application of such loci (Craig *et al.* 1988).

These advantages notwithstanding, concerns have been raised with regard to the population genetic characteristics of such polymorphisms (Lander, 1989; Cohen, 1990). In part this is caused by limited population data of VNTR polymorphisms from genetically well-characterized populations. Five studies in this regard are noteworthy. Baird *et al.* (1986) have shown extensive variation at two VNTR loci, HRAS-1 and D14S1, in three major ethnic groups. Using such data, Clark (1987) postulated that allelic variation at these loci follow the mutation-drift model of genetic variation (Kimura and Crow, 1964). Jeffreys *et al.* (1988) entertained other models of allele frequency distribution such as a finite allele mutation model or a stepwise mutation model to explain the relationship between heterozygosity and mutation rate at such loci. Flint *et al.* (1989) showed that the differences of heterozygosity levels of specific VNTR loci across populations can be used to postulate whether or not events such as a population bottleneck could have occurred during the geographic dispersal of humans. Chakraborty (1990a, b) argued that even a single VNTR locus can provide information concerning substructuring within a population with a statistical power far greater than several classical genetic markers studied simultaneously.

The well known advantages of VNTR loci such as their high heterozygosity and a large number of alleles also poses problems in the statistical interpretation of population data. For example, the presence of a large number of alleles increases sample size requirements for population survey studies, since even with respectable sample sizes all possible genotypes can not be generally detected. As a result, precisions of allele frequency estimates are compromised which subsequently hinders statistical calculations based on classical text book methods of analysis. Since all genotypes are not observed, and even the ones observed occur in low frequencies, the application of the large sample theory of chi-square goodness-of-fit test is not valid for checking whether or not the observed genotypic distribution conforms to their Hardy-Weinberg expectations. Similarly, the association among alleles at multiple loci cannot be adequately determined from their genotypic distributions by standard summary measures such as linkage disequilibrium. In addition, a single VNTR locus may exhibit more than one allele of similar size and incomplete resolution of distinct alleles by Southern gel electrophoresis is not uncommon. Such inescapable laboratory phenomena can not be overlooked in the statistical interpretation of VNTR population data (Devlin *et al.*, 1990). There are also claims that the hypervariability at these loci is caused by 'mutational' changes whose rate is high in comparison with other classical markers. Hence, when used singly, a VNTR locus may lead to a wrong conclusion regarding biological relationships between individuals (Odelberg *et al.*, 1989). Furthermore, very little is known about the molecular mechanism of production of new alleles at VNTR loci, although it has been suggested that replication slippage, sister chromatid exchange, and/or unequal recombination may be involved in this process. Virtually nothing is known regarding the functional requirements of such loci in general, even though DNA-binding proteins which appear to bind specifically to VNTR loci are known to exist (Collick and Jeffreys, 1990). In view of these uncertainties, it is difficult to determine the mode and rate of evolution of genetic variation at VNTR loci.

In spite of these difficulties, we believe that the statistical interpretation of VNTR polymorphism data is not an insurmountable problem. Certain modifications of classic population genetic methods of data analysis can be introduced which lead to rigorous and legitimate estimation and hypothesis-testing principles for analyzing such data. Such methods also may suggest molecular mechanisms that generate and maintain genetic variation at these loci. We have initiated a series of studies on the population genetics of VNTR polymorphism at our Center, and this presentation is a preliminary summary of several results from these studies. Two VNTR loci have been typed by PCR-based experimental protocols which can be employed for population surveys at such loci and which minimize the problem of incomplete resolution of alleles (Boerwinkle *et al.*, 1989). Here we describe the genotype and allele frequency distributions at two VNTR loci; one locus is 3′ to the apolipoprotein-B (ApoB) gene and the other is within the ApoC-II gene, which have been scored on individuals belonging to 121 nuclear families. Several alternative methods are suggested for examining the conformity of the genotypic distribution of VNTR data with HWE and for testing independent segregation of alleles at unlinked loci in the presence of large number of alleles. Futhermore, the allele frequency distributions of these loci are used to predict possible molecular mechanisms that generate and maintain such genetic variation. Characterization of such population genetic features implies that VNTR polymorphisms detected through PCR-based studies conform to classic population genetic principles, and hence are useful in human genetic research and forensic applications.

Materials and Methods

The genotype data analyzed here were obtained from a random sample of 121 nuclear families taking part in routine health examinations at the Center for Preventive Medicine in Nancy, France. Genomic DNA was isolated by phenol/chloroform extraction of proteinase K treated crude buffy coat preparations. Two VNTR loci with different core sequences were selected for the present analyses. The first is an AT-rich VNTR 3′ to the human apolipoprotein B gene on chromosome 2 (Huang and Breslow, 1987). Detailed PCR-based methods for typing this VNTR have been previously presented (Boerwinkle *et al.*, 1989). Our previous results indicate that this locus differs in the number of copies (from 29 to 51) of a conserved core sequence 14 or 15 base pair long (Boerwinkle *et al.*, 1989). The second VNTR locus is a microsatellite located in the first intron of the human apolipoprotein C-II gene on chromosome 19 (Fojo *et al.*, 1987). This ApoC-II VNTR consists of a $(TG)_n(AG)_m$ core motif repeated from 16 to 34 times. The oligonucleotides used for priming the PCR bind immediately adjacent to the $(TG)_n(AG)_m$ block. One member of the pair of primers was 5′ end-labeled using bacteriophage T4 kinase. The PCR was carried out in a 50 μl volume containing approximately 0.5 μg of genomic DNA and samples were processed through 30 temperature cycles consisting of 1 minute at 92°C (denaturation), 1 minute at 55°C (annealing), and 1.5 minutes at 72°C (elongation). After 10 cycles with only cold oligonucleotide the reaction mixture was spiked with the end-labeled oligonucleotide for the remaining 20 cycles. The amplified DNA was analyzed after being electrophoresed on 8% denaturing polyacrylamide gels and exposed to X-ray film for 20 hours. Size standards were created by dideoxy sequencing using M13 mp18 as a template. The size of the PCR products were directly determined from the size of the co-migrating M13 fragment in the ladder (data not shown).

The family data were used to verify Mendelian segregation of the identified alleles at the ApoB and ApoC-II VNTR loci. The parents of these families represent a sample of unrelated individuals and were used for allele frequency estimation and other calculations. Therefore, these data can be regarded as from a random sample of Caucasian individuals of French ancestry. Genotype data on 240 individuals for the ApoB locus and 162 individuals for the ApoC-II locus were used for the analyses presented here. Two locus genotype data were available for 160 individuals.

Because one purpose of this paper is to describe the analytical tools for the analysis of allele and genotype frequency data at VNTR loci, the statistical methods are not presented in this section but rather are given along with a corresponding question and resulting inference in the next section. Statistical analyses consist of: (1) analysis of genotype and allele frequency distribution for each locus individually to test whether or not HWE predictions hold for these two loci, (2) joint analysis of two-locus genotype data to determine independent segregation of alleles at these two unlinked loci; (3) examination of the relationship between heterozygosity and the number of alleles at each locus to determine the underlying mechanism of production of new alleles at these loci, and finally (4) to postulate possible reasons for the shape of the allele size distributions at these loci.

Results

Single-Locus Genotypic Distributions

Typically VNTR locus variation is codominant and multi-allelic. Letting k be the number of segregating alleles, there are $k(k + 1)/2$ possible genotypes at a locus, k of which are homozygous $A_i A_i$ ($i = 1, 2, \ldots, k$) and $k(k - 1)/2$ are heterozygous $A_i A_j$ ($i < j = 2, 3, \ldots, k$). It should be possible to observe all possible $k(k + 1)/2$ genotypes in any given sample. However, the sample size needed to observe all possible genotypes is generally large when k is large.

We observed 12 different alleles at the ApoB VNTR locus and 11 at the ApoC-II VNTR locus in a sample of 240 and 162 individuals, respectively. The number of possible genotypes ($k(k + 1)/2$), therefore, are 78 and 66, respectively. Table 1 shows the observed genotype and allele frequency distributions at the ApoB locus, demonstrating that even though the sample size ($n = 240$) is much larger than the number of possible genotypes, we observed only 42 distinct genotypes. Of the possible 12 homozygote genotypes only 8 are observed and of the possible 66 heterozygote genotypes only 34 are observed. A large fraction of the observed genotypes have frequencies below 5 individuals per genotype.

Obviously, such data cannot be subjected to a classical goodness-of-fit test examining whether or not the genotypic distribution conforms to HWE predictions. This is so because a chi-square approximation is inaccurate when the observed frequencies are small, and also we cannot assign any well-defined degrees of freedom because of the absence of a large fraction of genotypes in the sample. Three alternative test statistics are used for this purpose. First, grouping the entire data of the genotype table (Tab. 1) into two classes, heterozygotes and homozygotes (irrespective of the allele types, observed), we found that there are 59 individuals that are homozygous at this locus and 181 heterozygous. The expected numbers of individuals in these two categories, based on the observed allele frequencies (shown on the last column of Tab. 1), are 55.38 and 184.62, respectively, and their difference from the observed ones is not significant by a chi-square test with 1 d.f. ($P > 0.55$). Second, using the theory of Chakraborty et al. (1988) and Chakraborty (1991), we also determined the expectations (and their standard errors) of the number of distinct heterozygous and homozygous genotypes that can be observed in a sample of 240 individuals at the ApoB VNTR locus when the alleles unite at random to form the genotypes of these individuals. The sampling distributions of these statistics, under the HWE assumption, can be evaluated exactly (Chakraborty, 1991). In particular, denoting the expected genotype frequencies (under HWE) by $Q_1, Q_1, \ldots, Q_K$, the mean and variance of the number of distinct

Table 1. Genotype and allele frequency distributions at the ApoB VNTR Locus in a random sample of 240 individuals from Nancy, France

	Alleles												Freq.
	29	31	33	35	37	39	41	43	45	47	49	51	
29	—	1	1	—	—	—	—	—	—	—	—	—	2
31		3	2	7	6	1	2	1	—	2	2	1	31
33			2	2	10	2	—	—	—	1	3	1	26
35				12	50	1	2	—	—	10	6	2	104
37					36	11	4	—	2	16	17	6	194
39						2	1	—	—	—	3	2	25
41							1	—	—	—	—	—	11
43								—	—	—	—	—	1
45									—	—	—	—	2
47										1	2	1	34
49											2	—	37
51												—	13

genotypes observed in a sample of n individuals are given by

$$\mu = K - T_1, \quad \text{and} \quad \sigma^2 = T_1(1 - T_1) + 2T_2, \qquad (1a)$$

respectively, where

$$T_1 = \sum_{i=1}^{K} (1 - Q_i)^n, \quad \text{and} \quad T_2 = \sum_{i>j=1}^{K} (1 - Q_i - Q_j)^n. \qquad (1b)$$

Note that in equations (1) and (2) $K = k$ for the homozygous and $Q_i = p_i^2$, the square of the i-th allele frequency, and $K = k(k-1)/2$ for the heterozygous genotypes, with Q_i's being an array of $k(k-1)/2$ probabilities representing each heterozygote genotypes ($2p_ip_j$). Application of this method to the present data showed that the observed numbers of distinct homozygous and heterozygous genotypes for the ApoB locus, 8 and 34, are not significantly different from their expectations, 6.37 and 32.98. Therefore, we conclude that the observed numbers of distinct genotypes are also in concordance with HWE predictions.

Since all of these statistics disregard the specific allele types observed in the sample, and hence these tests do not detect deviations of each specific genotype frequency from its HWE prediction, we used a third test criterion which does not have this limitation. This is the G-statistic (Sokal and Rohlf, 1969), a likelihood ratio, which should not be significant if the HWE prediction is correct. Unfortunately, although the G-statistic is a contrast of every observed genotype frequencies with their respective expectations, no standard statistical distribution can be applied to determine the significance level of the G-statistic because of the absence (or small frequencies) of several genotypes. We employed a shuffling algorithm to determine the empirical distribution of the G-statistic, by randomly permuting the 480 allele labels (12 of them, since there are 12 different observed alleles) and reconstructing genotypes by pairing the shuffled alleles at random. The observed value of G in the given data (of Tab. 1) was 60.18, and its empirical probability level was 0.62, suggesting that by chance 62% times we could have observed G-values larger than the one observed under the assumption of HWE. Therefore, none of the three alternative tests offered any suggestion of non-random association of alleles at the ApoB VNTR locus. In conclusion, the genotype distribution at this locus among the French Caucasians can be assumed to satisfy the HWE predictions. The observed heterozygosity at this locus is 0.745 ± 0.028, and its expectation based on the estimated allele frequencies is 0.769 ± 0.014.

Table 2 shows the observed genotypic and allele frequency distributions at the ApoC-II VNTR locus. Eleven segregating alleles are found at this locus, but of the possible 66 genotypes only 42 are observed. As in the case of the ApoB VNTR, each of the three alternative test criteria shows that the observed genotypic distribution is in accordance with the

Table 2. Genotype and allele frequency distributions at the ApoC-II VNTR Locus in a random sample of 162 individuals from Nancy, France

	Alleles											
	16	20	24	25	26	27	28	29	30	31	34	Freq.
16	3	3	3	1	1	6	6	5	7	1	1	40
20		2	1	—	—	6	7	4	8	—	1	34
24			—	—	—	1	1	5	—	1	—	12
25				2	—	—	—	1	1	—	—	7
26					—	4	1	2	—	—	—	8
27						13	14	10	4	1	—	72
28							5	3	10	1	2	55
29								4	7	1	—	46
30									2	—	—	41
31										—	—	5
34											—	4

HWE predictions, based on the observed allele frequencies. The overall heterozygosity at this locus is 0.809 ± 0.031, while its expectation based on the allele frequencies is 0.859 ± 0.007.

Table 3 summarizes each of the three test statistics mentioned above. Even though the statistics based on the total number of heterozygotes or homozygotes or the number of distinct genotypes of these two categories disregard each specific genotypic combinations, their use in testing departures from HWE predictions is not invalid when the number of alleles is large and the sample size is inadequate to observe

Table 3. Tests for Hardy-Weinberg Equilibrium (HWE) of genotype frequencies at the ApoB and ApoC-II VNTR Loci

Locus and statistics	Observed value	Expected $\pm$s.e. (under HWE)	P
ApoB VNTR:			
Number of			
heterozygotes	181	184.62 ± 6.53	>0.55
homozygotes	59	55.38 ± 6.53	>0.55
Number of distinct			
heterozygote genotypes	34	32.98 ± 2.56	0.689
homozygote genotypes	8	6.37 ± 1.17	0.162
Likelihood ratio	60.18	—	0.621
ApoC-II VNTR:			
Number of			
heterozygotes	131	139.19 ± 4.43	>0.55
homozygotes	31	22.81 ± 4.43	>0.55
Number of distinct			
heterozygote genotypes	35	34.89 ± 2.61	0.968
homozygote genotypes	7	6.06 ± 0.83	0.258
Likelihood ratio	65.38	—	0.234

all possible genotypes in a given sample. The inference regarding the fit of the data to HWE prediction is identical when these simple test statistics are contrasted with the more complex likelihood ratio test (G-statistics). The latter is presumably the most powerful statistical test because no data summarization is involved in evaluating this statistic nor in computing its empirical significance level. The first two summary statistics have the advantage in the sense that standard large sample theory can be invoked in judging whether or not they reflect a departure from HWE, whereas tedious permutation tests are needed to determine the significance level of the G-statistic.

Two-Locus Genotypic Distribution

As mentioned earlier, information on the joint distribution of genotypes at the ApoB and ApoC-II VNTR loci is available for 160 unrelated individuals in the present sample. A complete tabulation of this joint distribution was made to determine whether or not there is any evidence of non-random association of alleles at these loci. We expect no association, since these two loci are not syntenic, and hence they should segregate independently of each other. Evidence of non-random association, on the contrary, would signify that the population from which this sample is derived is heterogeneous, since it is known that a pseudo-linkage disequilibrium can be generated due to mixture of two or more populations (Nei and Li, 1973).

In principle, the two-locus genotype data is a contingency table of categorical data. But, due to the sparse nature of data, the traditional large sample contingency chi-square test cannot be applied since many of the classes are not represented in the sample. Among the 160 individuals for which two-locus genotypic information is available, there are 31 different genotypes at the ApoB VNTR locus and 41 different genotypes at the ApoC-II VNTR locus, giving a total of 1271 possible gentoypes that could have been observed. We observed only 132 different two-locus genotypes among the 160 individuals. The likelihood ratio test statistic, G, for the observed genotypic combinations is 227.86, which is not significant ($P = 0.428$), after 1000 random permutations. Two alternative statistics are also computed to illustrate that some particular summary of such data can be used to check independence of their segregations. Individuals may be classified into heterozygous or homozygous types at each locus to form a standard 2×2 contigency table (Tab. 4). The expected frequencies of each of the four classes can be obtained from the heterozygosity values of each locus, under the assumption of independent segregation. These are shown in the third column of Tab. 4. Clearly, the observed frequencies are in agreement with the expected ones (χ^2 with 1 d.f. is 0.34, $P > 0.55$),

Table 4. Tests for independence of genotypic frequencies at the ApoB and ApoC-II VNTR Loci

(A) Test based on two-locus homozygosity/heterozygosity:

	ApoC-II Locus	
ApoB Locus	Homozygous	Heterozygous
Homozygous obs	9	32
exp	7.52	31.81
Heterozygous obs	22	97
exp	23.09	97.58

χ^2 with 1 d.f. $= 0.34$ $(P > 0.55)$

(B) Test based on variance of number of heterozygous loci:

	Observed	Expected
Mean:	1.55 ± 0.05	1.56 ± 0.05
Variance:	0.36	0.34^+

$^+$ 95% Confidence interval for variance is (0.27–0.41)

(C) Test based on likelihood ratio test criterion:

$$-2 \ln(L_0/L_1) = 227.86 \text{ (Empirical probability } = 0.428)$$

are in agreement with the expected ones (χ^2 with 1 d.f. is 0.34, $P > 0.55$), suggesting again that there is no evidence of allelic association between these two unlinked loci.

Data from Tab. 4 can also be subjected to an alternative test, originally suggested by Brown et al. (1980) and examined in further detail by Chakraborty (1984). In this test, a variable representing the number of loci for which the individual is heterozygous is defined from the two-locus genotype data for each individual. In this specific case, this variable can take values 0, 1, or 2, corresponding to homozygosity at both loci, homozygosity for one locus and heterozygosity at the other, or heterozygosity at both loci, respectively. From the observed distribution of this statistic, we can determine the mean and variance of number of heterozygous loci. Brown et al. (1980) suggested that the variance of this statistic can be used to test whether or not the two loci are in linkage equilibrium. In the middle panel (B) of Tab. 4 we present the observed mean and variance of this statistic, and their expectations based on the hypothesis of linkage equilibrium. Also shown is the 95% confidence limit of the variance. Clearly, there is no departure from the expectation under the null hypothesis of linkage equilibrium. Therefore, we conclude that the ApoB and ApoC-II VNTR loci are at linkage equilibrium in this French Caucasian population. Intuitively, this result is expected because the two loci are unlinked. Indirectly, this also establishes that the population is homogeneous and there is no evidence of internal substructuring large enough to produce departure from HWE or linkage equilibrium.

Allele Size Distributions and Relationship Between Heterozygosity and Number of Alleles

Having shown that the genotype distributions at the ApoB and ApoC-II VNTR loci conform to their Hardy-Weinberg equilibrium predictions and these two loci are at linkage equilibrium, some additional information regarding the production of new alleles may be extracted from the allele size distributions. Figures 1 and 2 show the size distributions of alleles for the ApoB and ApoC-II VNTR, respectively. Alleles are designated by the copy numbers of their respective core sequences of length 14 or 15 bp for the ApoB locus and 2 for the ApoC-II VNTR locus. Both distributions show multiple modes; the ApoB distribution is bimodal, while there are apparently three modal classes for the Apoc-II locus.

Size distributions of alleles at these and other VNTR loci show multiple modes. In the literature there is no clear indication as to how such multiple modes can be generated. Boerwinkle *et al.* (1989) argued that the presence of multiple modes cannot be readily explained by 'mutational events' such as unequal recombination resulting from mismatching of repeat units or from replication slippage. Two possible alternative explanations may be offered. First, presence of multiple modes may indicate some form of genetic heterogeneity within the

Apo B Alleles

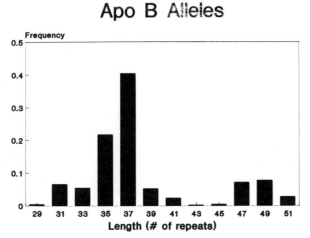

Figure 1. Size distributions of VNTR alleles at the ApoB Locus in the French Caucasian population (sizes are equivalent to the number of copies of a core sequence of length 14/15 bp)

138

ApoC-II (TG)n(AG)m Alleles

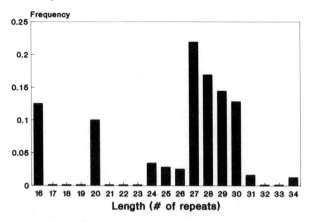

Figure 2. Size distributions of VNTR alleles at the ApoC-II Locus in the French Caucasian population (sizes are equivalent to the number of copies of a dinucleotide core sequence)

population. However, such heterogeneity would have produced significant departure from HWE predictions in the genotpye data analysis, and would have shown significant associations among alleles at the two loci. Since no such departure was found, we do not believe that population substructure is the cause of the observed multimodality. The second explanation is that the current distributions of alleles reflect their evolutionary antiquity, and therefore, it could be assumed that the modal classes reflect alleles that are older than the others. The suggestion of this possibility comes from the theory that under the infinite allele model, the age of an allele can be predicted from its frequency, in the sense that the most common allele is likely to be the oldest has a probability that equals its frequency (Watterson and Guess, 1977).

Such allele frequency profiles can be used to examine the relationship between heterozygosity and the observed number of alleles. In the context of electrophoretic loci, two mutation models have been proposed that can maintain genetic variation in a population. In one model, called the Infinite Allele Model (IAM), every mutational event is assumed to produce a new allele. When a population is at steady-state under the forces of such mutational events and random genetic drift, there is an expected relationship between heterozygosity and the observed number of alleles at a locus (Ewens, 1972; Chakraborty et al., 1978; Chakraborty and Griffiths, 1982). Chakraborty and Weiss (1991) also showed that the sampling distribution of the observed number of alleles can be analytically evaluated. Table 5 shows the summary results of such computations for both loci.

The ApoB VNTR locus has an expected heterozygosity of 76.9%. Given this heterozygosity, we expect to find 14.13 ± 2.05 alleles in a

Table 5. Relationship between heterozygosity and number of alleles at the ApoB and ApoC-II VNTR Loci

		ApoB Locus	ApoC-II Locus
Heterozygosity	obs[a]	0.754 ± 0.028	0.819 ± 0.030
	exp[b]	0.769 ± 0.014	0.859 ± 0.007
Sample size (n)[c]		480	324
Number of alleles	obs	12	11
	exp[d] (IAM)	14.13 ± 2.05	21.53 ± 1.79
	exp[e] (SMM)	5.88	8.71

[a]The observed (obs) heterozygosity is from the actual genotype counts;
[b]The expected (exp) heterozygosity is based on the estimated allele frequencies;
[c]The sample size (n) refers to the number of genes sampled;
[d,e]The expected (exp) number of alleles under the Infinite Allele Model (IAM) and Stepwise Mutation Model (SMM) are based on the expected heterozygosity and sample size, n.

sample of 480 genes (240 individuals), whereas the observed number of alleles at this locus is 12. The prediction of the Infinite Allele Model (IAM) is in statistical agreement with the observation; the probability of observing 12 or less alleles is 0.783 and the probability of observing 12 or more alleles is 0.321.

The expected heterozygosity at the ApoC-II VNTR locus is 85.9%. Given this level of heterozygosity, we would have expected 24.86 ± 1.73 alleles to be observed in the sample of 324 genes (162 individuals). We actually observed only 11 alleles. Since the probability of observing 11 or less alleles in such a sample under the Infinite Allele Model is 0.003, we infer that there are too few alleles observed at this locus for the given heterozygosity. Two possible reasons could explain this discrepancy. First, since this VNTR locus has a dinucleotide core repeat unit, similar sized alleles migrate close to one another on a gel. When copy numbers are large, some rare alleles appearing in heterozygous state in combination with more common and similar size alleles may have been erroneously neglected. Such individuals may easily be scored homozygous for the common type allele. This can account for the observed deficiency in the number of alleles, without markedly reducing the heterozygosity level of the locus, since such unscored alleles are rare in the population. This possibility should have resulted in a hetero- zygote deficiency of our HWE test procedure as well. Although we did not detect any significant departure of genotype frequencies at this locus from the HWE predictions, there is a slight indication that the observed number of heterozygotes is somewhat lower (131 versus 139.19) than its expectation. The second reason could be that the Infinite Allele Model may not apply to such VNTR loci. When the core sequence is small, every 'mutational' event may not necessarily yield a new allele. A form of forward-backward mutation, called Step-Wise Mutation Model, may be more relevant in such a case. In the context of electrophoretic studies, such a model has been proposed, where it is assumed that through a mutation the allelic state can either change by a single step in the forward or backward direction, or can keep the allelic state unaltered. Under such a model, Kimura and Ohta (1978) derived the relationship between number of alleles and heterozygosity. Applying their theory to the data on the ApoC-II VNTR locus, we found that for the given heterozygosity of 85.9%, we expect 8.71 alleles in a sample of 324 genes (162 individuals). The observed number of alleles, 11, is in between the expectations of the step-wise mutation model and the infinite allele model.

In summary, the relationship between heterozygosity and number of alleles at these two loci indicates that the genetic variation at such VNTR loci is maintained by joint effects of mutation and genetic drift, and the present population may be considered to be at a steady-state under these two counteracting forces.

Discussion and Conclusion

The above analyses of data on two VNTR loci performed on the same set of individuals from a genetically well-defined population showed that classic population genetic principles are applicable for understanding genetic characteristics of VNTR polymorphisms. The problems introduced by the large number of alleles can be circumvented by defining appropriate summary measures, such as the total number of heterozygotes, or the number of distinct genotypes observed in a sample. The sampling distributions of these summary measures are tractible and appropriate for hypothesis testing purposes. Alternatively, if one wishes to conduct genotype specific hypothesis testing, permutation tests can be performed on statistics relating expectations and observations of each specific genotype. Such permutation tests avoid problems inherent in sparse data (Efron, 1982). These alternative methods were shown here to result in identical conclusions.

Furthermore, our analyses also show that an apparent deficiency of observed heterozygosity should not be readily taken as evidence of substructuring within a population. This is so, because in the presence of substructuring we would have expected larger than expected number of alleles for the given value of heterozygosity (Chakraborty et al., 1988; Chakraborty, 1990a, b). On the contrary, if incomplete resolution of alleles is responsible for an observed deficiency of heterozygosity, then it is generally accompanied with a smaller observed numbers of alleles.

Lastly, we note an important difference between the allele frequency distributions at the ApoB and ApoC-II VNTR loci. For the ApoB locus, the allele frequency distribution is in agreement with the predictions of the infinite allele model, while this model does not apparently hold for the ApoC-II locus. The core sequence for the ApoB locus is substantially longer (14 or 15 bp) than that at the ApoC-II locus. When the core sequence is long, it may be true that replication slippage is relatively uncommon, while some form of unequal recombination or sister chromatid exchange may be the underlying mechanism of production of new alleles. In either of these two cases, recurrent mutations may not exactly revert allele sizes, because a fine tuning of such crossing-over events will be needed for generating an exact step-wise forward-backward form of mutation. Therefore, for VNTR loci characterized by relatively large core sequences, the infinite allele model may provide reasonable mathematical predictions of the allele frequency distribution, as in the case of the ApoB locus. On the other hand, when the core sequence is small, replication slippage can generate forward-backward mutations yielding several alleles of nearly similar sizes which can change from one to another. This process can occur in both a forward and backward fashion through recurrent mutational events. The large differences in some allele sizes at the ApoC-II locus may be produced by other mechanisms occurring at the same time. The observation that the observed number of alleles at the ApoC-II locus lies between the predictions of the Infinite Allele Model and the Step-wise Mutation Model indicates that at VNTR loci with a small repeat sequence, genetic variation may be generated by a mixture of two or more distinct molecular mechanisms. The first mechanism leads to new alleles not previously seen in a population and represents large differences of allele sizes, and the second produces small shifts of allele sizes in a forward-backward fashion. We speculate that the rate of occurrence of the first type of mutational changes is less than the second type. As a consequence of this, we observe a larger heterozygosity than expected at loci where step-wise changes are more common (reflected in larger heterozygosity at the ApoC-II locus compared to the ApoB locus).

A detailed mathematical study of such a mixed model of mutational changes is needed for a full understanding of the population dynamics of VNTR polymorphisms. Some initial attempt has been made in this direction. Li (1976) proposed a mixture model of mutation which incorporates the two types of mutations mentioned above. In his model, however, the step-wise changes were assumed to involve only one-step movements (forward or backward) in terms of allele states. Chakraborty and Nei (1982) proposed a step-mutation model where multiple step changes (in either direction) was introduced. Such models can be easily rationalized in the context of the molecular mechanisms of unequal recombination and replication slippage, and these should be examined in greater detail to study the evolutionary dynamics of VNTR polymorphism.

142

Acknowledgements

This work was supported by the grant GM-41399 from US National Institutes of Health and 90-IJ-CX-0038 from the National Institute of Justice. We thank Prof. A. J. Jeffreys for his constructive comments on the work and we are grateful to individuals from Nancy, France for their co-operation in our study.

References

Baird, M., Balazs, I., Giusti, A., Miyazaki, L., Wexler, K., Kanter, E., Glassberg, J., Rubinstin, P., and Sussman, L. (1986) Allele frequency distribution of two highly polymorphic DNA sequences in three ethnic groups and its application to the determination of paternity. *Am. J. Hum. Genet.* 39: 489–501.

Balazs, I., Baird, M., Clyne, M., and Meade, E. (1989) Human population genetic studies of five hypervariable loci. *Am. J. Hum. Genet.* 44: 182–190.

Bell, G. I., Selby, M. J., and Rutter, W. J. (1982) The highly polymorphic region near the insulin gene is composed of simple tandemly repeating sequences. *Nature* 295: 31–35.

Boerwinkle, E., Xiong, W., Fourest, E., and Chan, L. (1989) Rapid typing of tandemly repeated hypervariable loci by the polymerase chain reaction: Application to the apolipoprotein B 3′ hypervariable region. Proc. Natl. Acad. Sci. USA 86: 212–216.

Brown, A. H. D, Feldman, M. W., and Nevo, E. (1980) Multilocus structure of natural populations of *Hordeum spontaneum*. *Genetics* 96: 523–536.

Capon, D. J., Chen, E. Y., Levinson, A. D., Seeburg, P. H., and Goeddel, D. V. (1983) Complete nucleotide sequence of the T24 human bladder carcinoma oncogene and its normal homologue. *Nature* 302: 33–37.

Chakraborty, R. (1984) Detection of nonrandom association of alleles from the distribution of the number of heterozygous loci in a sample. *Genetics* 108: 719–731.

Chakraborty, R. (1990a) Mitochondrial DNA polymorphism reveals hidden heterogeneity within some Asian populations. *Am. J. Hum. Genet.* 47: 87–94.

Chakraborty, R. (1990b) Genetic profile of cosmopolitan populations: Effects of hidden subdivision. *Anthrop. Anz.* 48: 313–331.

Chakraborty, R. (1991) Generalized occupancy problem and its application in population genetics. In: Sing, C. F., and Hanism, C. L. (eds), Impact of Genetic Variation on Individuals, Families and Populations. Oxford University Press, New York (in press).

Chakraborty, R., Fuerst, P. A., and Nei, M. (1978) Statistical studies on protein polymorphism in natural populations. II. Gene differentiation between populations. *Genetics* 88: 367–390.

Chakraborty, R., and Griffiths, R. C. (1982) Correlation of heterozygosity and number of alleles in different frequency classes. *Theor. Pop. Biol.* 21: 205–218.

Chakraborty, R., and Nei, M. (1982) Genetic differentiation of quantitative traits between populations or species. I. Mutation and random genetic drift. *Genet. Res.* 39: 303–314.

Chakraborty, R., Smouse, P. E., and Neel, J. V. (1988) Population amalgamation and genetic variation: Observations on artificially agglomerated tribal populations of Central and South America. *Am. J. Hum. Genet.* 43: 709–725.

Chakraborty, R., and Weiss, K. M. (1991) Genetic variation of the mitochondrial DNA genome in American Indians is at mutation-drift equilibrium. *Am. J. Phys. Anthrop.* (in press).

Cohen, J. E. (1990) DNA fingerprinting for forensic identification: Potential effects on data interpretation of subpopulation heterogeneity and band number variability. *Am. J. Hum. Genet.* 46: 358–368.

Collick, A., and Jeffreys, A.J. (1990) Detection of a novel minisatellite-specific DNA-binding protein. *Nucleic Acid Res.* 18: 625–629.

Clark, A. G. (1987) Neutrality tests of highly polymorphic restriction fragment length polymorphisms. *Am. J. Hum. Genet.* 41: 948–956.

Craig, J., Fowler, S., Burgoyne, L. A., Scott, A. C., and Harding, H. W. J. (1988) Repetitive deoxyribonucleic acid (DNA) and human genome variation: A concise review relevant to forensic biology. *J. Forensic Sci.* 33: 1111–1126.

Devlin, B., Risch, N., and Roeder, K. (1990) No excess homozygosity at loci used for DNA fingerprinting, *Science* 249: 1416–1420.

Edwards, A., Hammond, H. A., Caskey, C. T., and Chakraborty, R. (1991) Population genetics of trimeric and tetrameric tandem repeats in four human ethnic groups. *Genomics* (in press).

Ewens, W. J. (1972) The sampling theory of selectively neutral alleles. *Theor. Pop. Biol.* 3: 87–112.

Efron, B. (1982) The Jackknife, the Bootstrap and Other Resampling plans. CBMS-NSF Regional Conference Series in Applied Mathematics, Monograph 38. SIAM, Philadelphia.

Flint, J., Boyce, A.J., Martinson, J.J., and Clegg, J.B. (1989) Population bottlenecks in Polynesia revealed by minisatellites. *Hum. Genet.* 83: 257–263.

Fojo, S., Law, S., and Brewer, H. B. (1987) The human preapolipoprotein C-II gene complete nucleic acid sequence and genomic organization. *FEBS Letters* 213: 221–226.

Goodbourn, S. E. Y., Higgs, D. R., Clegg, J. B., and Weatherall, D. J. (1983) Molecular basis of length polymorphism in the human zeta-globin complex. Proc. Natl. Acad. Sci. USA 80: 5022–5026.

Huang, L. S., and Breslow, J. L. (1987) A unique AT-rich hypervariable minisatellite 3' to the ApoB gene defines a high information restriction length polymorphism. *J. Biol. Chem.* 262: 8952–8955.

Jeffreys, A. J., Royle, V., Wilson, V., and Wong, Z. (1988) Spontaneous mutation rates to new length alleles at tandem-repetitive hypervariable loci in human DNA. *Nature* 332: 278–281.

Jeffreys, A. J., Wilson, V., and Thein, S. L. (1985) Hypervariable 'minisatellite' regions in human DNA. *Nature* 314: 67–73.

Kidd, K. K., Bowcock, A. M., Schmidtke, J., Track, R. K., Ricciuti, F., Hutchings, G., Bale, A., Perason, P., and Willard, H. F. (1989) Report of the DNA committee and catalogs of cloned and mapped genes and DNA polymorphisms. *Cytogenet. Cell Genet.* 51: 622–947.

Kimura, M., and Crow, J. F. (1964) The number of alleles that can be maintained in a finite population. *Genetics* 49: 725–738.

Kimura, M., and Ohta, T. (1978) Stepwise mutation model and distribution of allelic frequencies in a finite population. Proc. Natl. Acad. Sci. USA 75: 2868–2872.

Lander, E. S. (1989) DNA fingerprinting on trial. *Nature* 339: 501–505.

Li, W. H. (1976) A mixed model of mutation for electrophoretic identity of proteins within and between populations. *Genetics* 83: 423–432.

Ludwig, E. H., Friedl, W., and McCarthy, B. J. (1989) High-resolution analysis of a hypervariable region in the human apolipoprotein B gene. *Am. J. Hum. Genet.* 45: 458–464.

Nakamura, Y., Leppert, M., O'Connell, P., Wolff, R., Holm, T., Culver, M., Martin, C., Fujimoto, E., Hoff, M., Kumlin, E., and White, R. (1987) Variable number of tandem repeat (VNTR) markers for human gene mapping. *Science* 235: 1616–1622.

Nei, M., and Li, W. H. (1973) Linkage disequilibrium in subdivided populations. *Genetics* 75: 213–219.

Odelberg, S. J., Platke, R., Eldridge, J. R., Ballard, L., O'Connell, P., Nakamura, Y., Leppert, M., Lalouel, J. M., and White, R. (1989) Characterization of eight VNTR loci by agarose gel electrophoresis. *Genomics* 5: 915–924.

Sokal, R. R., and Rohlf, J. F. (1969) Biometry, 2nd edition. Freeman, New York.

Watterson, G.A., and Guess, H.A. (1977) Is the most frequent allele the oldest? *Theor. Pop. Biol.* 11: 141–160.

Wyman, A. R., and White, R. (1980) A highly polymorphic locus in human DNA. Proc. Natl Acad. Sci. USA 77: 6754–6758.

DNA Fingerprinting: Approaches and Applications
ed. by T. Burke, G. Dolf, A. J. Jeffreys & R. Wolff
© 1991 Birkhäuser Verlag Basel/Switzerland

Population Genetic Data Determined For Five Different Single Locus Minisatellite Probes

L. Henke, S. Cleef, M. Zakrzewska, and J. Henke

Institut f. Blutgruppenforschung, Otto-Hahn-Str. 39, D 4000 Düsseldorf 13, Germany

Summary

We report on the population genetic data (frequencies of restriction fragments, heterozygosity rates, and mutation rates) obtained by analysis of approximately 1100 HinfI-digested DNAs from West Germans. Probe G3 detects a common 1.7 kb DNA fragment showing a population frequency of about 13%. All the other fragments detected with probes MS1, MS31, MS43, G3 and YNH24 show frequencies of less than 8%. These data suggest that single locus DNA probes can provide valuable information for parentage evaluation and individualisation.

Introduction

Since the first highly polymorphic locus in human DNA (D14S1) was reported (Wyman and White, 1980), numerous additional hypervariable loci have been discovered (e.g. Nakamura *et al.*, 1987; Wong *et al.*, 1986; Wong *et al.*, 1987). Estimates state that there may be at least 1500 highly variable loci. Because of the more clear-cut patterns of inheritance single-locus polymorphisms may be regarded as being superior to multi-locus polymorphisms in forensic haemogenetics. In addition, locus specific probes have great potential in many other fields because of their sensitivity and ease of use.

The aim of this paper is to present family and population genetic data concerning the HinfI polymorphisms of the loci D1S7, D7S21, D12S11, D7S22, and D2S44 as detected by probes MS1, MS31, MS43, G3, and YNH24 respectively (Wong *et al.*, 1986; Wong *et al.*, 1987). This report thus contributes to fulfilling the generally accepted requirement that basic reliability and biostatistical standards established in conventional blood group tests must also be applied to the forensic use of DNA polymorphisms.

Materials and Methods

We have tested single individuals, multi-member families (Henke *et al.*, 1990) and people involved in paternity cases. All of them were of

Caucasian origin. In parentage cases, paternity was established by conventional serological means ($W \geq 99.8\%$ at a given *à priori* probability of 0.5) (Essen-Möller, 1938a, b; Essen-Möller and Quensel, 1939), while exclusions from paternity were based on at least 2 contradictions to the genetic pathways of the respective blood group systems.

DNA Isolation and Digestion

Genomic DNA was isolated from peripheral blood according to the non-toxic extraction method described by Miller *et al.* (1988). DNA samples (3 μg) were digested with the restriction enzyme HinfI at a concentration of 6 U/μg DNA according to the manufacturer's specifications.

Electrophoresis and Hybridisation

Separation of DNA fragments was performed by electrophoresis in 0.7% agarose (Pharmacia) gels (20 × 20 cm) in TBE buffer (0.134 M Tris base, 74.9 mM boric acid, 2.55 mM EDTA-Na, pH 8.8) for approximately 24 hours at 0.9 V/cm. Following electrophoresis, DNA was transferred to nylon membranes (Hybond N, Amersham) by using the capillary transfer technique (Maniatis *et al.*, 1982; Southern, 1975). Radioactive DNA probe was prepared by random oligopriming (Feinberg and Vogelstein, 1984). The prehybridisation was performed for 1 hr at 62°C in 10 × Denhardt's solution (1 × Denhardt's: 0.02% bovine serum albumin, 0.02% Ficoll 400, 0.02% polyvinylpyrrolidone 44000) followed by a 3 hr step in hybridisation solution (6.0% polyethylene glycol 6000, 0.1% sodium dodecylsulfate, 50 μg/ml herring sperm DNA, 1 × Denhardt's solution) with a total volume of 10 ml in the circulating drum of a hybridisation oven. Hybridisation was carried out overnight at 62°C. The washing was performed at 20°C for 2 × 10 min (2 × SSC, 1.5% SDS) (1 × SSC = 0.15 M sodium chloride, 15 mM sodium citrate, pH 7.0) followed by two washing steps for 15 min at 62°C (0.1 × SSC, 1% SDS). Radioactive bands were visualized by autoradiography at −20°C with Fuji medical (Fuji) x-ray films and G12 (Fuji) or Lgy HS 600 (Dr. Goos) intensifying screens (Henke *et al.*, 1988).

Fragment Size Calculation

The sizes of the DNA fragments detected were calculated using EcoRI/HindIII and BstEII digested lambda DNA, a high molecular weight

146

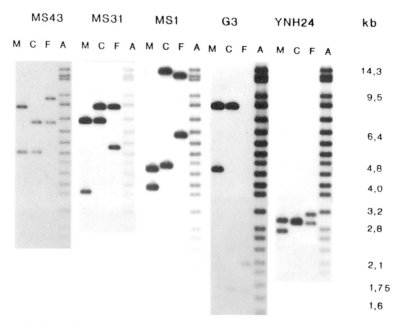

Figure 1. Examples of paternity cases analysed with a single locus minisatellite probe. 10 ng analytical marker (B) and 5 µg digested human DNA were loaded onto the gel. The membranes were hybridized with ^{32}P labeled probes MS31 and analytical marker. M = mother, C = child, A = alleged father, F = marker fragment size (kb).

marker (BRL) and the 1 kb ladder (BRL) as external markers and manual calculation by means of a plotted curve of the DNA size standards. Since a recently introduced "analytical marker" (Promega) turned out to show more and better distributed known fragments, the "analytical marker" was used additionally (Fig. 1).

Calculation of the Standard Deviations of Fragment Size Estimates

Genomic DNA from a single individual was used as an internal control. After digestion with HinfI it was hybridized with probes MS1, MS43, G3, and YNH24. Thus 10 different fragments ranging from 1.7 kb to 10.8 kb were observed on at least 23 different autoradiographs. This allowed multiple kb size measurements of the same fragment.

Results

1. Distribution of Restriction Fragments

Kb size measurements were carried out as described in Material and Methods. For practical reasons the scale was subdivided into steps of

Probe MS1 - Distribution of Fragments in West Germans

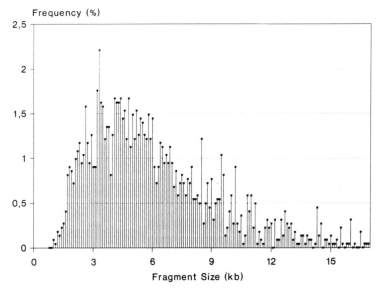

Figure 2. Size distribution of HinfI alleles at locus D1S7 as revealed by probe MS1. 2219 fragments are included. Number of fragments larger than 17 kb: 42 $\cong$ 1.89%.

Probe MS31 - Distribution of Fragments in West Germans

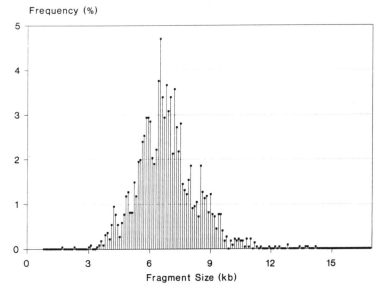

Figure 3. Size distribution of HinfI alleles at locus D7S21 as revealed by probe MS31. 2211 fragments are included.

148

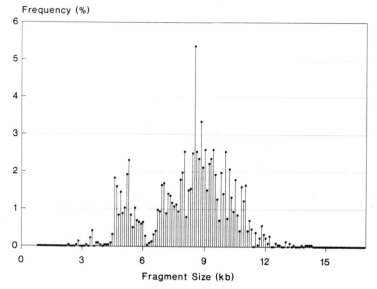

Probe MS43 - Distribution of Fragments in West Germans

Figure 4. Size distribution of HinfI alleles at locus D12S11 as revealed by probe MS43. 2108 fragments are included. Number of fragments larger than 17 kb: $4 \cong 0.19\%$.

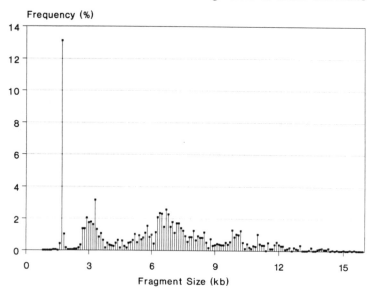

Probe G3 - Distribution of Fragments in West Germans

Figure 5. Size distribution of HinfI alleles at locus D7S22 as revealed by probe G3. 2108 fragments are included. Number of fragments larger than 17 kb: $\cong 0.14\%$.

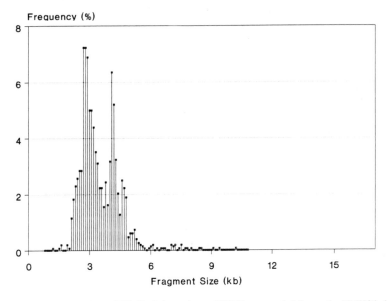

Figure 6. Size distribution of HinfI alleles at locus D2S44 as revealed by probe YNH24. 1408 fragments are included.

Table 1a. Variation in allele sizing in the laboratory standard human DNA by using lambda HindIII, lambda EcoRI/HindIII, lambda BstEII, and 1 kb ladder as sizing standards

Total number of measurements	63	65	67	65	61	67	24	23	59	40
Minimal fragment size	10.10	9.25	7.60	5.90	5.30	5.15	4.05	2.85	2.35	1.55
Maximal fragment size	11.50	10.00	8.80	6.30	5.80	5.80	4.30	3.10	2.70	1.85
Mean fragment size	10.86	9.58	7.96	6.11	5.60	5.35	4.20	2.98	2.50	1.70
Standard deviation	0.29	0.16	0.21	0.09	0.11	0.12	0.06	0.08	0.08	0.04
Variance	0.08	0.03	0.04	0.01	0.01	0.02	0.00	0.01	0.01	0.00

Table 1b. Variation in allele sizing in the laboratory standard human DNA by using the "analytical marker"

Total number of measurements	61	64	60	64	65	60	49	46	60	62
Minimal fragment size	10.50	9.50	7.70	6.00	5.50	5.30	4.10	3.00	2.40	1.65
Maximal fragment size	11.00	9.60	8.00	6.20	5.70	5.40	4.30	3.10	2.60	1.70
Mean fragment size	10.82	9.54	7.89	6.11	5.57	5.36	4.20	3.10	2.55	1.70
Standard deviation	0.10	0.05	0.07	0.07	0.06	0.05	0.03	0.03	0.05	0.01
Variance	0.01	0.00	0.00	0.00	0.00	0.00	0.00	0.00	0.00	0.00

100 bp. The odd numbers of fragments arose due to the inclusion of those paternal fragments which were found in children when the nominal alleged father was excluded from paternity. The size distribution of fragments is illustrated in Figs. 2–6.

The DNA of a single person was used as a laboratory standard for calculation of the standard deviation. After HinfI digestion, this DNA was hybridized with 5 single locus minisatellite probes (MS1, MS31, MS43, G3, and YNH24). Because of heterozygosity at all the appropriate loci involved, 10 fragments were observed. The results are compiled in Tabs. 1a and 1b.

2. Meioses

This study of mother-father-child trios comprises at least 646 meioses. The result of this evaluation is presented in Tab. 2.

3. Mutations

If in serologically established families one single restriction fragment of a child could not be visually attributed to either parent, we have interpreted this as a "mutational" event. Table 2, C1–C3, illustrates the results of this evaluation. Locus D1S7 shows the highest number of parent-child mismatches. Figure 7 presents the fragment pattern of a family showing the very rare but nevertheless expected event of a mutation in both parental meioses.

4. Heterozygosity

Heterozygosity reflects the degree of polymorphism. Our observed heterozygosity rates are listed in Tab. 2.

Table 2. Observed heterozygosity rates (A), numbers of meioses analysed (B1: total number, B2: paternal meioses, B3: maternal meioses), incidence of allele length mutation events (C1: total number of mutations, C2: paternal mutations, C3: maternal mutations)

Locus	Probe	A	B1	B2	B3	C1	C2	C3
D1S7	MS1	98.56%	989	424	565	38 (3.84%)	18 (4.25%)	20 (3.54%)
D7S21	MS31	94.08%	1008	432	576	10 (0.99%)	9 (2.08%)	1 (0.17%)
D12S11	MS43	94.57%	991	424	567	1 (0.10%)	1 (0.24%)	0
D7S22	G3	96.08%	977	419	558	3 (0.31%)	2 (0.48%)	1 (0.18%)
D2S44	YNH24	93.25%	646	272	374	3 (0.46%)	1 (0.27%)	2 (0.74%)

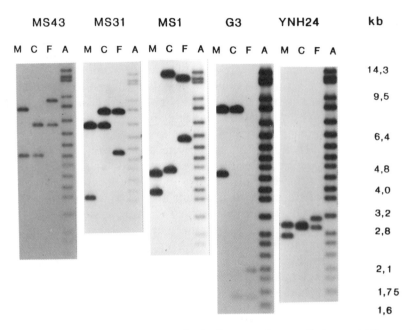

Figure 7. Example of a double mutant offspring in a paternity case. Both parental fragments at locus D1S7 (MS1) are mismatched, while the offspring fragments (C) at loci D7S21, D12S11, D7S22 and D2S44 match correctly the mother (M) and the alleged father (F). A, molecular weight standards.

5. Exclusion and False Inclusion Rates

In parentage testing it is of enormous interest to have "experimental data" on the exclusion rates (observed in true non-fathers), as well as data on false inclusions. Evaluations of paternity cases are presented in Tab. 3. The sequential use of the five probes would result in an average frequency of false inclusions of $1.93\% \times 7.98\% \times 8.17 \times 3.00\% \times 5.8\%$ $= 0.000022\%$ or 1 in about 4 500 000.

Table 3. Exclusion and false inclusion of true non-fathers. 138–213 cases of disputed paternity revealing exclusions were additionally subjected to DNA polymorphism analysis. The compilation illustrates how often the serological exclusions were confirmed

Locus	Probe	Total No.	Correct No.	Exclusion (%)	False No.	Inclusion (%)
D1S7	MS1	207	203	(98.07)	4	(1.93)
D7S21	MS31	213	196	(92.02)	17	(7.98)
D12S11	MS43	208	191	(91.83)	17	(8.17)
D7S22	G3	200	194	(97.00)	6	(3.00)
D2S44	YNH24	138	130	(94.20)	8	(5.80)

Discussion

Chromosomal localisation and repeat unit sequences of the loci D1S7, D7S21, D12S11, D7S22 and D2S44 are known (Nakamura *et al.*, 1987; Royle *et al.*, 1988; Wong *et al.*, 1987). The data suggest that restriction fragments of these minisatellite loci can be regarded as independent from each other. Thus, in paternity testing appropriate likelihood ratios can be simply summarized.

Sizes and frequencies of fragments were determined for five hyper-variable loci. The recent inclusion of the mentioned "analytical marker" led to a significant increase of accuracy in kb size measurements.

Probe G3 detects a common fragment at 1.7 kb, representing about 13% of the HinfI/G3 fragments observed in West Germans. The fragment frequency distribution has further been analysed by comparing recent data from Great Britain (Smith *et al.*, 1990) with ours. The similarity between both populations is astonishingly close, and would not have been expected from experiences in conventional blood group polymorphisms. The analysis of our data further reveals that the vast majority of fragments have frequencies below 5%.

Mendelian inheritance of fragments has been shown by many working groups, so that the application of DNA polymorphisms to parentage testing is underway. Preliminary evaluation of our data suggest that the investigated single locus polymorphisms show Hardy–Weinberg equilibrium. Like the interference of silent alleles in blood group systems, occurrence of spontaneous mutations at DNA minisatellite loci has to be taken into account. Our data, obtained by analyzing up to 975 meioses, show close similarity with those from Great Britain (Jeffreys *et al.*, 1988; Smith *et al.*, 1990). Thus, there can be no objection to deal with them in paternity likelihood calculations (Baur, pers comm.). The chance that 2 equidirectional mutations occur in one paternity case is 1 in 5000 at a given mutation rate of 2% for each locus (Balazs *et al.*, 1989).

The presented data on fragment frequencies and mutation rates provide a foundation for the application of these DNA polymorphisms to parentage testing as required by appropriate international organisations (American Association of Blood Banks, International Society of Forensic Haemogenetics).

The panel of locus-specific probes mentioned in this study has now been used along with conventional blood group tests in more than 200 cases of disputed paternity. Beside normal "trio cases", we have also dealt with prenatal paternity tests (fibroblast culture, fetal tissue), post mortem tests (either decayed or burnt bodies), tests with aged (deteriorated) blood samples, and tests in extremely complicated deficiency cases. The results of the comparisons will be published elsewhere and confirm the utility of locus-specific DNA polymorphisms in many aspects of individual identification, including for example, bone marrow transplantation.

References

Balazs, I., Baird, M., Clyne, M., and Meade, E. (1989) Human population genetic studies of five hypervariable DNA loci. *Am. J. Hum. Genet.* 44: 182–190.

Essen-Möller, E. (1938) Wie kann die Beweiskraft der Ähnlichkeit im Vaterschaftsnachweis in Zahlen gefaßt werden? *Verh. Dtsch. Ges. Rassenforsch.* 9: 76–78.

Essen-Möller, E. (1938) Die Beweiskraft der Ähnlichkeit im Vaterschaftsnachweis. Theoretische Grundlagen. *Mitt. Anthrop. Ges. Wien* 68: 9–53.

Essen-Möller, E., and Quensel, C. E. (1939) Zur Theorie des Vaterschaftsnachweises aufgrund von Ähnlichkeitsbefunden. *Dtsch. Z. Gerichtl. Med.* 31: 70–96.

Feinberg, A., and Vogelstein, B. (1984) A technique for radiolabelling DNA restriction endonuclease fragments to high specific activity. *Anal. Biochem.* 132: 6–13.

Henke, J., Henke, L., and Cleef, S. (1988) Comparison of different X-ray films for ^{32}P autoradiography using various intensifying screens at $-20°C$ and $-70°C$. *J. Clin. Chem. Clin. Biochem.* 26: 467–468.

Henke, L., Henke, J., Cleef, S., Zakrzewska, M., and Sander, K. (1990) Segregation of single-locus DNA fragments in a large family, in: Polesky H. F. and Mayr W. R. (eds.) Adv. Forens. Haemogenet 3: 75–76; Springer, Berlin, Heidelberg, New York.

Jeffreys, A. J., Wilson, V., and Thein, S. L. (1985) Hypervariable "minisatellite" regions in human DNA. *Nature* 314: 67–73.

Jeffreys, A. J., Royle, N. J., Wilson. V., and Wong. Z. (1988) Spontaneous mutation rates to new length alleles at tandem repetitive hypervariable loci in human DNA. *Nature* 332: 278–281.

Maniatis, T., Fritsch, E. F., and Sambrook, J. (1982) Molecular cloning: A Laboratory Manual, Cold Spring Harbor Lab., Cold Spring Harbor, N.Y.

Miller, S., Dykes, D., and Polesky, H. (1988) A simple salting out procedure for extracting DNA from human nucleated cells. *Nucleic Acids Res.* 16: 1215.

Nakamura, Y., Leppert, M., O'Connell, P., Wolff, R., Holm, T., Culver, M., Martin, C., Fujimoto, E., Kumlin, E., and White. R. (1987) Variable Number of Tandem Repeat (VNTR) markers for human gene mapping. *Science* 235: 1616–1622.

Royle, N. J., Clarkson, R. E., Wong, Z., Jeffreys, A. J. (1988) Clustering of hypervariable minisatellites in the proterminal regions of human autosomes. *Genomics* 3: 352–360.

Smith, J. C., Anwar, R., Riley, J., Jenner, D., Markham, A. F., and Jeffreys, A. J. (1990) Highly polymorphic minisatellite sequences: allele frequencies and mutation rates for five locus specific probes in a Causasian population. *J. For. Sci. Soc.* 30: 19–32.

Southern, E. (1975) Detection of specific sequences among DNA fragments separated by gel electrophoresis. *J. Mol. Biol.* 98: 503–517.

Wong, Z., Wilson, V., Jeffreys, A. J., and Thein, S. L. (1986) Cloning a selected segment from a human DNA "fingerprint": isolation of an extremely polymorphic minisatellite. *Nucleic Acids Res.* 14: 4605–4616.

Wong, Z., Wilson, V., Patel, I., Povey, S., and Jeffreys, A. J. (1987) Characterisation of a panel of highly variable minisatellites cloned from human DNA. *Ann. Hum. Genet.* 51: 269–200.

Wyman, A. R., and White, R. (1980) A highly polymorphic locus in human DNA. *Proc. Natl. Acad. Sci. USA* 77: 6754–6758.

DNA Fingerprinting: Approaches and Applications
ed. by T. Burke, G. Dolf, A. J. Jeffreys & R. Wolff
© 1991 Birkhäuser Verlag Basel/Switzerland

Multilocus and Single Locus Minisatellite Analysis in Population Biological Studies

T. Burke[a], O. Hanotte[a,b], M. W. Bruford[a,c], and E. Cairns[a]

[a]*Department of Zoology, University of Leicester, Leicester LE1 7RH, Great Britain;*
[b]*Service de Biochimie Moléculaire, Université de Mons-Hainaut, 7000 Mons, Belgium;*
[c]*Present address: Zoological Society of London, Institute of Zoology, Regent's Park, London NW1 4RY, Great Britain*

Summary
In this review we describe the situations in which minisatellite analysis is of value to studies of population and evolutionary biology. Evolutionary and population biologists need to be able to quantify genetic relationships among individual organisms at many different levels, from close familial relationships to evolutionarily distant phylogenetic ones. The use of minisatellite markers is put into this context and compared with the other molecular biological techniques. Examples of the use of multilocus minisatellite analysis in population biology are described. The limitations of multilocus fingerprinting are presented, together with the potential advantages of locus-specific probes. The use of locus-specific probes in population biology is now often feasible due to the recent development of a cloning system which allows their efficient isolation. The availability of locus-specific probes should significantly expand the role of minisatellite markers in population biology.

Introduction

Evolutionary and population biologists need to be able to quantify genetic relationships among individual organisms at many different levels, from close familial relationships to evolutionarily distant phylogenetic ones. The development of molecular biological techniques is allowing the quantification of variation within and between the DNAs of different individuals at many distinct levels across this broad range. The application of molecular methods to the study of systematics has been reviewed elsewhere (e.g. Hillis and Moritz, 1990; Hewitt, 1991). In this paper we will focus our attention on the development and application of methods for the analysis of highly polymorphic minisatellite DNA sequences. In particular, we will discuss separately the two alternative methods of minisatellite analysis: the multilocus and the locus-specific approaches. The locus-specific approach has only recently become generally feasible and seems likely to greatly expand the role of DNA fingerprinting methods in population biology. Highly polymorphic sequences are of most value in studies of close genetic relationships, but may nevertheless prove invaluable to some studies of more distant relationships, including population comparisons, especially for populations where their recent evolutionary history has resulted in relatively low heterozygosity.

Background

The measurement of close genetic relationships is of particular importance in behavioural ecology. Behavioural ecologists wish to understand the adaptive significance of the different behaviours and other characteristics displayed by the animals that they study, usually in the wild. Ideally, this understanding would come from the assessment and comparison of the inclusive fitnesses of different individuals. The precise measurement of absolute inclusive fitness is probably impossible (Mock, 1983), but a useful approximation requires reproductive output to be measured with a reasonable accuracy and – where apparent altruism and the possibility of kin selection occur – the estimation of genetic relatedness among socially interacting individuals.

Our own work has been concerned mainly with studies of birds. Birds are popular study organisms among behavioural ecologists as they are relatively observable, can often be easily trapped for individual marking, and they have obvious social behaviour, including parental care. The problems that need to be addressed in birds are typical of those of interest in other vertebrate taxa. For example, in recent years two types of behaviour have been discovered that potentially confound the measurement of individual reproductive success: extra-pair copulations, where a female is inseminated by a male other than the male with which she is paired, and intraspecific brood parasitism, where a female deposits an egg in the nest of another (see Birkhead *et al.*, 1990). Both these behaviours have on occasion, at least, now been observed in many different species, but they are very difficult to quantify from field observations alone. In general, an assessment of their adaptive significance requires the use of genetic methods to confirm the parentage of each adults' apparent offspring.

A related area where paternity analysis would be invaluable is in the study of sexual selection, especially in species with lek mating systems. In these species, which include Darwin's classical example of the peacock, *Pavo cristatus*, males display at a communal area (lek) where females visit to copulate with them. Females may mate with a number of different males and, as mating success among the males is highly skewed, a male might potentially be the father of many females' offspring.

As a final example, some species show cooperative behaviour, where they assist with the rearing of apparent non-kin. Several hypotheses involving inclusive fitness have been proposed to explain the evolution of this behaviour, and tests of these hypotheses require the assessment of both genetic relatedness *per se* and parentage (see Jones *et al.*, 1991; Rabenold *et al.*, 1990).

In principle, any Mendelian trait can provide information about relatedness. A variety of different kinds of genetic marker have been employed for this purpose (see Burke, 1989), but prior to the development of hypervariable DNA markers all suffered from a deficiency of variability. For example, in paternity analysis the probability of extra-pair paternity being detected through a putative father not carrying the offspring's paternal allele (the exclusion probability) is highly dependent on the number of detectable alleles at a locus (Fig. 1). The most variable markers available were until recently allozyme loci detected through starch gel electrophoresis. Allozyme loci typically show a

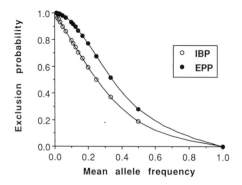

Figure 1. The probability of excluding an offspring (i) incorrectly assigned to a male, when correct maternity is known (as in extra-pair paternity, EPP) or (ii) incorrectly assigned to a pair (as in intra-specific brood parasitism, IBP), using loci with different mean allele frequencies. It is assumed that all alleles at a locus are equally frequent.

Table 1. Genetic markers in population biology

	Allozymes	RFLPs	System MLPs	SLPs	mtDNA/PCR
Applications					
Pedigree analysis	+	+	+ + +[5]	+ + +[5]	(+)[1]
Comparing genotypes (e.g. finding fathers)	+	+	+ +[5]	+ + +	(+)[1]
Population comparison	+	+	(+)[2]	+ +	+ + +
Phylogenetic studies	+	+	—	?	+ + +
QTL detection	+	+	+[5]	+ + +[5]	—
Molecular evolution	—	+	—	+	+ + +
Practicalities					
Robustness/ ease of use	+ + +	+ +	+	+ + +	+ +
Availability of probe/stain/ primers	+ + +	+[3]	+ + +	(+)[4]	+ + +

RFLP = restriction fragment length polymorphism; MLP = multilocus probe (DNA fingerprinting); SLP = single locus probe (minisatellite); mtDNA/PCR = use of PCR to obtain mtDNA sequence.
[1]May have some utility in identifying matrilines; [2]Of value in inbred populations only; [3]Refers to use of homologous coding sequences; [4]Have to be cloned, [5]Of less value than this in inbred populations.

handful of detected alleles and consequently, even when information concerning several such loci is available, exclusion probabilities rarely exceed 60%.

In this article we will concentrate on genomic minisatellite studies, though a qualitative comparison of the applicability to evolutionary biology of different types of genetic marker is presented in Tab. 1.

Minisatellite loci can be highly polymorphic due to the presence in a population of large numbers of alleles containing a different number of minisatellite repeat units (Jeffreys *et al.*, 1985a). This variability is detectable as electrophoretically distinguishable restriction fragments which contain minisatellites. There are sets of minisatellite loci in which the repeat units contain a common "core" sequence. Consequently, a poly-core probe can be used to simultaneously detect many such loci (Jeffreys *et al.*, 1985a).

The hypervariability at minisatellite loci makes such loci ideal for the analysis of parentage and the detection of close levels of relationship (Fig. 1; Tab. 1) (Jeffreys *et al.*, 1985b, c).

Fingerprinting Methods

The methods used in DNA fingerprinting have been described in detail elsewhere (e.g. Jeffreys *et al.*, 1985a, b, 1986; Burke, 1989; Birkhead *et al.*, 1990). A description of our overall approach and some of our experiences will be provided here.

DNA from any tissue can be used in a fingerprint analysis, though some tissues are easier to handle and different tissues may be subjected to different degrees of methylation. The problem of methylation can be of particular significance when comparing fingerprints from DNA obtained from different tissues. Comparison of fingerprint patterns obtained from house sparrow, *Passer domesticus*, blood DNA digested separately with MspI and HpaII showed that a proportion of DNA fragments were methylated (T.B. and E.C., unpublished). The problems of methylation can be minimized by the use of endonucleases such as AluI and HaeIII which do not contain the commonly methylated CpG dinucleotide in their recognition sequence (note that when a HinfI site is followed by a G, as GANTCG, methylation can occur and lead to inhibition, Kessler and Holtke, 1986).

In non-mammalian vertebrates blood is often the most convenient source of tissue as the red cells are nucleated and it can be obtained non-destructively. About 50 times less DNA per volume of blood is obtained from mammals as compared with birds and so skin, hair roots or other tissues may be more convenient.

When beginning studies on a new species we usually compare the fingerprints obtained using at least two multilocus probes from the

DNAs of two or more unrelated individuals digested with a panel of different 4-base recognition sequence endonucleases. This allows us to select the enzyme which produces the best compromise between having large numbers of resolved bands >2 kb and minimizing the proportion of shared bands. Obviously, for a given electrophoretic migration distance, as the number of bands increases the band sharing will also increase; in wild populations the overall band sharing value for the selected enzyme is typically around 20%. Band sharing is also greatly influenced by electrophoretic separation, and a compromise has to be made to avoid using very long gels and, as low voltages are used to minimize field distortion across the gel, to avoid having inconveniently long electrophoretic separation times.

The problems of field distortion and consequent variability in DNA migration rates across gels have been of particular concern to forensic biologists. However, even if these are minimized the very nature of the Southern blotting process (or gel drying if hybridization is to be carried out directly in gels, Tsao *et al.*, 1983), where a flexible and slightly elastic gel is transferred between supports, is such that some distortion is inevitably introduced. The use of direct measurement to compare the mobility of DNA fragments in different lanes is therefore of severely limited value and usually impractical when samples are run on different gels, even if gels include lanes containing standard marker fragments. However, fragments run in lanes that are adjacent, or nearly so, can be compared directly. Consequently, if many adjacent pairs of samples contain shared fragments, as they will in a pedigree analysis or if the population contains low levels of variability (Gilbert *et al.*, 1990), fingerprints can by extrapolation be compared across many lanes.

Attempts are underway to develop methods which will allow the direct comparison of any pair of fingerprints. In our own laboratory, we include a cocktail of standard size marker fragments at low concentration in every lane (Birkhead *et al.*, 1990). This cocktail can be revealed on an autoradiograph by reprobing with ^{32}P-labelled marker fragments. We have found that the interpolation of fragment sizes (Plikaytis *et al.*, 1986) can be achieved with a high degree of accuracy across the size range 2–20 kb provided that, under the conditions we use (Birkhead *et al.*, 1990), the agarose concentration is not more than 0.8%. One problem lies in achieving the accurate alignment of different autoradiographs, but this can probably be solved by using a probe cocktail which includes some sequence present only in separate size marker lanes, and using the marker lanes to align the separate exposures of the minisatellite and internal marker patterns. The presence in some species (see below) of apparently monomorphic satellite DNA fragments for which specific probes have been isolated may provide a better alternative to the use of internal marker fragments.

Why a DNA "Fingerprint"?

The probability of two individuals sharing the same set of minisatellite fragments may be conservatively found as x^n, where x is the probability of band sharing and n the number of resolved bands (see Brookfield, 1991), assuming that all bands are in linkage equilibrium. In humans x^n was found to be many orders of magnitude smaller than the reciprocal of the world human population size, and so this individual uniqueness (other than between identical twins) was encapsulated by analogy in the term "DNA fingerprint" (Jeffreys et al., 1985a, b).

The assumption of linkage equilibrium is extremely difficult to test directly because of the difficulty, referred to above, of quantifying and cross-comparing population samples of multilocus profiles. However, the most likely source of disequilibria – linkage among the minisatellite loci detected in a fingerprint – can be excluded through the completion of an analysis, specific to a probe/endonuclease combination, of the degree of cosegregation among parental fingerprint bands in a moderately large family. Such an analysis has been carried out in human (Jeffreys et al., 1986), Mus (Jeffreys et al., 1987), house sparrow (Burke and Bruford, 1987), dunnock, Prunella modularis (Burke et al., 1989), chicken, Gallus gallus (Bruford et al., 1990), zebra finch, Taeniopygia guttata (Birkhead et al., 1990), and in many other species (Burke et al., unpublished). In evolutionary biological studies the uniqueness of "fingerprint" patterns is not of direct concern, but the calculation of expected band sharing values when testing hypotheses about relatedness (e.g. Birkhead et al., 1990) relies equally upon the assumption of linkage equilibrium. The mean level of nonindependence detected in segregation analyses can be taken into consideration when hypothesis testing (Birkhead et al., 1990). However, in some species the "fingerprints" revealed by some enzyme/probe combinations are found to comprise a small number of sets of linked bands. For example, in such an analysis in the red grouse Lagopus lagopus we found that one individual's fingerprint contained 16 bands of which 12 represented only one locus (in two allelic cosegregating haplotypes) (O.H. and T.B., in preparation). It is therefore important that if the assumption of independence is made that it should actually be tested; there are already several published examples where statistics have been quoted without reference to this critical assumption (e.g. Georges et al., 1988).

In many situations it is impractical to obtain large families for a segregation analysis. Two studies have therefore used the observed distributions of band sharing values between known relatives and non-relatives to test hypotheses concerning the relatedness of other individuals (Westneat, 1990; Hunter et al., submitted). However, the statistical analysis used in these cases makes parametric assumptions concerning the distribution of band sharing values for a particular level of relatedness, but this assumption has not as yet been investigated.

Examples of Studies Using Multilocus Analysis

In population biology, DNA fingerprinting has so far been of most immediate value in studies of the reproductive behaviour of birds. Parentage analysis has allowed the quantification of extra-pair paternity and conspecific nest parasitism in population studies of several species. Published and preliminary data from several studies are summarized in Tab. 2. The frequency with which birds raise non-kin as a result of extra-pair mating and nest parasitism varies widely among species and probably even among and within populations.

The most beneficial studies are those where detailed observations of the behaviour of individuals are combined with a DNA fingerprinting analysis. One such study, of the dunnock, sought to understand why both the males in polyandrously mated trios (two males mated to one female) do not always make an equal contribution to feeding the female's brood (Burke et al., 1989). The analysis of paternity by DNA fingerprinting demonstrated that there was a highly significant association between whether the subordinate male fed the brood and whether he had fathered any of the nestlings. The males did not, however, feed their own offspring preferentially, and their relative feeding effort was apparently determined by their amount of exclusive access to (and therefore opportunities to copulate with) the female.

Table 2. Data from DNA fingerprinting studies

Species	Offspring Tested	Frequency EPP	IBP
House sparrow[1]	321	10%	0%
European bee-eater[2]	100	1%	0%
Dunnock[3]	133	1%	0%
Willow warbler[4]	120	0%	0%
Wood warbler[4]	56	0%	0%
Zebra finch[5]	92	2%	11%
Fulmar[9]	91	0%	0%
Oystercatcher[1]	23	5%	0%
Indigo bunting[6]	63	35%	0%
Purple martin[7]	52	18%	17%
Red-winged blackbird[8]	111	28%	0%

EPP = extra-pair paternity; IBP = intra-specific brood parasitism.
[1]Our laboratory, unpublished; [2]Jones et al., 1991; [3]Burke et al., 1989; [4]Gyllensten et al., 1990; [5]Birkhead et al., 1990; [6]Westneat, 1990; [7]Morton et al., 1990; [8]Gibbs et al., 1990; [9]Hunter et al., submitted.

In another study, of extra-pair copulation behaviour in the fulmar, *Fulmarus glacialis*, it was possible to record every copulation made by the individuals within the study sample (Hunter *et al.*, submitted). It was therefore possible to determine the rate of extra-pair paternity from a relatively small number of DNA fingerprints as it was not necessary to check the paternity of the offspring of all the many females which had not taken part in extra-pair copulations.

In studies of cooperative behaviour it is often desirable to be able to measure the relatedness of pairs of individuals directly. While fingerprint similarity between first order relatives can usually be distinguished from that of non-relatives with a high probability, overlap among the band sharing distributions for lower levels of relatedness makes tests of relatedness progressively less powerful. Two studies of avian helper-at-the-nest systems which have employed DNA fingerprinting have now been carried out (of bee-eaters, *Merops apiaster*, Jones *et al.*, 1991 and T.B. and M.W.B., unpublished, and of stripe-backed wrens, *Campylorhynchus nuchalis*, Rabenold *et al.*, 1990). In each case hypotheses have been tested which relate to explanations of the evolution of this cooperative behaviour: (i) do helpers assist because of the possibility of direct fitness benefits through copulations with the nesting female(s) or laying their own eggs in the nest (detectable through parentage analysis) or (ii) do they assist because of the inclusive fitness benefits of increasing a relative's reproductive success (detectable through relatedness analysis). In both the studies reported so far there was more evidence for (ii) than for (i). Though we have distinguished here between "parentage" analysis and "relatedness" analysis, both usually require the same hypothesis testing approach, where observed and expected band-sharing values are compared (Birkhead *et al.*, 1990). This can be avoided only in paternity testing and only then when maternity is certain, the mother is available and either if all possible fathers can be tested or that it has been shown that only the true father would contain all the paternal-specific bands.

Studies of two other species have used fingerprint data to estimate relatedness (naked mole-rat, *Heterocephalus glaber*, Reeve *et al.*, 1990, cf. Faulkes *et al.*, 1990; chicken, Kuhnlein *et al.*, 1989). Naked mole-rats are apparently unique among mammals in having a social structure that resembles that of the social insects: only one female and her 1–3 mates reproduce within each colony and their offspring help to raise the offspring in subsequent litters. Pamilo and Crozier's (1982) method of estimating relatedness from marker genotypes was adapted for use on fingerprint data and the resulting mean relatedness estimate of 0.81 was considered to result from consanguineous mating. Kunhlein *et al.*, (1989) measured fingerprint similarities within experimental chicken lines of known pedigree and were able to demonstrate a close and linear relationship between the mean band frequency and the mean inbreeding coefficient.

DNA fingerprinting is not normally the method of choice when examining genetic relationships among populations (Tab. 1). However, the method may be appropriate when, for unusual historical reasons, populations contain relatively low levels of genetic variation. An example of such a situation, in which DNA fingerprinting has been found to be appropriate, is provided by the Channel Island fox populations off the coast of California (Gilbert et al., 1990). In terms of the evolutionary history of mammals, these populations are young and their fingerprints show relatively low levels of variability within populations and a high degree of differentiation among populations. A phylogenetic tree constructed to demonstrate the evolutionary relationships among the populations was found to agree with that expected from the known geological history of the islands.

Limitations of the Multilocus Approach

The situations in which multilocus DNA fingerprinting provides, in comparison with other methods, an appropriate degree of genetic resolution are summarised in Tab. 1. However, as referred to above and discussed by Lynch (1988), though the system is the best that we have for examining close degrees of relatedness, it cannot be used to estimate the relatedness of two particular individuals with very great precision. The variance of the mean fingerprint similarity appropriate to any specific level of relatedness includes components due to the background population band-sharing among nonrelatives and variation in the number of loci that are represented in the scorable region of a fingerprint. These two components are major contributors to the lack of precision. In a sense, the problem is that the minisatellite alleles scored in a fingerprint cannot be attributed to specific loci (except for a very few loci in rare situations; Reeve et al., 1990, Amos et al., 1990).

One way around these problems is to develop probes that only hybridize to a single minisatellite locus. The statistical power achieved through the use of such probes will, of course, depend upon the heterozygosity at the detected loci, but it seems that typically just a few such probes used in succession, or even together as a cocktail (Wong et al., 1987), will be as powerful as a single multilocus analysis. The genotypes obtained from several such probes should allow the more precise estimation of relatedness than does multilocus fingerprint data. Queller and Goodnight (1989; cf. Pamilo, 1989) describe a method for the estimation of relatedness from diploid genotypes at single loci.

A major advantage of multilocus analysis is that there are now many probes available which will hybridize to minisatellites in species belonging to very diverse taxa. As predicted from neutral theory, the degree of polymorphism at a minisatellite locus is related to the mutation rate (Jeffreys *et al.*, 1988). The great majority of minisatellites are non-coding and seem unlikely to be significantly influenced by natural selection. It therefore seems likely that hypervariable minisatellite loci will evolve rapidly, leading to dramatic variation in the level of heterozygosity at an homologous locus even between closely related species, and even the frequent loss or gain of a locus. Evolutionary studies now provide some support for this scenario (Gray and Jeffreys, 1991). This implies that for a locus-specific analysis it will usually be necessary to clone minisatellite loci in the species of interest (but see below).

The Development of Locus-specific Minisatellite Probes

Until very recently, minisatellite loci were very difficult to isolate and the major effort that was required seemed justifiable only in humans and organisms of economic importance. Recently, however, a cloning system has been identified which allows the very efficient isolation of minisatellite loci in humans (Armour *et al.*, 1990), and this system has now been successfully applied in our laboratory to obtain locus-specific minisatellite probes from several species including peafowl (Hanotte *et al.*, 1991a, b), chicken (Bruford *et al.*, 1990, Bruford and Burke 1991), ruff, *Philomachus pugnax* (T.B., O.H., S. Bailey and E. Needham, unpublished), house sparrow (T.B. and E.C., unpublished) and pied flycatcher, *Ficedula hypoleuca* (T.B., M.B. and L. Jenkins, unpublished). As in some earlier studies (Wong *et al.*, 1986, 1987), the new protocol involves the cloning of a size-selected fraction of a restriction digested pool of DNA from several individuals. This fraction is therefore greatly enriched for the presence of minisatellites. The new method, however, takes advantage of a cosmid vector, known as a charomid (Saito and Stark, 1986), which can accept a wide size range of inserts before being packaged and efficiently introduced into a suitable host. The use of a cosmid avoids the problem which occurs with lambda hosts when minisatellites lose repeat units and the recombinant molecule becomes too small to repackage and reinfect further host cells. Also, a host *E. coli* has been identified in which the loss of repeat units is minimized (NM554; Raleigh *et al.*, 1988).

The charomid libraries are screened using a variety of multilocus probes. Our experiences with libraries from five avian species are summarized in Tab. 3. Only relatively small numbers of colonies have to be screened to find useful numbers of polymorphic probes. In principle, this procedure could be applied in any species in which variable, multilocus fingerprints can be obtained using the same restriction enzyme to be used in the cloning. Locus-specific probes have the practical advantage that they are used at high stringency and they make it easier, compared with multilocus probes, to obtain a good hybridization signal with minimal background. When used either singly or in cocktails they provide patterns which are much more suitable for analysis by scanning and densitometry methods than are multilocus fingerprints. They are also more sensitive than the multilocus probes.

The probes that we have isolated are now being applied to studies of the molecular evolution of minisatellites, genomic mapping, paternity analysis, relatedness and population structure. With respect to the latter, the loci that we have so far isolated from wild populations (ruff, house sparrow and pied flycatcher) typically have heterozygosities around 90%, which probably makes them too variable to be useful for comparing populations. However, there appears to be a relationship between mutation rate, minisatellite length and, consequently, heterozygosity (Jeffreys et al., 1988). We therefore anticipate that cloning different size fractions should provide minisatellites having average heterozygosities appropriate to different levels of population genetic problem. It seems unlikely, however, that minisatellite loci will be useful to phylogenetic analyses at the species level in the way that the more slowly evolving allozyme markers have been (see Tab. 1).

Studies of the cross-hybridization of minisatellites among related species indicate that loci vary considerably in their rates of evolution. For example, some peafowl probes appear to be specific to peafowl genera, whilst others hybridize at high stringency to a locus, which may be homologous, in species as evolutionarily distant as the grey par-

Table 3. Cloning summary

Species	Fraction ($\neq$)	Total Recombinants	Number Plated	Positives .6	.15	M13	$3'\alpha$	(Minipreps) $N \geq 3$	↑Alleles
Peafowl	6.6–16	7.2×10^4	1520	133	222			(50)	10
Chicken	11–21	4.5×10^4	1700	9	20	5	18	(26)	7
	2–7	1.2×10^6	4400	56	72	17	22	(21)	14
Ruff	4.4–9.6	4.4×10^5	2400	84	70			(15)	7
House sparrow	4.4–9.6	3.2×10^5	1200	142	64			(15)	7
Pied flycatcher	4.4–9.6	1.0×10^6	1800	51	47	64	171	(8)	5
TOTAL								(133)	57

Summary of minisatellite cloning in five species of bird. In the chicken, two libraries were constructed from different restriction fragment size fractions. For the colonies transferred into microtitre plates (number plated), the number of positives detected using up to four multilocus probes (.6: 33.6 and .15: 33.15, Jeffreys et al., 1985a; M13: the M13 minisatellite, Vassart et al., 1987; $3'\alpha$: the $3'\alpha$-globin hypervariable region, Fowler et al., 1988). For each library, the number N of different probes which detected different loci having at least three alleles among a test panel of four unrelated individuals is indicated. The number of different recombinant clones that were isolated as minipreps and probed against the test panels is shown in parentheses.

tridge, *Perdix perdix* (Hanotte *et al.*, 1991*a*, *b*). This suggests the possibility that minisatellites cloned from a limited set of species might provide useful locus-specific (and, at low stringency, multilocus) probes for studies of several other species. This seems likely to apply especially to more speciose taxa such as the Passerine birds.

Two of the species in which we have cloned minisatellites are of interest because they have a lek mating system. We anticipate that the use of locus-specific probes, in combination with internal size markers as described above, will greatly assist the classification of male genotypes and, in particular, allow cross-comparisons between gels when searching for fathers. In some species, including ruff, we have isolated a probe which hybridizes to a regularly spaced satellite fragment "ladder", and this may provide a useful alternative as a size marker to the use of internal standards.

Locus-specific minisatellite probes are likely to be of particular value in animal and plant breeding as markers potentially linked to quantitative trait loci and in genomic mapping projects (see Bruford and Burke, 1991). In the very long term, we can expect that such studies will contribute to the isolation of some of these quantitative trait loci and that these loci will be homologous across many species. This will ultimately enable the direct study of the selective processes acting on such loci under natural conditions.

Alternative Approaches

During our studies of many bird species and a number of other vertebrates we have encountered the occasional one in which none of our standard panel of multilocus probes hybridizes to a useful (i.e. reliably detectable and sufficiently variable) fingerprint. In such circumstances it may be necessary to try an alternative to minisatellites. One approach, used by Quinn *et al.* (1987) in studies of lesser snow geese, *Chen c. caerulescens*, is to clone random RFLP loci. Another possibility might be to develop polymorphic microsatellite markers (Tautz, 1989; Love *et al.*, 1990), which seem likely to be present in all vertebrates but are expensive to isolate. Alternatively, it may be possible to employ the available probes from the highly polymorphic major histocompatibility complex as single locus probes to homologous loci in other species. Mouse MHC probes have been successfully used for this purpose in studies of mating success in the red-winged blackbird *Agelaius phoeniceus* (Gibbs *et al.*, 1990). A final possibility is to use (TTAGGG)n probes to detect hypervariable telomeric sequences separated in pulsed field gels (Kipling and Cooke 1990).

166

Acknowledgements
We thank our colleagues S. Bailey, J. Dann, J. Gardner, L. Jenkins, A. Krupa, E. Needham and I. van Pijlen for their enthusiastic participation in our minisatellite cloning studies. We are collaborating with the following in unpublished studies of various species mentioned in the text or Tab. 1: C. M. Lessels and J. R. Krebs (bee-eater), B. Ens (oystercatcher), M. Petrie and H. Budgey (peafowl), A. Lundberg (pied flycatcher), D. B. Lank (ruff). We thank I. P. F. Owens for comments on the manuscript.

References

Amos, W., and Dover, G. A. (1990) DNA fingerprinting and the uniqueness of whales. *Mamm. Rev.* 20: 23–30.

Armour, J. A. L., Povey, S., Jeremiah, S., and Jeffreys, A. J. (1990) Systematic cloning of human minisatellites from ordered array charomid libraries. *Genomics*, 8: 501–512.

Birkhead, T. R., Burke, T., Zann, R., Hunter, F. M., and Krupa, A. P. (1990) Extra-pair paternity and intraspecific brood parasitism in wild zebra finches, *Taeniopygia guttata*, revealed by DNA fingerprinting. *Behav. Ecol. Sociobiol.* 27: 315–324.

Brookfield, J. F. Y. (1991) The statistical interpretation of hypervariable DNAs In: Hewitt, G. M. (ed.) Molecular Techniques in Taxonomy. Springer, Berlin. pp 159–170.

Bruford, M. W., Burke, T., Hanotte, O., and Smiley, M. B. (1990) Hypervariable markers in the chicken genome. Proceedings of the 4th World Congress on Genetics Applied to Livestock Production XIII: 139–142.

Bruford, M. W., and Burke, T. (1991) Hypervariable DNA markers and their applications in the chicken. In: Burke, T., Dolf, G., Jeffreys, A. J., and Wolff, R. (eds) DNA Fingerprinting: Approaches and Applications. Birkhäuser, Basel. pp 230–242. (This volume).

Burke, T. (1989) DNA fingerprinting and other methods for the study of mating success. *Trends. Ecol. Evol.* 4: 139–144.

Burke, T., and Bruford, M. W. (1987) DNA fingerprinting in birds. *Nature* 327: 149–152.

Burke, T., Davies, N. B., Bruford, M. W., and Hatchwell, B. J. (1989) Parental care and mating behaviour of polyandrous dunnocks *Prunella modularis* related to paternity by DNA fingerprinting. *Nature* 338: 249–251.

Faulkes, C. G., Abbott, D. H., and Mellor, A. L. (1990) Investigation of genetic diversity in wild colonies of naked mole-rats (*Heterocephalus glaber*) by DNA fingerprinting. *J. Zool.* 221: 87–97.

Fowler, S. J., Gill, P., Werrett, D. J., and Higgs, D. R. (1988) Individual-specific DNA fingerprints from a hypervariable region probe: alpha-globin 3'HVR. *Hum. Genet.* 79: 142–146.

Georges, M., Lequarre, A.-S., Castelli, M., Hanset, R., and Vassart, G. (1988) DNA fingerprinting in domestic animals using four different minisatellite probes. *Cytogenet. Cell Genet.* 47: 127–131.

Gibbs, H. L., Weatherhead, P. J., Boag, P. T., White, B. N., Tabak, L. M., and Hoysak, D. J. (1990) Realised reproductive success of polygynous red-winged blackbirds revealed by DNA markers. *Science* 250: 1394–1397.

Gilbert, D. A., Lehman, N., O'Brien, S. J., and Wayne, R. K. (1990) Genetic fingerprinting reflects population differentiation in the California Channel Island fox. *Nature* 344: 764–767.

Gray, I. C., and Jeffreys, A. J. (1991) Evolutionary transience of hypervariable minisatellite loci in man and the primates. *Proc. Roy. Soc. Series B*, 243: 241–253.

Gyllensten, U. B., Jakobsson, S., and Temrin, H. (1990) No evidence for illegitimate young in monogamous and polygynous warblers. *Nature* 343: 168–170.

Gyllensten, U. B., Jakobsson, S., Temrin, H., and Wilson, A. C. (1989) Nucleotide sequence and genomic organisation of bird minisatellites. *Nucleic Acids Res.* 17: 2203–2214.

Hanotte, O., Burke, T., Armour, J. A. L., and Jeffreys, A. J. (1991a) Hypervariable minisatellite DNA sequences in the Indian peafowl, *Pavo cristatus*. *Genomics* 9: 587–597.

Hanotte, O., Burke, T., Armour, J. A. L., and Jeffreys, A. J. (1991b) Cloning, characterization and evolution of Indian peafowl *Pavo cristatus* minisatellites. In: Burke, T., Dolf, G., Jeffreys, A. J., and Wolff, R., (eds) DNA Fingerprinting: Approaches and Applications. Birkhäuser, Basel. pp. 193–216 (This volume).

Hewitt, G. M. (ed.) (1991) Molecular Tehniques in Taxonomy. Springer, Berlin.

Hillis, D. M., and Moritz, C. (eds) (1990) Molecular Systematics. Sinauer, Sunderland, MA.

Hunter, F. M., Burke, T., and Watts, S. E. (submitted) Paternity determined by frequent copulation in the northern fulmar.

Jeffreys, A. J., Wilson, V., and Thein, S. L. (1985a) Hypervariable 'minisatellite' regions in human DNA. *Nature* 314: 67–73.

Jeffreys, A. J., Wilson, V., and Thein, S. L. (1985b) Individual-specific 'fingerprints' of human DNA. *Nature* 316: 76–79.

Jeffreys, A. J., Brookfield, J. F. Y., and Semeonoff, R. (1985c) Positive identification of an immigration test-case using human DNA fingerprints. *Nature* 317: 818–819.

Jeffreys, A. J., Wilson, V., Thein, S. L., Weatherall, D. J., and Ponder, B. A. J. (1986) DNA "fingerprints" and segregation analysis of multiple markers in human pedigrees. *Am. J. Hum. Genet.* 39: 11–24.

Jeffreys, A. J., Wilson, V., Kelly, R., Taylor, B. A., and Bulfield, G. (1987) Mouse DNA 'fingerprints': analysis of chromosome localisation and germ-line stability of hypervariable loci in recombinant inbred strains. *Nucleic Acids Res.* 15: 2823–2836.

Jeffreys, A. J., Royle, N., Wilson, V., and Wong, Z. (1988) Spontaneous mutation rates to new length alleles at tandem repetitive hypervariable loci in human DNA. *Nature* 332: 278–281.

Jones, C. S., Lessells, C. M., and Krebs, J. R. (1991) Helpers-at-the-nest in European bee-eaters (*Merops apiaster*): a genetic analysis. In: Burke, T., Dolf, G., Jeffreys, A. J., and Wolff, R. (eds) DNA Fingerprinting: Approaches and Applications. Birkhäuser, Basel. pp 169–192 (This volume).

Kipling, D., and Cooke, H. J. (1990) Hypervariable ultra-long telomeres in mice. *Nature* 347: 400–402.

Kessler, C., and Holtke, H.-J. (1986) Specificity of restriction endonucleases and methylases – a review (Edition 2). *Gene* 47: 1–153.

Kuhnlein, U., Dawe, Y., Zadworny, D., and Gavora, J. S. (1989) DNA fingerprinting: a tool for determining genetic distances between strains of poultry. *Theoret. Appl. Genet.* 77: 669–672.

Love, J. M., Knight, A. M., McAleer, M. A., and Todd, J. A. (1990) Towards construction of a high resolution map of the mouse genome using PCR-analysed microsatellites. *Nucleic Acids Res.* 18: 4123–4130.

Lynch, M. (1988) Estimation of relatedness by DNA fingerprinting. *Mol. Biol. Evol.* 5: 584–599.

Mock, D. G. (1983) On the study of avian mating systems. In: Brush, A. H., and Clark, G. A. (eds), Perspectives in Ornithology. Cambridge University Press, Cambridge, pp. 55–85.

Morton, E. S., Forman, L., and Braun, M. (1990) Extra-pair fertilisations and the evolution of colonial bredding in purple martins. *Auk* 107: 275–283.

Pamilo, P. (1989) Estimating relatedness in social groups. *Trends Ecol. Evol.* 4: 353–355.

Pamilo, P., and Crozier, R. H. (1982) Measuring genetic relatedness in natural populations: methodology. *Theor. Pop. Biol.* 21: 171–193.

Plikaytis, B. D., Carlowe, G. M., Edmonds, P., and Mayer, L. W. (1986) Robust estimation of standard curves for protein molecular weight and linear duplex DNA base pair number after gel electrophoresis. *Anal. Biochem.* 152: 346–364.

Queller, D. C., and Goodnight, K. F. (1989) Estimation of relatedness using genetic markers. *Evolution* 23: 258–275.

Quinn, T. W., Quinn, J. S., Cooke, F., and White, B. N. (1987). DNA marker analysis detects multiple maternity and paternity in single broods of the lesser snow goose. *Nature*, 362: 392–394.

Rabenold, P. P., Rabenold, K. N., Piper, W. H., Haydock, J., and Zack, S. W. (1990) Shared paternity revealed by genetic analysis in cooperatively breeding tropical wrens. *Nature* 348: 538–540.

Raleigh, E. A., Murray, N. E., Revel, H. Blumenthal, R. M., Westanay, D., Reith, A. D., Rigby, P. W. J., Elhai, J., and Hanahan, D. (1988) *Mcr*A and *mcr*B restriction phenotypes of some *E. coli* strains and implications for gene cloning. *Nucleic Acids. Res.* 16: 1563–1575.

Reeve H. K., Westneat D. F., Noon, W. A., Sherman, P. W., and Aquadro, C. F. (1990) DNA "fingerprinting" reveals high levels of inbreeding in colonies of the eusocial naked male rat. *Proc. Natl. Acad. Sci. USA* 87: 2496–2500.

Saito, I., and Stark, G. R. (1986) Charomids: cosmid vectors for efficient cloning and mapping of large or small restriction fragments. *Proc. Acad. Natl. Sci. USA* 83: 8664–8668.

Tautz, D. (1989) Hypervariability of simple sequences as a general source for polymorphic DNA markers. *Nucleic Acids Res.* 17: 6463–6471.

Tsao, S. G. S., Brunk, C. F., and Perlman, R. E. (1983) Hybridization of nucleic acids directly in agarose gels. *Anal. Biochem.* 131: 365–372.

Vassart, G., Georges, M., Monsieur, R., Brocas, H., Lequarre, A. S., and Christophe, D. (1987) A sequence in M13 phage detects hypervariable minisatellites in human and animal DNA. *Science* 235: 683–684.

Westneat, D. F. (1990) Genetic parentage in the indigo bunting: a study using DNA fingerprinting. *Behav. Ecol. Sociobiol.* 27: 67–76.

Wong, Z., Wilson, V., Jeffreys, A. J., and Thein, S. L. (1986) Cloning a selected fragment from a human DNA "fingerprint": isolation of a highly polymorphic minisatellite. *Nucleic Acids. Res.* 14: 4605–4616.

Wong, Z., Wilson, V., Patel I., Povey, S., and Jeffreys, A. J. (1987) Characterization of highly variable minisatellites cloned from human DNA. *Ann. Hum. Genet.* 51: 269–288.

DNA Fingerprinting: Approaches and Applications
ed. by T. Burke, G. Dolf, A. J. Jeffreys & R. Wolff

Helpers-at-the-nest in European Bee-eaters (*Merops apiaster*): a Genetic Analysis

C. S. Jones[a], C. M. Lessells[b] and J. R. Krebs[a]

[a]*The Edward Grey Institute of Field Ornithology, Department of Zoology, Oxford University, South Parks Road, Oxford, OX1 3PS, Great Britain;* [b]*Department of Animal and Plant Sciences, Sheffield University, Sheffield, S10 2TN, Great Britain*

Summary
Helping-at-the-nest provides a classic example for the analysis of the costs and benefits of altruism in relation to kin selection. Most studies to date have relied on genealogical data obtained by ringing parents and offspring. However, growing evidence shows that both extra-pair copulations and intra-specific nest parasitism are sufficiently frequent in birds to produce patterns of relatedness that differ appreciably from those inferred from ringing studies. This could have major repercussions for calculations of the genetic costs and benefits of helping. We therefore used DNA fingerprinting, in conjunction with detailed field observations over eight years (1983–1990) to analyse the helping behaviour of European bee-eaters (*Merops apiaster*). Fingerprints of 11 unhelped (60 chicks) and 8 helped (40 chicks) families using 33.6, 33.15 and M13 probes showed that: (a) every chick from unhelped families could be correctly assigned to their putative parents; (b) helpers are invariably genetic relatives ($r = 0.5$) of one or both of the breeding pair they help; and (c) in a preliminary analysis of parentage in helped families, about 2% (1/40) chicks in helped nests are not sired by the putative father. The helper in this case did not appear to be the father, but the illegitimate offspring may have been sired by a close relative of the helper.

Introduction

The evolutionary maintenance of altruistic behaviour is a central issue in behavioural ecology. The problem, put simply, is "how can genes that cause their bearers to sacrifice reproductive success or survival, whilst benefiting others, be maintained by selection?" A variety of theories have been proposed including kin selection (Hamilton, 1964), reciprocity (Trivers, 1971), manipulation (Alexander, 1974), and mutualism. "Helpers-at-the-nest" in birds have proved fertile ground for testing these ideas. For reviews see Brown, 1987; Stacey and Koenig, 1990; and Emlen, 1991.

In species with helpers, individuals appear to forego the opportunity to breed themselves and help others to breed instead. There is a considerable diversity of helper systems, but a common pattern is for one or both sexes to remain near their natal site (most helper species are non-migratory) and help their parents to raise young in subsequent breeding attempts.

The question of why (in the evolutionary sense of costs and benefits) helpers help is often divided into two (Emlen, 1991). (*a*) Why stay at home instead of breeding? This is usually answered in terms of "ecological constraints" that preclude or reduce any benefits of dispersal from home. These include shortage of mates, territories, nest sites, food supplies and the disadvantages of solitary (versus group) living. (*b*) Given that birds stay at home, why do they help? Here two hypotheses are currently favoured; either individuals gain some "direct" long-term benefit from helping (increased experience, increased survival as a result of being in a larger group, increased chance of inheriting a territory) or they gain an "indirect" benefit as a result of helping to raise collateral kin (nephews, nieces or siblings) or direct descendents (helping their own offspring to raise chicks). These two ideas are not mutually exclusive. Where it is possible to estimate the relative importance of the two effects, Emlen (1991) concludes that in 5 out of 7 species for which data are available, both effects are essential to maintain helping. In 2 out of 7 (primary helpers in pied kingfishers – Reyer, 1984; and white-fronted bee-eaters – Emlen and Wrege, 1988; Emlen, 1990) kin selection plays a major role. This is not to say that in other species helpers do not enhance the breeding success of the birds they help – there is almost invariably a positive correlation between the presence of helpers and breeding success of the recipients of this help (Stacey and Koenig, 1990 – 15/17 species).

European bee-eaters (*Merops apiaster*) are unusual among "helper" species in that they are migratory and colonial (non-territorial); they have no feeding territory, but do defend the nest hole. Helpers do not remain at their natal site in the same way as territorial species such as the Florida scrub jay (Woolfenden and Fitzpatrick, 1984), in which helpers "stay at home" in their natal territory. Helpers, which are predominantly male, are often birds that have returned to breed in their natal colony and failed. Following the failure of a nesting attempt, birds that become helpers move to the nest of a relative and contribute to feeding the young. Although there is no apparent "ecological constraint" to prevent birds from re-nesting instead of helping, Lessells (1990) argues that the seasonal decline in breeding success makes re-nesting a relatively poor option for birds who fail late in the breeding season. Ringing provides data on the costs and benefits of helping. It appears that helpers enhance the success (reproduction and survival) of the breeding pair that they help and that the recipients of this help are close relatives of the helper. However, ringing data cannot provide a conclusive evaluation of the costs and benefits of helping because extra-pair copulations (EPCs) and intra-specific nest parasitism (ISNP) may alter the patterns of relatedness between helpers and the recipients of help. Furthermore, our ringing records are incomplete (approximately 30% of helpers are known to be close relatives of the breeding pair from ringing genealogies, but because they nest down metre-long burrows, only a small sample of chicks are ringed each year), and the exact genealogies of many birds are not known. For this reason, a genetic analysis is essential. Other studies have successfully applied DNA finger-printing to the study of helping behaviour including in dunnocks (Burke et al., 1989), and stripe-backed wrens (Rabenold et al., 1990).

In this paper we use DNA fingerprinting to answer the following questions. (a) Are helpers always related to the birds they help. If so, in what way? Or, more pertinently, are helpers ever genetically unrelated to the pairs they help? Ideally, we require an unbiased estimate of propor-tions of different relatives helped. (b) Do helpers ever contribute to paternity (most helpers are male) of the brood they help rear? The incidence of copulations outside the pair-bond is well documented in many avian species (Birkhead, 1987; Westneat, 1987; Westneat et al., 1990). Indeed, EPCs comprise 4% of copulations in the colony and 14% of copulations at a feeding site in European bee-eaters and ISNP is also known to occur in this species (Lessells and Ovenden, 1989). Yet the degree to which these EPCs actually result in extra-pair fertilisations (offspring) is relatively unexplored (Gyllensten et al., 1990). However, a number of studies to date suggest that both EPCs and ISNP are sufficiently frequent in birds to produce patterns of relatedness that differ considerably from those inferred from ringing and behavioural observa-tions (Burke and Bruford, 1987; Wetton et al., 1987; Westneat, 1990). Preliminary analysis of both helped and unhelped families were carried out to assess this, to calculate the parameters necessary to evaluate kin selection as an explanation of helping behaviour in bee-eaters.

European Bee-eaters: the Study Species

The European bee-eater is a colonially breeding migratory species. We have studied a colony of about 100 pairs since 1983 at a colony in the Camargue, southern France, where they build their metre-long nest burrows in an earth spoil bank. We have obtained genealogical data and information about the identities of helpers and helped birds by marking both nestlings and full-grown birds with metal leg rings (which last throughout an individual's lifetime) and painting full-grown birds on the tail (which lasts one breeding season) for field identification. These ringing data, in conjunction with behavioural observations, show that one-fifth of the nests that fledge chicks have from one (75% of cases) to four helpers. Helpers arrive at nests after the chicks have hatched and help provision the chicks.

Helped broods are fed more in total than unhelped broods and the female, but not the male, parent brings less food if she is helped. Helping apparently increases slightly the number of chicks fledged and decreases the overwinter mortality of the parents. The majority (93%) of helpers are male, and part of this sex bias is accounted for by a male bias in natal philopatry. Individuals start breeding at one year of age, and helpers are one or more years old and are generally birds who have failed at breeding earlier in the season. Further details of the study methods and helping behaviour are given by Lessells (1990).

DNA Fingerprinting Methodologies

Bee-eater blood (approximately 100–400 μl) collected by venipuncture, from both adults and nestlings, was suspended in at least one volume of $1 \times$ SSC (0.15M NaCl, 15 mM trisodium citrate), 10 mM EDTA, pH 7.0, and all blood was immediately put onto icepacks in a cool box (4°C) and frozen (-20°C) within 12 hours.

For DNA extractions, approximately 100 μl of blood solution was resuspended in 500 μl of STE buffer (0.1 M Tris-HCl, pH 8.0, 0.1 M NaCl, 1 mM EDTA), 10 μl of proteinase K (10 mg/ml) and 20 μl of 25% SDS (sodium dodecyl sulphate), and incubated overnight at 55°C. This solution was extracted twice with phenol/chloroform, once with chloroform/isoamyl alcohol (24:1) and the DNA precipitated with absolute ethanol, washed with 70% ethanol, vacuum dried and dissolved overnight in 500 μl of TE buffer (10 mM Tris-HCl, pH 7.5, 1 mM EDTA).

Total genomic DNA (5–7 μg; concentrations determined by ethidium bromide fluorescence by comparison with known standards) was digested with an excess of HaeIII restriction endonuclease (20 units), in the presence of 4 mM spermidine trichloride (Sigma) to facilitate complete digestion, for between 6–16 hours at 37°C. Digested DNA was phenol extracted, and recovered by ethanol precipitation, washed with 70% ethanol and dissolved in $1 \times$ ficoll loading buffer (Maniatis et al., 1982). Each sample was electrophoresed through 30 cm long 1% agarose gels in $1 \times$ TBE (0.089 M Tris-base 0.089 M boric acid, 0.002 M EDTA, pH 8.3), run at approximately 1.5 V/cm for at least 48 hours until the 2 kilobase molecular weight marker (lambda HindIII) was about 2 cm from the end of the gel. These conditions generally produce gels with relatively straight tracks, giving between 30–40 reasonably spaced, clearly resolved bands, scorable down to at least 2 kilobases. However, despite standardised running conditions, some tracks run better than others for various reasons, including under- or over-loading, partially degraded or contaminated samples or varying salt concentrations. As a consequence of lanes being unequally scorable, estimates of band-sharing become ambiguous and inaccurate. To counter this we routinely add small quantities (5 ng) of internal DNA molecular size markers (lambda Hind III) to each sample and after obtaining the fingerprints with the hypervariable probes, the filters are reprobed with uncut lambda DNA to expose the molecular size markers in each track. Thus, by precisely aligning the marker autoradiograph with the fingerprint autoradiograph, the bands can be read from the midpoint of every track and accurate band interpolation achieved (Burke *et al.*, 1989).

DNA was then depurinated by treatment with 0.25 M HCl, denatured in 0.5 M NaOH, 1.5 M NaCl, neutralised in 1.5 M NaCl, 0.5 M Tris-HCl, pH 7.2, 0.001 M EDTA, and transferred by Southern blotting (Southern, 1975) to Hybond N+ nylon membranes (Amersham International). DNA was alkali fixed to the filter by soaking in 0.4 M NaOH and rinsing briefly in 5 × SSC. Filters were prehybridised in 0.26 M sodium phosphate, pH 7.2, 1 mM EDTA, 7% SDS, and 1% BSA, fraction V (Sigma), for 3 hours at 60°C. Hybridisations were carried out in the same solution, with the addition of radioactively labelled probes, incubated in a shaking waterbath at 60°C for 24–48 hours. Single-stranded 280 base-pair HaeIII- Cla I restriction fragment from wildtype M13 bacteriophage (mp8, Amersham International) (Vassart *et al.*, 1987) obtained after agarose gel electrophoresis and purification with Gene Clean (Stratech Ltd) or single-stranded M13 DNA with human minisatellite clones with either 33.15 or 33.6 inserts (Jeffreys *et al.*, 1985a), was labelled with ^{32}P by random priming (Feinberg and Vogelstein, 1983, 1984) to high specific activities ($>1 \times 10^9$ cpm/μg DNA). Wash conditions are as given in Westneat *et al.* (1988). Probed filters were placed on Kodak X-Omat film in X-ray cassettes and exposed for 1–14 days at −70°C, with 1 or 2 intensifying screens. Each filter was then stripped by immersing in boiling 0.5% SDS, and allowed to cool to room temperature and reprobed as required.

Results

Variability of DNA Fingerprints

Hybridisation patterns produced by each of the three hypervariable probes, 33.15, 33.6 and M13, from unrelated European bee-eaters display considerable variability between individuals. When autoradiographs from each of the three probes for a single blot were superimposed, no significant overlap in fragments was observed, consistent with the findings of other studies using a similar set of probes (Georges *et al.*, 1988; Westneat, 1990); each probe presumably detects a slightly different subset of minisatellites (Jeffreys *et al.*, 1985b, 1986). A mean (±SD) of 23.7 (±2.8), 24.2 (±0.8), and 18.9 (±1.1) fragments were scored per individual for the probes 33.15, 33.6, and M13, respectively. They were detected in molecular weight size range of 2–30 kb for the former two probes, and 1–30 kb for M13. The latter scorable range was extended because this probe detected significantly fewer fragments than the other two probes.

Band-sharing probabilities, x, defined as the mean of each pairwise comparison of the proportion of one individual's bands that were shared by those in another individual (a band was only scored if it had the same position and intensity), were calculated from presumed unrelated adults (see Tab. 1), representing the background level of band-sharing between individuals chosen at random from the colony. It is important to note that we cannot conclusively state that these individuals are all unrelated and some of the variability in band-sharing in this group of individuals may be the result of related individuals being included in the sample. Each of the three probes gave similar band-sharing estimates for both the overall mean and the within and between

Table 1. Band-sharing between apparently unrelated European bee-eaters. DNA fingerprints were scored for each of 10 one year old and older individuals

	33.15	33.6	M13	Average across all 3 probes
Size range (kb)	2–30	2–30	1–30	—
Band-sharing $(x)^1 \pm$ SD comparison:				
(i) male–male ($n = 21$ pairwise comparisons)	0.169 ± 0.082	0.198 ± 0.074	0.180 ± 0.075	0.182 ± 0.050
(ii) male-female ($n = 21$)	0.212 ± 0.076	0.210 ± 0.067	0.165 ± 0.095	0.196 ± 0.054
(iii) female–female ($n = 3$)	0.227 ± 0.052	0.204 ± 0.096	0.115 ± 0.095	0.182 ± 0.018
(iv) overall mean ($n = 45$)	0.193 ± 0.079	0.204 ± 0.071	0.169 ± 0.085	0.189 ± 0.055
Mean no. fragments $\pm$ SD	23.7 ± 2.8	24.2 ± 0.8	18.9 ± 1.1	
Mean allele frequency $(q)^2$	0.102	0.108	0.088	
Probability two fingerprints are identical $(Pf)^3$	1.2×10^{-20}	5.9×10^{-21}	9.7×10^{-17}	
Probability two fingerprints from siblings are identical $(Ps)^4$	5.1×10^{-10}	3.9×10^{-10}	2.9×10^{-8}	

[1]:$x = [(N_{AB}/N_A) + (N_{AB}/N_B)]/2$, where N_A and N_B are the number of bands in individual A and B, and N_{AB} is the number shared by both (Jeffreys et al., 1985b).

[2]:$q = 1 - (1 - x)^{0.5}$ (assuming that $x = 2q - q^2$ and that any shared bands are always derived from identical alleles) (Jeffreys et al., 1985b).

[3]:$Pf = (1 - 2x + 2x^2)^{n/x}$, where n is the number of bands in a typical DNA fingerprint (Jeffreys et al., 1985b).

[4]:$Ps = [1 - 0.5q(1 - q)^2 (4 - q)]^{n/x}$ (Jeffreys and Morton, 1987).

sex comparisons, although, generally, the M13 probe gave slightly lower values, in particular in the female comparisons, probably due to the fact that there were fewer scorable fragments in the detectable range. The combined mean band-sharing probability ($\pm$SD) was 0.189 ($\pm$0.055), similar to that reported for many avian species (Burke and Bruford, 1987; Wetton *et al.*, 1987; Gyllensten *et al.*, 1990; Morton *et al.*, 1990; Westneat, 1990).

Frequency distributions of band-sharing estimates between unrelated individuals for each probe, and the combined average are given in Fig. 1a–d. The probability (Pf) that two bee-eaters have identical fingerprints (assuming an independent assortment of alleles and no significant linkage between loci, supported by analysis of one sibship of 12 offspring, Burke pers. comm), using the combined average of the three probes, is extremely small, approximately 10^{-19}; similarly, if the individuals compared are siblings, the probability (Ps) is approximately 10^{-9} (see Tab. 1 for calculation details and separate estimates for each probe). These probabilities confirm that the bee-eater fingerprints are individual-specific and as such can be used in analyses of parentage and relatedness.

Parentage Analysis

Segregation of bands (linkage and allelism) was generally unknown in the bee-eater families analysed here due to the small sibships involved. However, relationships can still be determined because (*i*) all bands present in an offspring must be present in one or other of its parents and (*ii*) each chick inherits half of its autosomal DNA from each parent, and is, therefore, expected to share approximately 50% of its bands with each parent. Below we have used band-matching, i.e. checking for unassigned bands (not present in either putative parent) in the chicks, for maternity and paternity exclusion. Only relationships confirmed in this way have been included in analyses of band-sharing.

Analysis of Unhelped Families

Paternity and maternity was assessed using DNA fingerprinting in 60 chicks from 11 unhelped families. As a preliminary analysis of parentage, genomic blots of HaeIII digests were hybridised with the minisatellite probe 33.15 in all 11 families, and in a few cases (3 families) results were confirmed with probes 33.6 and M13. Bands which were exclusive to each adult in the breeding pair (resolved maternal and paternal specific fragments) were scored on an acetate overlay of the autoradiograph, and their presence in the offspring tracks documented. DNA fingerprints were available for both putative parents for only 9 out of 11

176

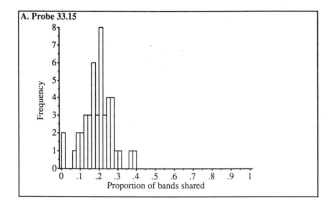

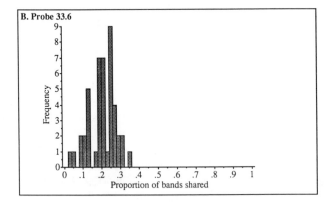

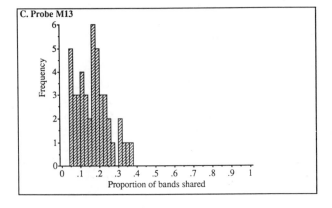

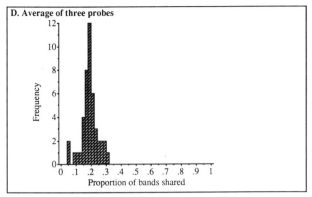

Figure 1. Frequency distributions of band-sharing probabilities from 10 presumably unrelated adults (background bandsharing, x) using probes a) 33.15 b) 33.6 c) M13 and d) band-sharing averaged over all three probes.

families. Of the 51 chicks in these 9 families (e.g. Fig. 2), 46 had no unassigned bands. The remaining 5 chicks had one novel (unassigned) band each. We consider these novel bands to be mutations due to the high expected mutation rates in these minisatellites (Jeffreys et al., 1985b). These novel bands allow us to estimate the mutation rate, m, as $M/(N.n)$, where M is the number of novel bands, N is the number of chicks examined and n is the mean number of bands per fingerprint. The mutation rate for probe 33.15 was 0.0041 [$5/(51 \times 23.7)$] mutation/band/generation. This value is comparable to mutation rates found in other species including humans (Jeffreys et al., 1986), naked mole-rats (Reeve et al., 1990) and birds (Burke and Bruford, 1989; Westneat, 1990).

In the remaining two unhelped families either sufficient good quality, restrictable DNA could not be extracted or the putative father was missing. However, the paternal profile could be reconstructed from fragments occurring in the chicks' fingerprints, but not present in the mother's, using a simplified technique derived from that of Tegelstrom et al. (1991) (Fig. 3). The total number of male-specific bands, divided by the total number of resolvable bands in the chicks was 0.507 and 0.471 for the two families. The first of these values is slightly higher than expected if all the chicks were fathered by a single male (unpubl. analysis). (If background band-sharing, x, = 0, the expected value in the absence of mutation is 0.5. As x increases, the expected value decreases.) However, further analysis of this family (Fig. 3) shows that no chick had more than one band which was not either shared by at least one other chick or present in the faint paternal fingerprint. We provisionally suggest that all chicks from each of the two families were sired by single fathers.

178

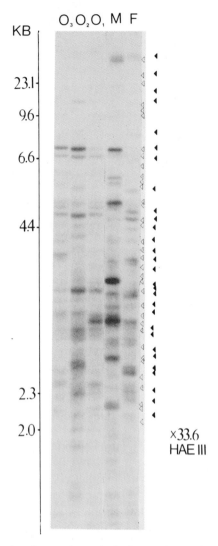

Figure 2. DNA fingerprint of genomic DNAs from an unhelped family (nest F24, 1987) of European bee-eaters, digested with HaeIII and probed with the minisatellite probe 33.6 (short exposure; 2 days with one intensifying screen). M – putative mother; F – putative father; O_{1-3} offspring 1–3. All three chicks can be assigned to the putative parents. Solid arrows indicate resolved paternal bands; open arrows maternal bands; all bands were interpreted directly from short, medium and long exposures of the original autoradiographs. Only those bands whose relative position and intensity matched those in the putative parents were scored as matching in the offspring. Size of fragments (in kilobases, kb) are indicated to the left, determined using the molecular weight marker lambda HindIII.

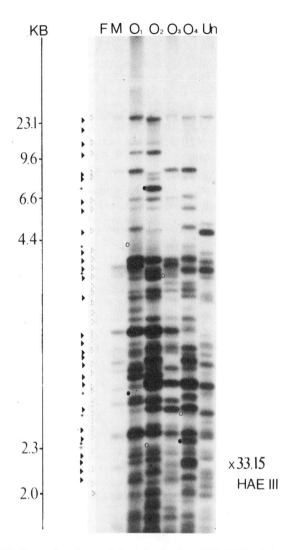

KB   F M O₁ O₂ O₃ O₄ Un

x33.15
HAE III

Figure 3. DNA fingerprint of an unhelped family (nest F14, 1987) of European bee-eaters. Offspring (O₁₋₄), their putative mother (M) and putative father (F – severely under-loaded) and an unrelated individual (Un) for comparison. Total genomic DNA was digested with HaeIII, and the filter probed with Jeffreys's probe 33.15. Scored maternal bands are indicated by open arrows. The mother can be assigned in all four chicks from band-sharing probabilities. Of the fragments present in the chicks, but not in the mother, 28 were present in two or more chicks (solid arrows) and 7 in only one chick (asterisks). A very long exposure identified three of the latter bands as paternal (solid circles), whilst the other four remained unassigned (open circles). These four bands could be due to mutation or chance transmission to a single chick.

180

In conclusion, each of the 60 chicks from 11 unhelped families could be assigned to their putative parents.

Estimates of band-sharing between each parent and offspring were made for probe 33.15. Mean band-sharing for parent to offspring is 0.620 ± 0.070 ($\pm$ SD) (range 0.50–0.75), and was significantly lower between father and offspring (0.607 ± 0.054; range 0.50–0.71) than mother and offspring (0.633 ± 0.050; range 0.53–0.75. $F_{1,\,108} = 6.62$, $p = 0.011$). The reason for this difference is unknown, but it should be borne in mind that multiple band-sharing estimates within families between a parent and each offspring may not be independent, and hence the statistical significance may be inflated. Band-sharing between the breeding male and breeding female was 0.172 ± 0.034, not significantly different from the background band-sharing estimate of 0.193 ± 0.079.

Analysis of Helped Families

The breeding pair within each of the 8 helped families analysed proved to be the parents by band-matching in all but one of the chicks ($n = 40$), using probe 33.15. These results were confirmed by band-matching using the additional probes 33.6 and M13 for four of the families. Chick 2 in family 2 caught in 1989 from nest M4 (track 12 in Fig. 4b and c) had a total of 19 unassigned bands from the three probes,

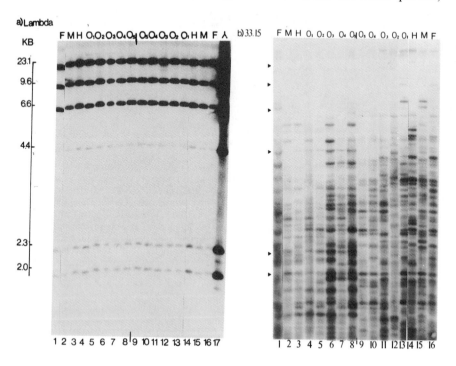

c)MI3   F M H O₁ O₂ O₃ O₄ O₄O₅ O₄ O₃O₂ O₁ H M F

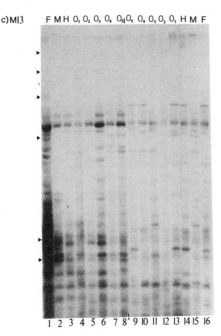

1 2 3 4 5 6 7 8 9 10 11 12 13 14 15 16

Figure 4. Examples of DNA fingerprints from two helped European bee-eater families, obtained by digesting nuclear DNA with HaeIII and hybridising with b) Jeffreys's minisatellite probe 33.15 and c) wild type M13 mp8 bacteriophage DNA. Small quantities (5 ng) of lambda HindIII were added to each sample DNA before electrophoresis to allow accurate interpolation of fragment sizes, which were detected by stripping the filter of previous hypervariable probes and reprobing with a) radioactively labelled, uncut lambda DNA. Track 17 in a) contains lambda HindIII at a higher concentration (0.6 μg) used to visualise the markers by ethidium bromide staining to allow the correct length of the electrophoresis run to be determined. Arrow-heads in b) & c) indicate the size of fragments (kb) using the molecular weight marker lambda HindIII. Tracks 1–8 show helped family one (nest M8, 1989); putative father (F; overloaded sample which is running slightly faster than the rest of the tracks in the family, the profile can be scored from a very short exposure); putative mother (M); the helper (H); and offspring 1–5 (O_{1-5}). Tracks 9–16 illustrate family two (nest M4, 1989), the same letter designations depict the relationships. The helper in family one (track 3) is 0.5 related to both members of the breeding pair, and is deduced to be their offspring from the previous year's brood. The helper in family two (track 14) is deduced (from ringing and behavioural data) to be the brother of the breeding male; this is confirmed by band-sharing estimates. The helper is 0.5 related to the breeding male but unrelated to the breeding female. All chicks in these two helped families can be assigned to the putative parents, with the exception of chick 2 (O_2) in the second family (track 12). Maternity could be assigned to chick 2 by band sharing. However, numerous bands not present in either the putative father or the helper were detected (a few obvious ones are highlighted by arrows). This suggests that this offspring is the result of an extra-pair copulation. A couple of helper specific bands were detected (open circles), indicating that the illegitimate offspring may have been sired by a close relative of the helper (and breeding male).

33.15, 33.6 and M13 (6, 9 and 6, respectively). Approximately 50% of the bands in the chick were maternal in origin, and a higher proportion of bands were shared with the putative mother (0.58, using an average of three probes) than with the putative father (0.29), suggesting that this chick was not the offspring of the breeding male.

Calculating Kinship (Coefficient of Relatedness, r) from Band-sharing Estimates – a Calibration

The theoretical relationship between band-sharing and relatedness can be derived as follows. Unrelated individuals have a proportion $(1 - x)$ of unshared bands, where x is defined as the background band-sharing. Among related individuals a proportion r of these bands ($r(1 - x)$ of all bands) are shared by descent, where r, the coefficient of relatedness, is defined as the expected fraction of genes in the genome of one individual that is identical by descent from a common ancestor with genes in the genome of another individual (Hamilton, 1972; Brown, 1987). Hence, observed band-sharing, $b = x + r(1 - x)$, comprising background plus non-background band-sharing due to relatedness. This is a linear function which has the value x when $r = 0$ (unrelated), and the value 1 when $r = 1$. Although this relationship is theoretically linear, empirical measurements (Packer et al., 1991; Gilbert et al., pers comm.) have shown the relationship to be non-linear in some species, so the shape of the relationship must be empirically determined in bee-eaters.

A calibration curve of band-sharing in relation to relatedness can be constructed where genealogies have been independently established from ringing data and behavioural observations. Band-sharing data for such individuals potentially suffer from three forms of non-independence: first, background band-sharing is often estimated from multiple comparisons between a group of individuals. For instance a sample of 10 birds yields 45 estimates of band-sharing, but these 45 values are not all independent. Second, the conditions under which DNA fingerprints are produced may vary between gels in such a way as to produce consistent differences in band-sharing between gels. Thus multiple estimates of band-sharing from one gel are not independent. Third, estimates of band-sharing between one individual and several of its relatives may not be independent (e.g. between a parent and each of its offspring). We have selected band-sharing estimates for use in the construction of the calibration curve in order to minimise these sources of non-independence (see caption to Fig. 5). However, it must be emphasised that we consider this calibration curve to be provisional, and to be in need of confirmation based on a much larger sample.

The calibration curves for each of the three probes, and the combined average of the three, is given in Fig. 5a–d. The relationship of band-sharing to relatedness is significant ($p < 0.001$) for all three probes, and

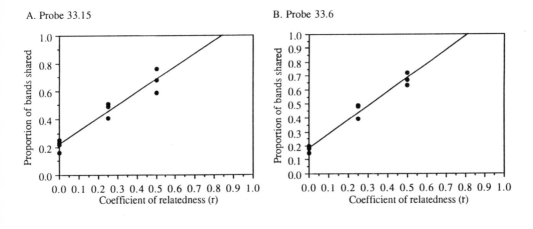

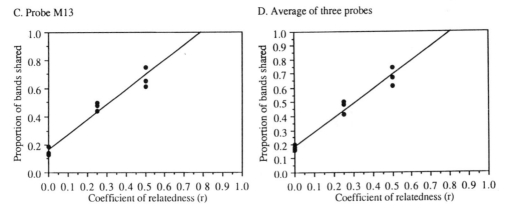

Figure 5. Calibration curve of band-sharing in relation to the coefficient of relatedness (r) for probes a) 33.15, b) 33.6, c) M13, and d) average of the three probes. Band-sharing estimates were obtained as follows: for $r = 0$, from band-sharing between the male and female of 7 (33.15, or 4 for the other 2 probes) helped breeding pairs. (These estimates did not differ significantly ($p \geq 0.4$) from the estimate of background band-sharing obtained from 10 unknown (presumably unrelated) breeding individuals (Tab. 1). Values of band-sharing for $r = 0.25$ and $r = 0.5$ were obtained from three families in which the helper was known from ringing to have a coefficient of relatedness of 0.5 to one of the breeding pair, and hence 0.25 to the chicks (all probes). These DNA fingerprints were scored blind, without prior knowledge of the relationships. For $r = 0.25$, only the three mean values (for each of the helpers) were used in the analysis. (a) Probe 33.15: band-sharing, $b = 0.222 + 0.925$ (r, coefficient of relatedness). $F_{1, 11} = 194.6$, $P < 0.001$. Departure from linearity: $F_{1, 10} = 0.35$, $P = 0.57$. Departure of line from point 1.0, 1.0: $t_{11} = 2.59$, $P < 0.05$. (b) Probe 33,6: $b = 0.188 + 0.992\,r$. $F_{1, 8} = 257.2$, $P < 0.001$. Departure from linearity: $F_{1, 7} = 1.29$, $P = 0.29$. Departure of line from point 1.0, 1.0: $t_8 = 3.69$, $P < 0.01$. (c) Probe M13: $b = 0.164 + 1.056\,r$. $F_{1, 8} = 168.3$, $P < 0.001$. Departure from linearity: $F_{1, 7} = 3.80$, $P = 0.09$. Departure of line from point 1.0, 1.0: $t_8 = 3.62$, $P < 0.01$. (d) Average of all three probes: $b = 0.188 + 0.999\,r$. $F_{1, 8} = 197.5$, $P < 0.001$. Departure from linearity: $F_{1, 7} = 1.64$, $P = 0.24$. Departure of line from point 1.0, 1.0: $t_8 = 3.36$, $P < 0.01$.

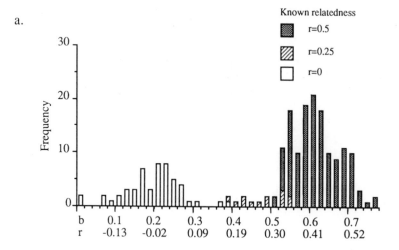

a.

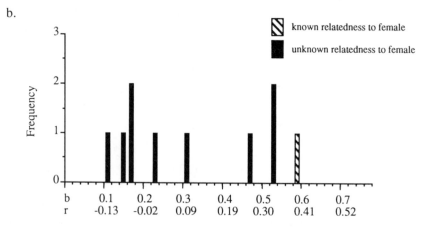

b.

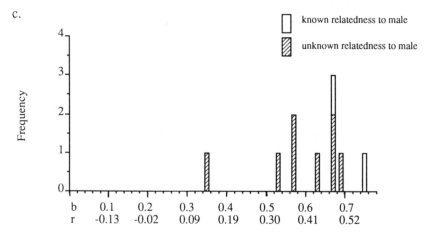

c.

the average of the three probes. Departure from linearity was tested by fitting a quadratic term; in none of the four cases was this significant. However, the theoretical relationship passes through the point 1, 1, but all four of the empirical lines pass significantly above this point. Thus, although no significant non-linearity could be detected over the range of relatedness examined ($r = 0.0-0.5$), the relationship cannot be linear over the entire range of $r = 0.0-1.0$. A series of analyses using various sub-sets of the data available (including the 45 pair-wise estimates of background band-sharing and parent-offspring relationships in both helped and unhelped families), in every case yielded either a linear relationship that passed significantly above the point 1,1, or a significantly curved relationship. Thus, the conclusion that the relationship is curvilinear over some part of the range does not depend on the sub-set of data analysed here. This curvilinearity implies that accurate prediction of relatedness from band-sharing will be more difficult for higher values of r; however, for lower degrees of relatedness ($r \leq 0.5$) there is little overlap in the band-sharing distributions for values of r of 0.0, 0.25 and 0.5 (Fig. 6a), so these levels of relatedness may be distinguished with a fair degree of certainty.

The calibration line for probe 33.15 is: band-sharing = $0.222 + 0.925r$, which may be rearranged (Snedecor and Cochran, 1967, page 150) to predict r from band-sharing: $r = (\text{band-sharing} - 0.222)/0.925$. Having established this relationship using individuals of known relationship, we used this calibration to estimate the relationships of helpers to each of the breeding pair (Fig. 6b and c). The mean estimated relatedness of the helper to the breeding male (including two helpers used to construct the calibration curve) was 0.42 ± 0.13 (SD), and of the helper to the breeding female 0.12 ± 0.20 (including 1 helper used to construct the calibration curve) (see Tab. 2 for estimates for individual helpers). The mean relatedness to the breeding pair, and hence to the helped offspring, was 0.27. The estimates of relatedness suggest that helpers were generally 0.5 related to the breeding male, but that the relationship to the breeding female was more variable. No helper

---

Figure 6a–c. Frequency distributions of the proportion of bands shared (b) and the estimated coefficient of relatedness (r) among helped European bee-eater families (using probe 33.15 only). Estimated coefficients of relatedness were obtained using the equation given in the caption to Fig. 5a. (a) Birds of known relatedness (from ringing and behavioural data). The sample for $r = 0$ includes 45 pairwise comparisons between 10 birds, and 7 comparisons within helped breeding pairs, for $r = 0.25$ includes 13 comparisons between 3 helpers and the chicks which they helped and for $r = 0.5$ includes 3 comparisons between helpers and one of the breeding pair that they helped, 60 mother-offspring and 51 father-offspring comparisons in unhelped families and 13 mother-offspring and 13 father-offspring comparisons in helped families. (b) Estimated coefficient of relatedness between the helper and female of the breeding pair. (The stippled bar indicates a relationship used in generating the calibration curve.) (c) Estimated coefficient of relatedness between the helper and male of the breeding pair. (The unshaded bars indicate two relationships used in generating the calibration curve.)

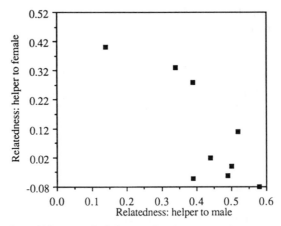

Figure 7. Relatedness (*r*) between the helpers and each member of the breeding pair that they helped. Relatedness was estimated from band-sharing using probe 33.15 (see caption to Fig. 5a for calibration equation).

was unrelated to both of the breeding pair (Fig. 7; Tab. 2). Using the relatedness estimates from DNA fingerprinting, in combination with the ages of the helper and helped pair (determined from plumage characteristics; Lessells and Krebs, 1989) and available ringing data, we were able to determine the actual relationships between the helpers and helped pairs (Tab. 2). The distribution of types of relationship between helpers and the breeding pair determined from DNA fingerprinting in combination with ringing data is similar to that determined from ringing alone (Tab. 3).

Do Helpers Ever Contribute to Paternity of the Brood They Help to Rear?

Chick 2 in the helped family two (nest M4, 1989) (Fig. 4b and c) was attributable to an EPC. Mean band-sharing between putative mother and this chick was 0.58, averaged over three probes, which gives an *r* value of 0.39, and thus maternity could be assigned to the putative mother. However, mean band-sharing (using an average of three probes) of the chick with the putative father (breeding male) and with the helper was 0.29 and 0.30, respectively, giving *r* values of 0.10 and 0.11, indicating that both were much less than 0.25 related to the chick. Ringing and behavioural data showed that the helper was the brother of the breeding male, but unrelated to the breeding female. To summarise, paternity could not be assigned to either the breeding male or the helper, but could be the result of an extra-pair copulation by a close relative.

Table 2. The relationship between helpers and breeding pairs in European bee-eaters as determined from DNA fingerprinting and ringing

Year-Nest	Helper	Age[1]/Sex[2]	Coefficient of relatedness from band-sharing Probe 33.15 only (Average of 3 probes)		Relationships known from ringing information[3]	Relationships known from fingerprinting + ringing[4]
			To breeding male	To breeding female		
89-M4	AX22487	JM	0.58 (0.55)	−0.08 (−0.06)	brother of male	brother of male
89-M8	AX27722	JM	0.34 (0.34)	0.33 (0.39)		son of pair
89-M22	AX27773	JM	0.44	0.22		brother of male
89-M22	AX22479	AM	0.39	0.28		father of male
89-M22	AX22432	AF	0.52	0.11		mother of male
89-N47	AX21872	AM	0.50 (0.47)	−0.11 (−0.07)	father of male	father of male
89-N35	AX27675	JM	0.39	−0.05		son of male, not of female
89-M2B	AX22303	AM	—	0.33	son of female (unknown to male)	son of female (unknown to male)
88-M6	AX22355	JM	0.14 (0.13)	0.40 (0.42)	son of female, not of male	son of female, not of male
88-M2	AX22169	AM	0.49	−0.04	son of male, not of female	son of male, not of female

[1]:J, juveniles hatched the previous calender year. A, adults hatched before the previous calender year.

[2]:M, male; F, female.

[3]:Relationships based on the identification of breeding adults and ringing chicks in the nest.

[4]:As 3, but ringing information is used to exclude certain relationships. In addition, juveniles may be separated from adults in the hand on the basis of plumage criteria (Lessells and Krebs, 1989). It has been assumed that (a) birds help siblings only from the same brood and (b) incestuous matings do not occur.

Table 3. The frequency of different relationships between helpers and breeding pairs in European bee-eaters: a comparison of ringing and DNA fingerprinting data. 'Ringing data' refers to relationships which were determined by identifying breeding adults and ringing chicks in the nest ($n = 26$). The relationships of a further 55 helpers could not be determined from ringing. 'Ringing + fingerprinting data' refers to the 10 helpers in Tab. 2. In all cases the exact relationship, as well as the coefficient of relatedness could be determined

Relative(s) helped	Ringing data (1983–90) Sex of helper		Ringing + fingerprinting data Sex of helper	
	Male	Female	Male	Female
Both parents	6	1	1	
Father, not mother	1		2	
Father, possibly mother	1			
Mother, not father	5		1	
Mother, possibly father	2		1	
Brother	6		2	
Son	3	1	2	1

Discussion

I. The Application of Band-sharing Estimates to Calculating Relatedness

DNA fingerprinting has proved to be an extremely powerful tool for paternity analyses (Burke and Bruford, 1987; Jeffreys et al., 1987; Wetton et al., 1987; Morton et al., 1990; Westneat, 1990). However, its discriminatory power for obtaining measures of biological relatedness per se is somewhat less well known and is often controversal (Lynch, 1988; Burke, 1989; Lewin, 1989).

An empirical calibration using helped families, including three of known kinship obtained from ringing data and behavioural observations, demonstrated that there was a linear relationship between the proportion of bands shared and the degree of relatedness over the range $r = 0.0$ to 0.5. In their study of inbred chickens, Kuhnlein et al. (1990) show a linear relationship (their Fig. 2) but comment that there is a non-linear relationship between band-sharing and inbreeding. In bee-eaters, kinship categories much beyond $r = 0.5$ become difficult to define as the regression lines do not go through the point 1, 1, indicating that the relationship between band-sharing and relatedness is not linear over some part of the range $0.0–1.0$. A similar curvilinear relationship has been found in free-ranging lion populations from the Serengeti (Packer et al., 1991; Gilbert et al., pers. comm.). The fact that the band-sharing distributions for $r = 0$, 0.25, and 0.5 overlap (Fig. 6a) implies that relatedness cannot always be determined from band-sharing with complete certainty, but these distributions suggest that first and second degree relatives and unrelated individuals for which no accurate demographic data exist will generally be correctly determined from band-sharing estimates. In addition, the precise biological relationships of helpers could be determined by combining genetic estimates with information on the age and genealogies of birds obtained from ringing. Our results suggest that DNA fingerprinting in this colony of bee-eaters might not discriminate more finely between different degrees of relatedness, for example between cousins ($r = 0.125$) and half-siblings ($r = 0.25$). Nevertheless, DNA fingerprinting offers an estimate of the relatedness in these cases and could, in combination with genealogical data from other sources, be informative. Moreover, such limits to the capabilities of DNA fingerprinting for distinguishing some categories of relatedness (Lewin, 1989) may be extended by increasing the number of loci scored (Lynch, 1988; Gilbert et al., pers comm.).

Estimates of coefficients of relatedness from band-sharing suggest that all helpers examined ($n = 10$) have r values of 0.5 to at least one of the breeding pair that they help and that helpers are more likely to be closely related to the breeding male (89%; 8 out of 9 helpers) than to the breeding female (30%; 3 out of 10 helpers). Ninety-three percent of helpers are male, and hence are more likely to have male relations in the colony due to juvenile female-biased dispersal and male natal philopatry. Using a larger sample based on ringing data alone, 73% of the helpers were related to the breeding male and 54% to the breeding female (Tab. 3).

Generally the relationship of the helper to the breeding pair and to the offspring was more variable in the European bee-eater than in many other helper species. This may provide sufficient scope for investigating whether helping varies (provisioning rates vary enormously among helpers) predictably with the degree of relatedness between the donor and the recipients.

II. Extra-pair Paternity and Maternity

Unhelped families (representing 80% of nests fledging chicks in the colony) were examined to investigate whether the number of chicks in each nest actually represents the breeding pair's reproductive success, and as such can be used to estimate the costs or benefits of raising young unaided, for comparison with helped families. All chicks in unhelped families, both from complete families and those where the father is absent, could be assigned to their putative parents. Indeed, even in the absence of a paternal sample within the unhelped families, paternity can still be inferred as a partial paternal fingerprint can be reconstructed from the mother-offspring comparisons (Jeffreys et al., 1985; Burke, 1989; Tegelstrom et al., 1991). No extra-pair fertilisations or intra-specific nest parasitism were observed in any of the unhelped families examined. Thus, from the small number of nests examined in this analysis, the number of chicks represents an accurate indicator of reproductive success in unhelped families. The lack of extensive extra-pair paternity in European bee-eaters explains the similarity of estimates of the heritabilities of winglength based on offspring-father and offspring-mother regressions (Lessells and Ovenden, 1989).

Finally, preliminary analysis of parentage among 8 helped families indicates that only about 2% (1/40) chicks in helped nests are not sired by the putative father. As an alternative behavioural strategy, helpers may attempt to gain extra-pair copulations with the female of the breeding pair (or female helpers may acquire matings with the breeding male), to accrue direct benefits from helping instead of only indirect benefits through helping kin (Burke et al., 1989). However, the helper in the single detected case of an EPC did not appear to be the father, but band-sharing indicated that the illegitimate offspring may have been sired by a close relative of the helper (and of the breeding male). Too few families were examined to rule out this reproductive strategy in the European bee-eater, thus this needs to be tested further. Perhaps, breeding pairs allow extra-pair copulations by related males (or intraspecific nest parasitism by related females) to improve the chances of receiving the helping behaviour. Alternatively, forced or voluntary EPCs (and ISNP) may be a consequence of social living (Emlen and Wrege, 1986).

III. Kin Selection and Helping in the European Bee-eater

Helpers in the European bee-eater do not appear to gain any direct advantages from helping and preferentially direct help to relatives, suggesting that the major benefits of helping are accrued through enhanced production and/or survival of relatives other than offspring (kin selection) (Lessells, 1990). The genetic analysis reported here confirms that European bee-eaters invariably help relatives and that they do not contribute paternity or maternity to the brood that they help.

Acknowledgements
We would like to thank Professor A. J. Jeffreys for generously providing minisatellite probes. Dr's. L. R. Noble and C. Packer for reading through the manuscript, Mrs B. Wood for helping to type the tables and the Centre de Recherches sur la Biologie des Populations d'Oiseaux for permission to ring and bleed bee-eaters. C. S. J. was supported by a N.E.R.C. post-doctoral assistantship.

References

Alexander, R. D. (1974) The evolution of social behaviour. *Ann. Rev. Ecol. Sys.* 5: 325–383.
Birkhead, T. R. (1987) Sperm competition in birds. T.R.E.E. 2: 268–271.

Brown, J. L. (1987) Helping and communal breeding in birds: ecology and evolution. New Jersey, Princeton University Press.

Burke, T., and Bruford, M. W. (1987) DNA fingerprinting in birds. *Nature* 327: 149–152.

Burke, T., Davies, N. B., Bruford, M. W., and Hatchwell, B. J. (1989) Parental care and mating behaviour of polyandrous dunnocks *Prunella modularis* related to paternity by DNA fingerprinting. *Nature* 388: 249–251.

Burke, T. (1989) DNA fingerprinting and other methods for the study of mating success. T.R.E.E. 4: 139–144.

Emlen, S. T., and Wrege, P. H. (1986) Forced copulations and intra-specific parasitism: two costs of social living in the white-fronted bee-eater. *Ethology* 71: 2–29.

Emlen, S. T., and Wrege, P. H. (1988) The role of kinship in helping decisions among white-fronted bee-eaters. *Behav. Ecol. Sociobiol.* 23: 305–315.

Emlen, S. T. (1990) White-fronted bee-eaters: helping in a colonially nesting species. In: Cooperative Breeding in Birds: longterm studies of ecology and behaviour. (eds) Stacey, P. B., and Koenig, W. D. Cambridge University Press.

Emlen, S. T. (1991) Cooperative breeding in birds. In: Behavioural Ecology: An Evolutionary Approach. (Eds) J. R. Krebs & N. B. Davies, Third Edition. Blackwell Scientific Publications (in press).

Feinberg, A. P., and Vogelstein, B. (1983) A technique for radiolabelling DNA restriction endonuclease fragments to high specific acitvity. *Anal. Biochem.* 132: 6–13.

Feinberg, A. P., and Vogelstein, B. (1984) Addendum – a technique for radiolabelling DNA restriction endonuclease fragments to high specific activity. *Anal. Biochem.* 137: 266–67.

Georges, M., Lequarre, A.-S., Castelli, M., Hanset, R., and Vassart, G. (1988) DNA fingerprinting in domestic animals using four different minisatellite probes. *Cytogenet. Cell. Genet.* 47: 127–131.

Gyllensten, U. B., Jakobsson, S., and Temrin, H. (1990) No evidence for illegitimate young in monogamous and polygynous warblers. *Nature* 343: 168–170.

Hamilton, W. D. (1964) The genetical evolution of social behaviour. I & II *J. Theor. Biol.* 7: 1–52.

Hamilton, W. D. (1972) Altruism and related phenomena, mainly in social insects. *Ann. Rev. Ecol. Syst.* 3: 193–232.

Jeffreys, A. J., Wilson, V., and Thein, S. L. (1985a) Hypervariable 'minisatellite' regions in human DNA. *Nature* 314: 67–73.

Jeffreys, A. J., Wilson, V., and Thein, S. L. (1985b) Individual-specific 'fingerprints' of human DNA. *Nature* 316: 76–79.

Jeffreys, A. J., Wilson, V., Thein, S. L., Weatherall, D. L., and Ponder, B. A. J. (1986) DNA 'fingerprints' and segregation analysis of multiple markers in human pedigrees. *Am. J. Hum. Genet.* 39: 11–24.

Jeffreys, A. J., and Morton, D. B. (1987) DNA fingerprints of dogs and cats. *Anim. Genet.* 18: 1–15.

Kuhnlein, U., Zadworny, D., Dawe, Y., Fairfull, R. W., and Gavora, J. S. (1990) Assessment of inbreeding by DNA fingerprinting: Development of a calibration curve using defined strains of chickens. *Genetics* 125: 161–165.

Lessells, C. M., and Krebs, J. R. (1989) Age and breeding performance of European bee-eaters. *The Auk* 106: 375–382.

Lessells, C. M., and Ovenden, G. N. (1989) Heritability of winglength and weight in European Bee-eaters *Merops apiaster*. *Condor* 91: 210–214.

Lessells, C. M. (1990) Helping at the nest in European bee-eaters: Who helps and why? In: Blondel, J., Gosler, A. G., Lebreton, J. D., and McCleery, R. H. Population Biology of Passerine birds: An integrated Approach. *NATO ASI Series* pp. 357–368.

Lewin, R. (1989) Limits to DNA fingerprinting. *Science* 243: 1549–1551.

Lynch, M. (1988) Estimation of relatedness by DNA fingerprinting. *Mol. Biol. Evol.* 5: 584–599.

Maniatis, T., Fritsch, E. F., and Sambrook, J. (1982) Molecular Cloning: A Laboratory manual. Cold Spring Habor Laboratory, N. Y.

Morton, E. S., Forman, L., and Braun, M. (1990) Extra-pair fertilisations and the evolution of colonial breeding in purple martins. *The Auk* 107: 275–283.

Packer, C., Gilbert, D. A., Pusey, A. E., and O'Brien, S. J. (1991) A molecular genetic analysis of kinship and cooperation in African lions. *Nature* (in press).

Rabenold, K. N. (1985) Cooperation in breeding by nonreproductive wrens: kinship, reciprocity, and demography. *Behav. Ecol. Sociobiol.* 17: 1–17.

Rabenold, K. N. (1990) Campylorhynchus wrens: the ecology of delayed dispersal and cooperation in Venezuelan savanna. In: Cooperative Breeding in Birds: longterm studies of ecology and behaviour. (eds) Stacey, P. B., and Koenig, W. D. Cambridge University Press.

Rabenold, P. P., Rabenold, K. N., Piper, W. H., Haydock, J., and Zack, S. W. (1990). Shared paternity revealed by genetic analysis in cooperatively breeding tropical wrens. *Nature* 348: 538–40.

Reeve, H. K., Westneat, D. F., Noon, W. A., Sherman, P. W., and Aquadro, C. F. (1990) DNA 'fingerprinting' reveals high levels of inbreeding in colonies of the eusocial naked mole-rat. Proc. Natl. Acad. Sci. USA. 87: 2496–2500.

Reyer, H. U. (1984) Investment and relatedness: a cost/benefit analysis of breeding and helping in the pied kingfisher (*Ceryle rudis*) *Anim. Beh.* 32: 1163–1178.

Snedcor, G. W., and Cochran, W. G. (1976) Statistical methods. Iowa State University Press, Ames, Iowa.

Southern, E. M. (1975) Detection of specific sequence among DNA fragments separated by gel electrophoresis. *J. Mol. Biol.* 98: 503–517.

Stacey, P. B., and Koenig, W. D. (1990) Cooperative Breeding in Birds: longterm studies of ecology and behaviour. Cambridge University Press.

Tegelstrom, H., Searle, J., Brookfield, J., and Mercer, S. (1991) Multiple paternity in wild caught common shrews (*Sorex araneus*) confirmed by DNA fingerprinting. *Heredity* (in press).

Trivers, R. L. (1971) The evolution of reciprocal altruism. *Q. Rev. Biol.* 46: 35–57.

Vassart, G., Georges, M., Monsieur, R., Brocas, H., Lequarre, A.-S., and Christophe, D. (1987) A sequence in M13 phage detects hypervariable minisatellites in human and animal DNA. *Science* 235: 683–4.

Westneat, D. F. Noon, W. A., Reeve, H. K., and Aquadro, C. F. (1987) Extra-pair fertilisations in a predominantly monogamous bird: genetic evidence. *Anim. Behav.* 35: 877–866.

Westneat, D. F. (1988) Improved hybridisation conditions for DNA 'fingerprints' probed with M13. *Nucl. Acids. Res.* 16: 4161.

Westneat, D. F. (1990) Genetic parentage in the indigo bunting: a study using DNA fingerprinting. *Behav. Ecol. Sociobiol.* 27: 67–76.

Westneat, D. F., Sherman, P. W., and Morton, M. L. (1990) The ecology and evolution of extra-pair copulation in birds. *Curr. Ornithol.* 7: 331–369.

Wetton, J. H., Carter, R. E., Parkin, D. T., and Walters, D. (1987) Demographic study of a wild house sparrow population by DNA fingerprinting. *Nature* 327: 147–149.

Woolfenden, G. E., and Fitzpatrick, J. W. (1984) The Florida scrub jay. Princeton University Press, Princeton, NJ.

DNA Fingerprinting: Approaches and Applications
ed. by T. Burke, G. Dolf, A. J. Jeffreys & R. Wolff
© 1991 Birkhäuser Verlag Basel/Switzerland

Cloning, Characterization and Evolution of Indian Peafowl *Pavo cristatus* Minisatellite Loci

O. Hanotte*, §, T. Burke*, J. A. L. Armour¶, and A. J. Jeffreys¶

Department of Zoology, University of Leicester, University Road, Leicester LE1 7RH, Great Britain; §Université de Mons-Hainaut, Faculté de Medecine, Service de Biochimie Moléculaire, 24 Av. du Champ de Mars, 7000 MONS, Belgium; ¶Department of Genetics, University of Leicester, University Road, Leicester LE1 7RH, Great Britain

Summary
We have recently described the large scale isolation of Indian peafowl *Pavo cristatus* minisatellite sequences and their potential applications in behavioural ecology (Hanotte *et al.*, 1991). We report here further details about our genomic DNA library, the characterization of the minisatellites and the evolutionary conservation of the cloned loci in the Phasianidae. The strategy adopted for the cloning has now been applied in several other species of birds and is reviewed in a broader context, taking into account our previous and new experiences.

Introduction

Multilocus DNA fingerprinting is now widely applied in birds for studies in behavioural ecology (Burke, 1989; Burke *et al.*, 1991) and poultry breeding programmes (Kuhnlein *et al.*, 1989; Bruford *et al.*, 1990; Bruford and Burke, 1991). However, its application to population genetic studies is limited as DNA fragments from homologous loci cannot be identified in unrelated individuals (Burke *et al.*, 1991). Also, while the study of reproductive success in socially "monogamous" species of birds has been revolutionized by this technique (Burke, 1989), the study of more complicated mating systems (polyandrous or polygynous species) usually requires the identification of the true mother or, more frequently, the true father among an often large number of possible parents. For example, in lekking species the number of males displaying together in one mating arena (lek) can be very large, as many as ten to fifteen in the Indian peafowl *Pavo cristatus* (M. Petrie, pers. comm.) or sometimes exceeding twenty birds in species such as the black grouse *Tetrao tetrix* (Cramp and Simmons, 1980); females might also visit more than one lek. To use multilocus fingerprinting in this kind of study is a relatively inefficient process; we are limited by the number of samples which can be loaded on one gel, and comparison of individual DNA fingerprints run on separate gels is difficult in the absence of established, precise methods to compare the migration of bands between different gels

(Burke *et al.*, 1991). The scoring of a large set of multilocus DNA fingerprints which includes many potential parents is also very time-consuming. Finally, the genomic mapping of avian species of commercial interest (chicken) requires the isolation of highly polymorphic DNA markers (Bruford and Burke, 1991).

The isolation of single locus probes detecting highly variable loci will consequently be very useful. Until recently, significant numbers of such probes had only been obtained from human and only a few hypervariable loci had been isolated from other species (see Hanotte *et al.*, 1991). Recently, using a cloning strategy initially developed in human (Armour *et al.*, 1990), we have been successful in the first large scale isolation of minisatellite sequences from a non-human species (Hanotte *et al.*, 1991).

In our first study, the strategy for cloning minisatellites was described and five polymorphic minisatellite loci were analysed in detail (Hanotte *et al.*, 1991). We report here further details of the characterization and evolution of these and other loci. The cloning strategy is reviewed taking into account both the overall results in the Indian peafowl and preliminary data obtained in other species of birds using the same methods.

Materials and Methods

The cloning procedure has been described elsewhere (Hanotte *et al.*, 1991) and only those parts of the protocol that are of special relevance to this paper, or have not previously been described in detail, will be presented.

Birds

Blood samples were collected from captive birds in various collections: Indian peafowl (from Mr. Q. Spratt, Norwich, UK; World Pheasant Association, Lower Basildon, UK or Whipsnade Zoo, Whipsnade, UK), green peafowl *Pavo muticus* (Mr. Q. Spratt), Congo peacock *Afropavo congensis* (Antwerpen Zoo, Antwerpen, Belgium), and golden pheasant *Chrysolophus pictus* (Mr. J. Corder, London, UK). The green peafowl is the species most closely related to the Indian peafowl. The two species hybridize easily, producing fertile hybrids of both sexes (Johnsgard, 1983). The Congo peafowl, an African species, is considered to belong to a peafowl lineage which has been separated from the Asian peafowl since at least the upper Miocene, more than 6 million years ago (Mourer-Chauviré, 1989). The divergence time between the pheasant and the peafowl lineages is more uncertain. Based on restriction site data from nuclear genes, the peafowl lineage is supposed to have separated from the rest of the Phasianidae about 20 million years ago (Helm-Bychowski and Wilson, 1986); however, DNA–DNA hybridization studies suggest a separation time between the pheasant and the peafowl lineages of up to twice this length (Sibley and Ahlquist, 1990).

Cloning and Characterization of the Genomic Library

200 ng of size-selected Indian peafowl DNA (6.4–16.0 kb) cut to completion with MboI was ligated into the BamHI site of 1.8 μg Charomid 9–36 (Saito and Stark, 1986). One-third of the ligated DNA was packaged *in vitro* (Gigapack Plus, Stratagene) and the packaged DNA was used to infect *Escherichia coli* NM554 cells (*recA, mcrA, mcrB*) (Raleigh *et al.*, 1988). Ampicillin-resistant clones were plated at low density and the library screened directly with multilocus probes 33.6 and 33.15, after duplication of the colonies onto nylon filters (Hybond-N, Amersham) or after transfer of the colonies into the wells of microtitre plates and subsequent transfer onto nylon filters. Positive recombinant Charomid DNAs were extracted (Birnboim and Doly, 1979), digested with Sau3AI and the cloned insert collected after gel electrophoresis by electroelution onto dialysis membrane. We observed that Charomid extracted from the bacterial host NM554 is particularly sensitive to nuclease, so, to avoid any degradation, the purification of the genomic insert was processed in the absence of RNase. Gel electrophoresis, Southern blotting, radioactive labelling of the probes and prehybridization, hybridization and washing of the filters were performed as described previously (Hanotte *et al.*, 1991). The ordered array Charomid library was also screened at high stringency with the first specific minisatellite sequences isolated; the aim was to detect new minisatellite sequences which had not been detected previously with the multilocus probes 33.6 and 33.15 ("probe-walking", Washio *et al.*, 1989).

Results

Cloning Strategy and Screening of the Libraries

Our first aim was to isolate hypervariable minisatellite sequences for use in behavioural studies. It was first necessary to obtain multilocus DNA fingerprints of the species (Indian peafowl) under study with the restriction enzyme (MboI) used in our cloning system. This enabled us to select a size fraction rich in highly variable minisatellite sequences for cloning purposes (Hanotte *et al.*, 1991). Also, from studies in humans, it is known that alleles of very different sizes may occur at a minisatellite locus (Wong et al., 1986, 1987). We therefore prepared our genomic DNA from fourteen unrelated individuals so that alleles from a large number of minisatellite loci would occur within the chosen size range (Wong et al., 1987). As the size-selected fraction represents only a small portion (1%) of the total peafowl DNA (Hanotte et al., 1991) and the cloned fragments are large, the library size required to include with a probability of 95% (Clarke and Carbon, 1976) at least one copy of a given DNA fragment is small: approximately 3000 colonies, corresponding to three genome equivalents (Hanotte et al., 1991). However, this supposes that for each minisatellite locus the MboI alleles always appear in the 6.4–16 kb size fraction; hypervariable loci at which MboI alleles only rarely entered the 6.4–16 kb size range would have a much lower representation in the library.

The results from screening the library with the multilocus probes 33.6 and 33.15 test the efficiency of the strategy described above. Our first

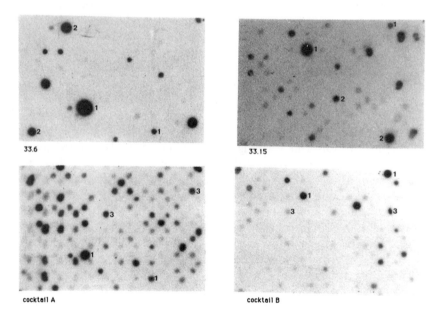

Figure 1. Screening of an ordered array Indian peafowl Charomid library, replica-plated from a microtitre plate, by hybridization at low stringency with the human polycore probes 33.6 or 33.15 and at high stringency with two cocktails of peafowl single locus probes (A: cPcr1, cPcr2, cPcr5, cPcr6; B: cPcr5, cPcr17). Examples of different classes of positively hybridizing colonies are indicated as follows: 1 examples of positives detected both with the polycore probes 33.6 or 33.15 and peafowl probe cocktails A or B, 2 polycore probe-specific positives and 3 peafowl probe-specific positives. The full library of 1520 colonies was replicated at two-fold density onto eight filters. Hybridization and washing at low or high stringency were done in the absence of any competitor DNA (Hanotte et al., 1991).

screening of a random spread of about 6000 colonies established that around 10% hybridized positively to each multilocus probe. The use of an ordered array library (1520 clones) allowed us to obtain more precise estimates: 222 clones (14%) were detected with 33.15 and 133 (8%) with 33.6, of which 71 (5%) were detected with both probes (Hanotte et al., 1991). Our probe-walking approach, in which we used, at high stringency without any vertebrate competitor DNA, two cocktails of the first peafowl minisatellite probes isolated (A: cPcr1, cPcr2, cPcr5, cPcr6 detected by 33.6; B: cPcr5, cPcr17 detected by 33.15; Tab. 1, Fig. 1), indicated that these probes collectively hybridized to 79% and 83% respectively of the positively hybridizing clones detected with 33.6 and 33.15. In addition, cocktail A hybridized to 340 new clones and cocktail B to more than 55, offering the opportunity to isolate hypervariable sequences previously undetected with the multilocus probes 33.6 and 33.15. The high number of new positives detected, in the absence of peafowl competitor DNA, with cocktail A suggests the possible presence in this cocktail of a sequence repeated frequently in the Indian peafowl genome (satellite or interspersed repeat sequences). It should be borne in mind that as the cocktails were found to detect additional minisatellite clones, at least some of these clones that hybridized to both 33.6 or 33.15 and the cocktails presumably also contained sequences from loci other than those represented in the cocktails. The probe-walking approach therefore increases the efficiency with which novel sequences are isolated, but can potentially lead to some clones that contain novel loci being ignored.

Identification and Characterization of the Positively Hybridizing Clones

Recombinant Charomid was extracted from positively hybridizing clones, and the genomic inserts used as specific probes against a standard panel of four unrelated Indian peafowl and one green peafowl (Hanotte et al., 1991). The first twenty-one probes tested were shown to originate from only five different loci, which seem to be predominant in the library (Hanotte et al., 1991; see above). More precisely, the probes isolated from eleven positively hybridizing clones detected by the multilocus probe 33.6 belong to only three different loci; two polymorphic ones having three or four different alleles were isolated five and three times respectively, and an apparently identical monomorphic locus was detected three times. All the probes isolated from seven positively hybridizing clones detected by the multilocus probe 33.15 hybridized to the same polymorphic locus having three alleles. Similarly, three positively hybridizing clones which hybridized to both multilocus probes 33.15 and 33.6 seem to originate from the same monomorphic locus. Details of all the probes tested are shown in Tab. 1.

198

Table 1. Identification and characterization of the recombinant peafowl Charomids detected by multilocus minisatellite probes

Probe	Detected by	Insert size kb[a]	Locus[b] Pattern & fragment size, kb[a]
cPcr1	33.6	6.6	poly. (4) 9.4–2.3
cPcr2	33.6	6.6–4.4	mono. 9.4–6.6
cPcr3	33.6	6.6–4.4	same locus as cPcr2
cPcr4	33.6	9.4–6.6	poly. (3)[c] 23.0–4.4
cPcr5	33.6 & 33.15	6.6–4.4	mono. 6.6
cPcr6	33.6	9.4–6.6	same locus as cPcr4
cPcr7	33.6	6.6–4.4	same locus as cPcr4
cPcr8	33.15	6.6–4.4	same locus as cPcr10[d]
cPcr9	33.15	6.6–4.4	same locus as cPcr10
cPcr10	33.15	6.6–4.4	poly. (3)[c] 9.4–0.5
cPcr11	33.6 & 33.15	6.6–4.4	same locus as cPcr10
cPcr12	33.6	6.6–4.4	same locus as cPcr10
cPcr13	33.6 & 33.15	6.6–4.4	same locus as cPcr5
cPcr14	33.15	6.6–4.4	same locus as cPcr10
cPcr15	33.15	6.6–4.4	same locus as cPcr10
cPcr16	33.15	6.6–4.4	same locus as cPcr10
cPcr17	33.15	6.6–4.4	same locus as cPcr10
cPcr18	33.6	6.6–4.4	same locus as cPcr4
cPcr19	33.6	6.6–4.4	? ?
cPcr20.1	33.6	6.6–4.4	same locus as cPcr2
cPcr20.2	33.6	4.4–2.3	? ?
cPcr21	33.6	6.6–4.4	same locus as cPcr1
cPcr22.1	33.6	6.6–4.4	same locus as cPcr1
cPcr22.2	33.6	4.4–2.36	? ?
cPcr23	cocktail B	6.6–9.4	poly. (2) 9.4–4.4
cPcr24	33.15	6.6–9.4	poly. (3) 23.0–9.4
cPcr25	cocktail B	6.6–4.4	poly. (4)[c] 9.4–2.3
cPcr26	cocktail B	6.6–4.4	poly. (2) 23.0–9.4
cPcr27	33.15	6.6–4.4	multiband[e] –
cPcr28	33.15	6.6–4.4	poly. (5) 9.4–6.6
cPcr29.1	33.15	4.4–2.3	? ?
cPcr29.2	33.15	4.4–2.3	poly. (2) 6.6
cPcr30	cocktail B	6.6–4.4	same locus as cPcr2
cPcr31	cocktail B	6.6	mono. 9.4–6.6
cPcr32	cocktail B	6.6–4.4	multiband[e] –
cPcr33	33.15	6.6–4.4	poly. (2) 6.6
cPcr34	cocktail B	6.6–4.4	poly. (4) 9.4–6.6
cPcr35	33.15	6.6–4.4	poly. (2) 6.6–4.4
cPcr36	33.6	6.6–4.4	mono. 6.6
cPcr37	33.6	6.6	same locus as cPcr36
cPcr38	33.6	6.6	mono. 9.4–6.6
cPcr39	33.6	6.6	multiband[e] –
cPcr40	cocktail A	6.6–4.4	poly.(3)[c] 23.0–6.6
cPcr41	cocktail A	4.4–2.3	multiband[e] –
cPcr42.1	cocktail A	6.6	multiband[e] –
cPcr42.2	cocktail A	6.6–4.4	? ?
cPcr43	33.6	6.6–4.4	mono. 6.6–4.4
cPcr44.1	cocktail A	6.6–4.4	poly. (3) 9.4–6.6
cPcr44.2	cocktail A	4.4–2.3	same locus as cPcr 44.1
cPcr45	33.6	6.6–4.4	poly.(2) 9.4–6.6
cPcr46	cocktail A	9.4–6.6	poly.(2)[c] 9.4–4.4
cPcr47	cocktail A	6.6–4.4	multiband[f] –
cPcr48	cocktail A	9.4–6.6	poly.(4) 23–6.6
cPcr49	cocktail A	9.4–6.6	poly.(2) 9.4–6.6

(a): estimated visually by comparison of the electrophoretic migration of the insert DNA fragments, in 0.8% agarose gel, with the migration of λ/HindIII fragments. (b): each probe was tested against the same panel of four unrelated Indian peafowl. mono.: monomorphic single locus pattern, one identical band detected in all four samples; poly.: polymorphic single locus, more than one band detected, the number of different alleles is indicated in parentheses; multiband: more than three bands in each individual; ?: not tested; –: not applicable. The fragment size refers to those alleles detected in the four Indian peafowl. (c): presence of three bands in some of the birds. (d): the locus cPcr10 is recognized within a multiband pattern. (e): presence of internal sites and/or recognition of several loci. (f): pedigree analysis shows the presence of one polymorphic locus with internal MboI restriction sites.

In order to avoid the repeated isolation of clones from the same loci, we screened our genomic library at high stringency with the first five probes isolated, as described above. Twenty-seven positively hybridizing clones were then characterized (cPcr23 to cPcr49, Tab. 1). Six positives had previously been detected with 33.15 but were not detected by cocktail B; they correspond to six new loci: three dimorphic loci

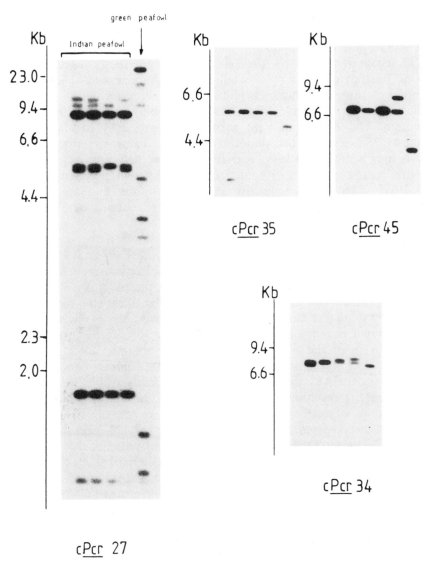

Figure 2. Examples of loci detected by Southern blot hybridization with cloned Indian peafowl minisatellites. The DNAs from four unrelated Indian peafowl (left lanes) and one green peafowl (right lanes) were digested with MboI. cPcr27: a multiband pattern: cPcr34: a polymorphic locus with four alleles; cPcr35, cPcr45: two dimorphic loci.

(including cPcr35, Fig. 2), one locus with three alleles, one highly polymorphic locus with five alleles, and a probe revealing a multiband pattern comprising DNA fragments of different intensities, suggesting the recognition of different loci or a large tandem array cleaved internally by MboI into a number of linked fragments (cPcr27, Fig. 2). Seven were "probe-walking positives" previously undetected by the multilocus probe 33.15 but detected by cocktail B; they correspond to one monomorphic locus, two dimorphic loci, two polymorphic loci with four alleles (including cPcr34, Fig. 2), a multiband pattern, and a monomorphic locus apparently identical to a locus previously isolated and recognized by 33.6. Also, six positives previously detected with 33.6, but not detected by cocktail A were characterized. They correspond to three new monomorphic loci, one being recognized twice; a dimorphic locus (cPcr45, Fig. 2), and a probe revealing a multiband pattern. The analysis of eight positives recognized by cocktail A but not by 33.6 revealed two dimorphic loci, one locus with three alleles, two loci with four alleles, and three probes detecting a multiband pattern. In conclusion, from the eighteen different polymorphic loci isolated (two or more alleles detected in four unrelated Indian peafowl), three were detected by 33.6, six were detected by 33.15, five by cocktail A, and four by cocktail B.

Tandemly repeated DNA sequences are known to be unstable in several cloning systems (Kelly et al., 1989; Neil et al., 1990). It is obvious from Tab. 1 that the size of the isolated insert tends to be smaller than the DNA fragments detected in the panel of four Indian peafowl (in 16 cases the insert fell within the allele size range, in 8 cases it was smaller and in no case was it larger). Also, we were unable to recover a genomic insert in approximately 10% of our extractions, in which cases the electrophoretic pattern obtained after digestion with Sau3AI was a smear without any clear genomic band. As the Charomid Sau3AI fragments were normally clearly visible, we interpret the smear as resulting from a particularly unstable genomic insert. In several cases, we electrophoretically recovered two distinct genomic inserts from a single positive clone (cPcr20, cPcr22, cPcr29, cPcr42, cPcr44; Tab. 1). Two possible explanations are the contamination of some positives with another bacterial colony containing a recombinant Charomid, as might have occurred during the transfer of the bacterial colonies from petri dishes to microtitre plates, or that a single clone contained two different Sau3AI inserts due to ligation between genomic fragments. However, in one case both inserts were tested (cPcr44, Tab. 1) and were shown to detect the same locus. This suggests that some of these double band patterns were the result of a single recombinant Charomid containing an unstable genomic insert from which two predominant differently sized recombinant Charomids arose.

Comparison of Multilocus DNA Fingerprints and Library Screening

The more polymorphic Indian peafowl minisatellite loci show only four or five different alleles among four unrelated birds (Tab. 1). The variability is lower than at several human minisatellite loci, which were found to have heterozygosities above 90% and more than 10 alleles (Wong *et al.*, 1987; Nakamura *et al.*, 1987). However, the multilocus DNA fingerprints of Indian peafowl show a high level of band sharing (around 70%) between unrelated birds (Tab. 2 and Fig. 3; Hanotte *et al.*, 1991). The relatively low heterozygosities among the isolated loci are therefore not surprising. All the peafowl studied originated from birds kept in captivity for many generations, during which time genetic variability might have been significantly reduced by founder effects and inbreeding (Hanotte *et al.*, 1991).

Screening of the library with multilocus probes 33.15 and 33.6 revealed some positives detected by both probes (Tab. 1). As in our genomic DNA library some minisatellite loci occur more commonly than others, it is difficult to estimate the exact proportion of loci recognized by both probes. Some degree of such identical hybridization was, however, expected as comparison of MboI multilocus DNA fingerprints obtained with 33.6 and 33.15 reveals that a proportion of bands were detected by both multilocus probes (Fig. 3; unpubl. data). More precisely, analysis of Fig. 3 indicates that fingerprints obtained using 33.6 and 33.15 have 22% of their bands in common (Tab. 2).

Table 2. Similarities of Indian peafowl DNA fingerprints produced by human and peafowl minisatellite probes

	33.6	(A) Between individuals 33.15	cPcr13	cPcr48
n:	34	39	24	25
p:	0.71	0.69	0.75	0.88

	33.6	(B) Between multilocus probes 33.15	cPcr13	cPcr48
33.6	—			
33.15	0.22	—		
cPcr13	0.36	0.33	—	
cPcr48	0.38	0.38	0.39	—

Similarities of DNA fingerprints (A) obtained from different individuals and (B) detected using different multilocus probes. The proportion of bands shared between two fingerprints was found as $p = 2n_{ab}/(n_a + n_b)$, where n_a and n_b are the numbers of resolved bands in the two fingerprints (mean n) and n_{ab} is the number of bands common to both. In (A), p was found by comparing the same two individuals for each probe. In (B), the matrix shows the mean of each pair of values for p obtained using the same pair of individuals as in (A).

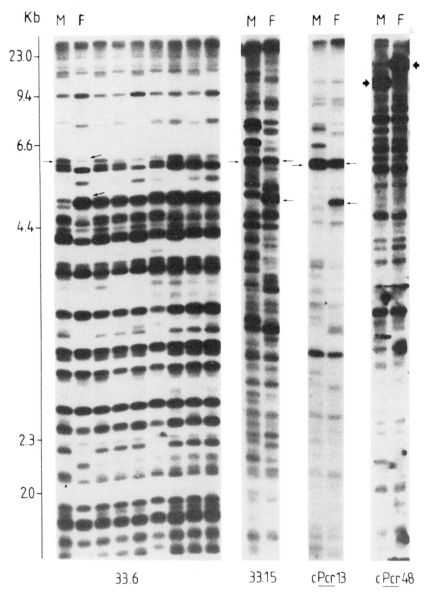

Figure 3. MboI multilocus DNA fingerprints of one pair of Indian peafowl (M: male, F: female) and seven offspring obtained with two human minisatellite probes (33.6, 33.15), and two peafowl minisatellite probes (cPcr13, cPcr48). 33.6: complete pedigree, 33.15, cPcr13, cPcr48: only the parents are shown. →, alleles homologous to cPcr13 (also present in 33.6 and in 33.15 fingerprints), and ♦, alleles homologous to cPcr48. All the hybridizations were done at 64°C, in the absence of any competitor DNA and the filters were washed at low stringency (1 × SSC, 0.1% SDS) (Hanotte et al., 1991).

These apparently similar bands detected by the two multilocus probes might reflect either the comigration of alleles from two different loci, separately detected by the two probes, or the recognition of the same locus. Indeed, when a band derives from a polymorphic locus it is highly unlikely that the same distribution among offspring would be obtained with both probes unless the same allele is being detected; however, in most cases here the same band was present in every individual and it seems probable that the shared bands mostly originate from less variable, or even monomorphic, loci.

We also compared the multilocus fingerprints obtained at low stringency using the Indian peafowl minisatellite probe cPcr13 (isolated from a clone positive for both multilocus probes 33.6 and 33.15, Tab. 1) with the multilocus fingerprints obtained by 33.6 and 33.15 (Tab. 2, Fig. 3). The monomorphic locus from which cPcr13 is presumed to have been isolated appears as a strong band in a cPcr13 fingerprint and is also apparently present in multilocus fingerprints obtained with both 33.6 and 33.15 (Fig. 3). The percentages of bands detected by both cPcr13 and 33.6 or both cPcr13 and 33.15 are similar (36% and 33% respectively, Tab. 2). The fingerprints obtained with cPcr13 show a high level of band sharing (75%, Tab. 2), as also observed in the multilocus fingerprints obtained with 33.15 and 33.6. From the 30 different bands scored in two unrelated individuals after hybridization with cPcr13, seven were also detected in 33.6 fingerprints, eight were detected in 33.15 fingerprints, six were present in both 33.6 and 33.15 multilocus fingerprints, and nine were new DNA fragments specific to the cPcr13 multilocus fingerprint.

Finally, we also used the minisatellite probe cPcr48 at low stringency (Fig. 3). At high stringency this probe detects a polymorphic locus with four alleles in four unrelated Indian peafowl (Tab. 1). Mendelian transmission at this minisatellite locus is illustrated in Fig. 4B (see also above). This clone was detected after probe-walking with cocktail A, and had not been detected after screening the library with multilocus probes 33.15 or 33.6; cPcr48 might therefore be expected to detect a new set of variable minisatellite sequences when used at low stringency. The alleles homologous to cPcr48, clearly visible in cPcr48 fingerprints, are absent from 33.6, 33.15 or cPcr13 fingerprints (Fig. 3). However, the percentages of bands detected by both cPcr48 and the other three multilocus probes (around 40%, Tab. 2) are slightly higher than the values obtained between pairs of the other probes. Despite the relatively high level of polymorphism at the cloned locus (Tab. 1), the band sharing value for the low stringency profiles between two unrelated birds is high (88%, Tab. 2) and pedigree analysis suggest that most of the bands detected represent monomorphic and/or homozygous loci (data not shown).

A

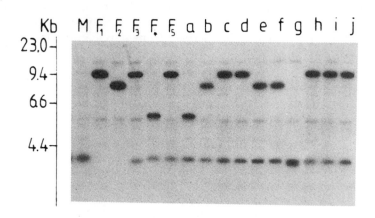

cPcr 1

B

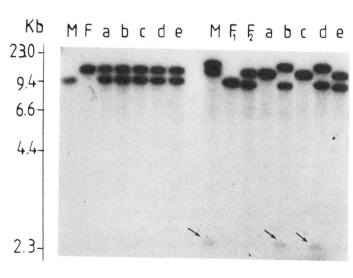

FAMILY I   FAMILY II

cPcr 48

Mendelian Inheritance and Linkage Analysis at Indian Peafowl
Minisatellite Loci

Examples of Mendelian inheritance at two polymorphic minisatellite loci are shown in Fig. 4A (cPcr1) and 4B (cPcr48). Figure 4A illustrates a maternity assignment problem. Each of the five different females might be the mother of any of the ten offspring; five different alleles can be recognized among the six supposedly unrelated adults (though only four different alleles were detected for this probe in the standard panel of four unrelated birds, Tab. 1). Three of the alleles are specific to individual females, one is present in three females and another in two. Using this single variable minisatellite probe, maternity can be assigned for five of the ten chicks (a, b, e, f, g) and we can exclude two females (F_2 and F_4) for the five other chicks (c, d, h, i, j). In Fig. 4B, two families were studied. In family I both parents are homozygous and all the offspring are heterozygous. In family II, precise maternity was unknown but cPcr48 allowed us to determine the maternity of offspring a and c which were each homozygous for an allele absent in female 1. Also, although the size difference between the two low molecular weight alleles of females 1 and 2 is small, offspring b and e appeared to inherit their maternal alleles from female 2, and only offspring d gets its allele from female 1. This result was confirmed using other polymorphic minisatellite probes (cPcr1, cPcr28, Hanotte *et al.*, 1991). The highest molecular weight allele at this locus (family II, male) cosegregates with a low molecular weight fragment (c. 2.3 kb) suggesting that this particular allele has an internal MboI site close to one end.

While the haploid genome size of birds is fairly constant, approximately 1.0×10^9 bp, the number of chromosomes varies between species (Shields, 1983). In the Indian peafowl, there are 18 macrochromosomes and 48 microchromosomes (Capanna *et al.*, 1987). The chromosomal location of minisatellites is not known in any bird species, but multilocus segregation analyses have shown in most of the cases investigated that minisatellite loci assort independently in birds, suggesting that

---

Figure 4. Mendelian inheritance of two variable Indian peafowl minisatellite loci. A DNA from one male, five possible mothers and ten offspring were digested with MboI and probed with cPcr1. Pedigree analysis indicates that: females 1, 3 and 5 could each be the mother of offspring, c, d, h, i and j; female 2 is the mother of offspring b, e, f; female 3 is the mother of offspring g; female 4 is the mother of offspring a. B: DNAs from two families of Indian peafowl (1: adult male, female and 5 offspring; II: adult male, female 1, female 2 and 5 offspring) were digested with MboI and probed with cPcr48. Maternity could be assigned in family II: only offspring d received its maternal allele from female 1 and offspring a, b, c and e received theirs from female 2. →: low molecular weight fragment cosegregating with the highest molecular weight allele of the male in family II. Hybridizations were done at 64°C in the presence of Indian peafowl competitor DNA (6 µg/ml) and the filters were washed at high stringency (0.1 × SSC, 0.01% SDS) (Hanotte *et al.*, 1991).

minisatellite loci are widely dispersed among the chromosomes (Burke and Bruford, 1987; Burke et al., this volume). We have previously assumed that all our cloned minisatellite loci segregate independently and estimated the probability of two unrelated Indian peafowl sharing the same genotype, and the probability of false paternal inclusion, for a combination of five polymorphic minisatellite loci (Hanotte et al., 1991). The possibility of chromosomal clustering among the cloned minisatellite loci can, however, be tested by linkage analysis among pairs of loci detected by probes in large pedigrees. For all but one pairwise comparison no evidence of linkage was found between the minisatellite loci cPcr1, cPcr24, cPcr25, cPcr28 and cPcr27 (data not shown). However, segregation analysis in two large Indian peafowl families, including 12 and 11 offspring, revealed significant linkage between the minisatellite loci detected by cPcr1 and cPcr25 ($\hat{z} = 4.53$, $\hat{\Theta} = 0.05$). Given the large number of chromosomes, this result is relatively surprising, and might suggest the existence in the avian genome, as in human (Royle et al., 1988) and in cattle (Georges et al., 1990), of some clusters of closely linked minisatellite loci. It emphasizes the need in both single locus and multilocus fingerprint analysis to demonstrate the independence of the scored loci prior to their use in behavioural or population genetic studies.

Molecular Structure of Indian Peafowl Minisatellite Loci

As no minisatellite loci have so far been sequenced in the Indian peafowl, the tandem repeat composition of the cloned loci can only be inferred indirectly, as follows. (i) They were selected after hybridization of our genomic library with multilocus minisatellite probes 33.6 and 33.15. (ii) The restriction enzyme used, MboI, is a four base pair recognition site enzyme which is expected to cut frequently in the avian genome. Very large MboI DNA fragments such as those that we cloned are therefore only likely to occur if they consist largely of a short, reasonably uniform tandemly repeated unit. (iii) A relatively high degree of instability of the cloned genomic inserts, resulting in multiple, smaller fragments, was detected with our cloning system, despite all the precautions taken (Tab. 1), as has been commonly observed during the cloning and analysis of other minisatellite sequences (Nakamura et al., 1987; Kelly et al., 1989). (iv) The presence of three or more alleles among unrelated birds, in particular where only a single restriction fragment is detected per allele, is more easily explained by variation in the number of copies of a repeat sequence rather than by restriction site polymorphism (single base variation). (v) The cloned minisatellites (including monomorphic ones) when used as probes at low stringency detect sets of polymorphic sequences (cPcr13, Fig.

3). (vi) In addition, a variable length polymorphism at a minisatellite locus can be visualized by several restriction enzymes, while a polymorphism resulting from a single base change is likely to be revealed by only one (Nakamura *et al.*, 1987).

Such multi-enzyme tests might reveal other information about the molecular structure of our minisatellite clones, including the size of the unique DNA sequences flanking the minisatellites or the detection of restriction enzymes cutting within each minisatellite repeat unit. Three examples are presented in Fig. 5. cPcr34 is a polymorphic locus (Tab. 1), and the individual shown is heterozygous; the two alleles are detected not only with MboI but also with AluI, HinfI, and MspI (Fig. 5). Depending on the restriction enzyme used, the size of the alleles varies but the size difference between the alleles remains similar. This implies that alleles of different size have identical restriction sites, probably in the DNA flanking the variable minisatellite repeat sequence. The total length of the flanking DNA is large in the MboI digest and can be estimated, from the difference in allele size between MboI and HinfI digests, to be at least 6 kb. If the DNA is cut with HaeIII no alleles are detected, indicating that most or all of the tandem repeats at this locus

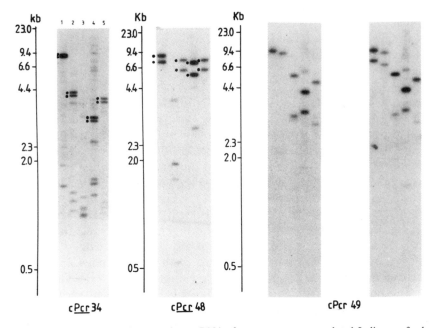

Figure 5. Multi-enzyme gel comparisons. DNAs from one or two unrelated Indian peafowl were digested with five different enzymes (1: MboI, 2: AluI, 3: HaeIII, 4: HinfI, 5: MspI) and, after electrophoresis into 0.8% agarose gel, probed at high stringency (0.1 × SSC, 0.01% SDS) with cPcr34, cPcr48 or cPcr49 (a homozygote and a heterozygous individual at this locus). Alleles at cPcr34 and cPcr48 are marked (●). See text for the interpretation of the electrophoretic pattern.

contain an internal HaeIII site. cPcr48 is from another polymorphic locus (Tab. 1) at which, as for cPcr34, two alleles having a constant size difference are detected with several restriction enzymes (MboI, HaeIII, HinfI, MspI). In contrast with cPcr34, HaeIII does not cleave inside the minisatellite locus, but the pattern obtained with AluI suggests occasional presence of AluI restriction sites inside the minisatellite.

cPcr49 detects a dimorphic locus (Tab. 1) and a single base pair substitution, resulting in a restriction site polymorphism, rather than variation in the minisatellite length might explain the presence of two alleles. Indeed, two allelic bands are only detected if the genomic DNA is cut with MboI or AluI. Restriction with the other enzymes (HaeIII, HinfI, and MspI) fails to detect a similar pattern; one, two or more bands being present in each lane (Fig. 5). Moreover, restriction digestion of an individual homozygous for the high molecular weight allele results in the same electrophoretic pattern (HaeIII, HinfI, and MspI digestions) as for a bird heterozygous at this locus (Fig. 5). The presence of a similar pattern with MboI and AluI digestion may be explained by two different restriction site polymorphisms in the flanking DNA or by a large DNA insertion in the high molecular weight alleles. However, the probability of either of these seems small. A more likely explanation would be the recognition by cPcr49 of a monomorphic locus next to a dimorphic minisatellite, where MboI and AluI restriction enzymes each excise the two minisatellites in a single fragment, while HaeIII, HinfI and PstI also cut within the sequence separating the two minisatellites.

Evolutionary Comparison of Indian Peafowl Minisatellite Loci

An unexpected observation was the apparent evolutionary conservation of Indian peafowl minisatellite loci in other species of Phasianidae (Hanotte et al., 1991). We investigated this point further by hybridizing several Indian peafowl minisatellite probes against three unrelated individuals of the three species of peafowl (Indian peafowl, green peafowl, Congo peacock) and one pheasant (golden pheasant). The results are summarised in Tab. 3, and illustrated in Fig. 6A and 6B. All the probes so far tested hybridized to the green peafowl and the Congo peacock genomic DNA. Two probes hybridized to the golden pheasant with a relatively weak but clear signal (Fig. 6A; Hanotte et al., 1991). Variability at a supposedly homologous locus may change between species even when they are closely related; some apparently monomorphic loci in the Indian peafowl were found to be polymorphic in other species (cPcr2, cPcr31). However, the two most conserved loci, in the sense that an apparently homologous locus was detected even in the most evolutionary distant of the species, were polymorphic in all four species (cPcr1, cPcr48).

Table 3. Cross-species comparisons of Indian peafowl minisatellite loci

Probe	*Pavo cristatus*	*Pavo muticus*	*Afropavo congensis*	*Chrysolophus pictus*
cPcr1	poly. (2)	poly. (2)	poly. (4)	poly. (4)
cPcr2	mono.	poly. (3)	poly. (2)	no hybridization
cPcr8	poly. (3)	multiband	multiband	no hybridization
cPcr28	poly. (4)	poly. (2)	poly. (3)	no hybridization
cPcr31	mono.	poly. (2)	poly. (3)	no prominent locus
cPcr33	mono.[a]	poly. (3)	mono.[b]	no hybridization
cPcr47	multiband[c]	multiband	multiband	no hybridization
cPcr48	poly. (4)	poly. (4)	poly. (3)	poly. (4)

The table describes the hybridization of probes derived from Indian peafowl against apparently homologous loci in related species. All the probes were tested against the same panel of three unrelated individuals of each species run in the same electrophoretic gel. See Tab. 1 legend for definitions. The numbers in parentheses indicate the number of different alleles present in three unrelated individuals. (a): dimorphic in the standard panel of four other unrelated Indian peafowl (Tab. 1). (b): several other bands present in the background. (c): pedigree analysis demonstrates, in the Indian peafowl, the presence of one polymorphic locus with an internal restriction site for MboI.

In Fig. 6B, probe cPcr1 has been hybridized to thirteen different male Congo peacocks, all captive birds descending from five wild birds, and two unrelated Indian peafowl. The intensity of the signal obtained in the Congo peacock and in the Indian peafowl is identical, suggesting a strong similarity between the cPcr1 probe and the locus recognized in the Congo peacock. The pedigree of all these birds is known (R. Van Bocxstaele, pers. comm.), and four different alleles can be recognized among nine different genotypes; four pairs of individuals show identical patterns. Three of the four alleles contain an internal HaeIII restriction site. These haplotypic patterns disappear if the DNA is cut with MboI, only one fragment being observed for each allele (data not shown).

It is known that if we use a tandemly repeated sequence as a single locus probe, a higher molecular weight allele will give a stronger intensity of signal than a smaller allele, reflecting the larger number of copies of repeat sequence (Nakamura *et al.*, 1987). This can, for example, be observed in Fig. 6B, where allele d is of the lowest intensity. Also, the low molecular weight fragments of alleles a and b are less intense than the high molecular weight fragments, suggesting the presence of the internal HaeIII restriction site towards one end of the tandemly repeated sequence. In the case of allele c the larger molecular weight fragment is less intense than the smaller one. This suggests that the internal HaeIII restriction site can vary in its position between alleles at the same locus or this might also be explained by the presence of unique sequences of different length flanking the minisatellite. The result is two DNA fragments with the high molecular weight fragment containing more tandem repeats but less DNA flanking sequence than the low molecular weight fragment. In allele d, the low molecular weight

A

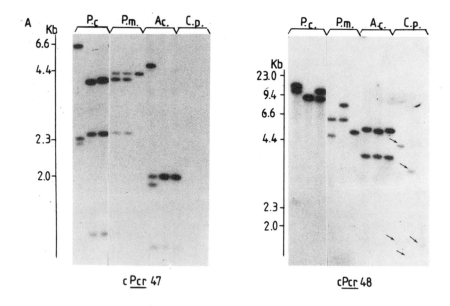

cPcr 47   cPcr 48

B

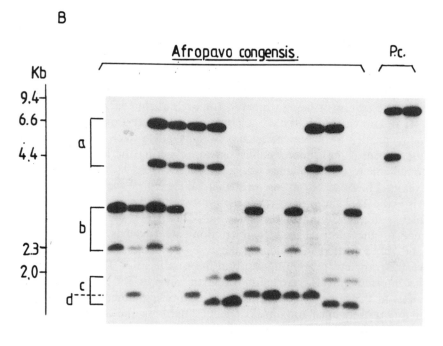

cPcr 1

fragment might have run out of the gel or this small allele might have lost, with other repeats, the tandem repeat containing the HaeIII restriction site.

Discussion

We have reported here and elsewhere (Hanotte *et al.*, 1991) the isolation and characterization of eighteen different polymorphic minisatellite loci in the Indian peafowl. We used a cloning strategy which has recently been shown to be very successful in the cloning and isolation of human minisatellites (Armour *et al.*, 1990). After extraction, the genomic DNA of several unrelated individuals is completely digested with the restriction enzyme MboI and a DNA fraction size known to be rich in hypervariable minisatellites is selected and ligated with the appropriate member of a family of cosmid-derived vectors known as Charomids (Saito and Stark, 1986). Charomid vectors contain a spacer region consisting of 1–23 tandemly repeated copies of a 2 kb DNA fragment which determines the total size of the vector between 7.3 kb (1 spacer) and 52 kb (23 spacers). Depending on the DNA fraction size selected for cloning, a different Charomid vector can be chosen to maximise the packaging *in vitro* of the recombinant molecules into the bacteriophage λ particles. To enrich our library for tandemly repeated sequences, our genomic DNA was digested to completion and a fraction size rich in minisatellite sequences was selected. The selected high molecular weight size fraction represents only a small proportion of the total avian genome and the library size required to include one haploid genome equivalent is greatly reduced. This allows the convenient transfer of our library into the wells of microtitre plates. Moreover, to minimize the instability or degradation of the minisatellite inserts, we used a recA⁻ *E. coli* host (NM554) which was also deficient for the two site-specific systems mcrA and mcrB known to restrict eukaryotic DNA containing methylated cytosine (Raleigh *et al.*, 1988).

Figure 6. Cross-species hybridization with specific Indian peafowl minisatellite probes. A: DNAs from three unrelated Indian peafowl (P.c.), green peafowl (P.m.), Congo peacock (A.c.) and golden pheasant (C.p.) were digested with MboI and probed with cPcr47 or cPcr48. For cPcr47 a multiband pattern is present in all the peafowl, but clearly absent in the golden pheasant. In contrast, with cPcr48 a polymorphic locus (faint, arrows) is also detected in the golden pheasant. B: DNAs from two unrelated Indian peafowl (P.c.) digested with MboI and 13 male Congo peacock digested with HaeIII were probed with cPcr1. In the Indian peafowl two different alleles can be recognized, and in the Congo peacock four different alleles with (a, b, c) or apparently without (d) internal HaeIII sites. See text for the interpretation of the electrophoretic pattern. After hybridization at 64 °C in the presence of Indian peafowl competitor DNA (6 µg/ml), the filters were washed at high stringency (0.1 × SSC, 0.1% SDS) (Hanotte *et al.*, 1991).

The transfer of our library into an ordered array in microtitre plates had several practical advantages. It allowed us to screen the library by a single round of hybridization, to compare easily the identities and signal intensities of the positives detected by different multilocus probes, and to apply a probe-walking strategy (Washio *et al.*, 1989) to avoid the repeated isolation of the same positives and to isolate new subsets of variable loci not detected by the original multilocus probes. Moreover, as the bacterial colonies were suspended in microtitre plate wells containing Luria broth medium with 40% glycerol, the ordered array library can be kept at $-20\ °C$ for at least several months for subsequent screening, in contrast to a library plated on petri dishes.

Other strategies have been used, nearly exclusively in the human species, to isolate minisatellite loci. Minisatellites were first discovered, and are still sometimes found, fortuitously (Wyman and White, 1980; Coppieters *et al.*, 1990). The systematic screening for polymorphisms among random segments of human DNA cloned in bacteriophage λ (Braman *et al.*, 1985) gave poor results with a detection rate of highly polymorphic loci of about 1%. Human λ libraries, constructed from size-selected fractions of pooled DNA from unrelated individuals were screened with multilocus probes 33.15 and 33.6 in the first selective isolation of minisatellites. This selective approach was more successful, resulting in the isolation of several highly variable human minisatellites (Wong *et al.*, 1986, 1987). Nakamura *et al.* (1987) showed that pre-screening human cosmid libraries with oligonucleotides, corresponding to the consensus sequences of known tandemly repeated sequences, increases by five- to tenfold the frequency of isolation of cosmids that will detect highly polymorphic loci. Two bird minisatellite loci have recently been isolated from a λ library (Gyllensten *et al.*, 1989). The genomic DNA of one willow warbler (*Phylloscopus trochilus*) was digested with EcoRI and ligated into vector λgt10. About 5×10^4 plaque forming units were plated and screened with human minisatellite multi-locus probes 33.6 and 33.15. Only twenty positives were detected from which the two minisatellite loci were characterized. The instability of tandem repeated sequences might explain the poor yield of minisatellites isolated from λ libraries as the loss of repeat units might lead to inviability of recombinant phage (Kelly *et al.*, 1989). This may explain why the screening of cosmid libraries grown in rec⁻ bacteria has been more successful. Nakamura *et al.* (1987) isolated a large number of highly polymorphic loci by screening human cosmid libraries with several distinct G-rich oligonucleotide probes. The procedure was, however, time-consuming as it involved the screening of large libraries which had not been enriched for minisatellites and the subcloning of the fragment responsible for the polymorphism from each selected clone.

None of these alternative approaches offers the advantages and efficiency of the system tested by Armour *et al.* (1990) in the human species. We show here and elsewhere (Hanotte *et al.*, 1991; Burke *et al.*, 1991) that the same system can be applied easily and efficiently to species other than the human, opening the way for the isolation of specific minisatellite loci and their application in behavioural and population genetic studies in any animal species which can be fingerprinted. The success of this strategy in the Indian peafowl can be measured by comparing the proportion of polymorphic loci isolated with the total number of positives tested. Eighteen (38%) of the 48 positives tested were polymorphic with two or more alleles detected in four unrelated individuals; 5 (10%) showed four or five alleles (Tab. 1). The numbers of alleles in Tab. 1 are minimum values, and new alleles are sometimes identified by the use of a higher resolution gel system (Hanotte et al., unpublished data) or through the screening of more birds. An example is shown at Fig. 4A; cPcr1 shows five different alleles but only four were detected in our standard panel of four unrelated Indian peafowl (Table 1; Hanotte *et al.*, 1991).

Recently, Armour *et al.* (1990) demonstrated the presence of a "null" allele in a minisatellite locus isolated in human. Moreover, segregation analyses of minisatellite loci isolated in chicken *Gallus gallus* show the existence of null alleles at several loci (M. W. Bruford, unpubl. data). Therefore, care should be taken when interpreting pedigree data. For example, in family II (cPcr48, Fig. 4B) we could not completely exclude the possibility that female 1 was heterozygous with a null allele (perhaps < 2.3 kb) and could therefore only assign maternity with complete certainty for offspring d (maternal allele from female 1) and b and e (maternal allele from female 2). However, we have not detected null alleles at this or any of the other isolated peafowl minisatellite loci in pedigree analyses, and the maternity of offspring a and c (assigned to female 2) was confirmed with other probes (Hanotte *et al.*, 1991).

We have already isolated several polymorphic loci in four other species of birds (Burke *et al.*, 1991) and although the results are preliminary, some general conclusions can be drawn. In most cases, screening of the Charomid library with two multilocus probes will be enough to find five to ten polymorphic loci (see Tab. 3 in Burke *et al.*, 1991). However, the results obtained also suggest a positive correlation between the variability of the multilocus fingerprints in the fraction size selected for cloning, and the mean heterozygosity of the isolated minisatellite loci. For example, in the Indian peafowl, the heterozygosity values of the cloned loci are small (Hanotte *et al.*, 1991) compared to the more polymorphic loci isolated in human (Wong *et al.*, 1986, 1987) or in mouse (Kelly *et al.*, 1987), but this is not surprising given the high band sharing values calculated from multilocus DNA fingerprints of unrelated Indian peafowl (Tab. 2). Also, the "probe-walking" approach will not always prove necessary, but has to be considered as a potentially invaluable adjunct to the system, especially when multilocus screening of the library reveals the predominant presence of a small number of minisatellite sequences or the presence of satellite sequences recognized by the multilocus probes. Apparently homologous loci in species more or less related to the species where the minisatellite loci have been cloned have also been detected in groups of birds other than the Phasianidae, including non-passerine families such as the Scolopacidae (T.B., O.H., S. Bailey, and E. Needham) and passerine families (T.B., E. Cairns, and L. Jenkins). We anticipate that in most cases minisatellite loci in an avian species will show enough sequence similarities to be detected by homologous minisatellites isolated in another species of the same genus. The use of these probes in species classified in other genera within the same family or belonging to other families within the same order is more uncertain and unpredictable but not impossible. However, each probe will need to be systematically tested in the species of interest. We already know that some probes cross-hybridize with species which have diverged > 20 million years ago (Fig. 6A and Tab. 3; Hanotte *et al.*, 1991).

PCR amplification of minisatellite sequences, using primers matching the unique flanking sequences of the minisatellite, has demonstrated in one case that the main locus detected by hybridization of a human locus-specific minisatellite probe in a related primate species is actually non-homologous (Gray and Jeffreys, 1991). We cannot at this stage exclude the possibility of non-homologous hybridization during our cross-species high stringency hybridization experiments. However, this seems likely to be a rare event. Moreover, the observation that the hybridization signal strength in the golden pheasant is weaker than in the peafowl lineage (Fig. 6A) is consistent with the recognition of a diverged, homologous locus. On the other hand, either independent and rapid evolution of the minisatellite locus or the recognition of a non-homologous locus might sometimes result in the signal obtained in a distantly related species being stronger in comparison with more related ones, or completely absent in closely related species but present in less related ones. The recognition of a non-homologous locus would not, in any case, prevent the use of a probe for behavioural or population genetic studies in a species other than the one from which it was isolated. This emphasises, however, that only the comparison of the unique DNA sequence flanking the minisatellite locus will confirm the homology of a minisatellite locus in different species, so allowing molecular evolutionary studies of minisatellites and phylogenetic reconstruction using minisatellite sequence data. Finally, Gray and Jeffreys (1991) have shown in primates that variability at a minisatellite locus can change very rapidly in a short period of evolutionary time. Our data indicate (Tab. 3) that this could also be true in birds.

Acknowledgements
This work was supported by grants from NERC and SERC to T.B. We thank Prof. A. O. A. Miller for his constant encouragement, Mr. R. Léonard for the preparation of the figures. Charomid vectors were generously supplied by the Japanese Cancer Research Resources Bank. Blood samples were kindly provided by the World Pheasant Association (Mr. T. Gardiner), Antwerpen Zoo (Mr. R. Van Bocxstaele), Mr. Q. Spratt and Mr. J. Corder. The human minisatellite probes 33.6 and 33.15 are the subject of Patent Applications. Commercial enquiries should be addressed to Cellmark Diagnostics, 8 Blacklands Way, Abingdon, Oxon., OX14 1DY, GB.

References

Armour, J. A. L., Povey, S., Jeremiah, S., and Jeffreys, A. J. (1990) Systematic cloning of human minisatellites from ordered array Charomid libraries. *Genomics* 8: 501–512.

Birnboim, H. C., and Doly, J. (1979) A rapid alkaline extraction procedure for screening recombinant plasmid DNA. *Nucleic Acids Res.* 7: 1513–1523.

Braman, J. C., Barker D., Schumm, J. W., Knowlton, R. G., and Donis-Keller, H. (1985) Characterization of very highly polymorphic RFLP probes. 8th Human Gene Mapping Workshop. *Cytogenet. Cell Genet.* 40: 589.

Bruford, M. W., Burke, T., Hanotte, O., and Smiley, M. (1990) Hypervariable markers in the chicken genome. *Proceedings of the 4th World Congress of Genetics to Livestock Production* XIII: 139–142.

Bruford, M. W., and Burke, T. (1991) Hypervariable DNA markers and their applications in the chicken. In: Burke, T., Dolf, G., Jeffreys, A. J., and Wolff, R. (eds), DNA Fingerprinting: Approaches and Applications. Birkhäuser, Basel. pp. 230–242. (This volume).

Burke, T., and Bruford, M. (1987) DNA fingerprinting in birds. *Nature* 327: 149–152.

Burke, T. (1989) DNA fingerprinting and other methods for the study of mating success. *Trends Ecol. Evol.* 4: 139–144.

Burke, T., Hanotte, O., Bruford, M. W., and Cairns, E. (1991) Multilocus and single locus minisatellite analysis in population biological studies. In: Burke, T., Dolf, G., Jeffreys, A. J., and Wolff, R. (eds), DNA Fingerprinting: Approaches and Applications. Birkhäuser, Basel. pp. 154–168. (This volume)

Capanna, E., Civitelli, M. V., and Martinico, E. (1987) I cromosomi degli uccelli Citotassonomia ed evoluzione cariotipica. *Avocetta* 11: 101–144.

Clarke, L., and Carbon, J. (1976) A colony bank containing synthetic ColE1 hybrid plasmids representative of the entire *E. coli* genome. *Cell* 9: 91–99.

Coppieters, W., Van de Weghe, A., Depicker, A., Bouquet, Y., and Van Everen, A. (1990) A hypervariable pig DNA fragment. *Anim. Genet.* 21: 29–38.

Cramp, S, and Simmons, K. E. L. (1980) The Birds of the Western Palearctic. Vol. II. Oxford University Press, Oxford, p. 420.

Georges, M., Lathrop, M., Hilbert, P., Marcotte, A., Schwers, A., Swillens, S., Vassart, G., and Hanset, R. (1990) On the use of DNA fingerprints for linkage studies in cattle. *Genomics* 6: 461–474.

Gray, I. C., and Jeffreys, A. J. (1991) Evolutionary transience of hypervariable minisatellite loci in man and the primates. *Proc. Roy. Soc. Series B*, 243: 241–253.

Gyllensten, U. B., Jakobsson, S., Temrin, H., and Wilson, A. C. (1989). Nucleotide sequence and genomic organisation of bird minisatellites. *Nucleic Acid Res.* 17: 2203–2214.

Hanotte, O., Burke, T., Armour, J. A. L., and Jeffreys, A. J. (1991) Hypervariable minisatellite DNA sequences in the Indian peafowl *Pavo cristatus*. *Genomics* 9: 587–597.

Helm-Bychowski, K. M., and Wilson, A. C. (1986) Rates of nuclear DNA evolution in pheasant-like birds: evidence from restriction maps. *Proc. Natl. Acad. Sci. USA* 83: 688–692.

Johnsgard, P. A. (1983) Hybridization and zoogeographic patterns in pheasants *Wld. Pheasant Ass. J.* 8: 89–98.

Kelly, R., Bulfield, G., Collick, A., Gibbs, M., and Jeffreys, A. J. (1989) Characterization of a highly unstable mouse minisatellite locus: evidence for somatic mutation during early development. *Genomics* 5: 844–856.

Kuhnlein, U., Dawe, Y., Zandworny, D., and Gavora, J. S. (1989) DNA fingerprinting: a tool for determining genetic distances between strains or poultry. *Theor. Appl. Genet.* 77: 669–672.

Mourer-Chauviré, C. (1989) A peafowl from the Pliocene of Perpignan, France. *Palaeontology* 32: 439–446.

Nakamura, Y., Leppert, M., O'Connell, P., Wolff, R., Holm, T., Culver, M., Martin, C., Fujimoto, E., Hoff, M., Kumlin, E., and White, R. (1987) Variable number of tandem repeat (VNTR) markers for human gene mapping. *Science* 235: 1616–1622.

Neil, D. L., Villasante, A., Fisher, R. B., Vetrie, D., Cox, B., and Tyler-Smith, C. (1990) Structural instability of human tandemly repeated DNA sequences cloned in yeast artificial chromosome vectors. *Nucleic Acids Res.* 18: 1421–1428.

Raleigh, E. A., Murray, N. E., Revel, H., Blumenthal, R. M., Westanay, D., Reith, A. D., Rigby, P. W. J., Elhai, J., and Hanahan, D. (1988) McrA and McrB restriction phenotypes of some *E. coli* strains and implications for gene cloning. *Nucleic Acids Res.* 16: 1563–1575.

Royle, N. J., Clarkson, R. E., Wong, Z., and Jeffreys, A. J. (1988) Clustering of hypervariable minisatellites in the proterminal regions of human automsomes. *Genomics* 3: 352–360.

Saito, I., and Stark, G. R. (1986) Charomids: cosmid vectors for efficient cloning and mapping of large or small restriction fragments. *Proc. Natl. Acad. Sci. USA* 83: 8664–8668.

Shields, G. F. (1983) Organization of the avian genome. In: Brush A. H., and Clark, G. A. (eds), Perspectives in ornithology. Cambridge University Press, Cambridge, pp. 271–290.

Sibley, C. G., and Ahlquist, J. E. (1990) Phylogeny and classification of birds: A study in molecular evolution. Yale University Press, New Haven, p. 299.

Washio, K., Misawa, S., and Ueda, S. (1989) Probe walking: development of novel probes for DNA fingerprinting. *Hum. Genet.* 83: 223–226.

Wong, Z., Wilson, V., Jeffreys, A. J., and Thein, S. L. (1986) Cloning a selected fragment from a human DNA "fingerprint": isolation of an extremely polymorphic minisatellite. *Nucleic Acids Res.* 14: 4605–4616.

Wong, Z., Wilson, V., Patel, I., Povey, S., and Jeffreys, A. J. (1987) Characterization of a panel of highly variable minisatellites cloned from human DNA. *Ann. Hum. Genet.* 51: 269–288.

Wyman, A. R., and White, R. (1980) A highly polymorphic locus in human DNA. *Proc. Natl. Acad. Sci. USA* 77: 6754–6758.

DNA Fingerprinting: Approaches and Applications
ed. by T. Burke, G. Dolf, A. J. Jeffreys & R. Wolff
© 1991 Birkhäuser Verlag Basel/Switzerland

Use of Sex-Linked Minisatellite Fragments to Investigate Genetic Differentiation and Migration of North American Populations of the Peregrine Falcon (*Falco peregrinus*)

J. L. Longmire[1], R. E. Ambrose[2], N. C. Brown[1], T. J. Cade[3],
T. L. Maechtle[3], W. S. Seegar[4], F. P. Ward[4], and C. M. White[5]

[1]*Genetics Group, M.S. 886, Los Alamos National Laboratory, Los Alamos, NM 87545, USA;* [2]*United States Fish and Wildlife Service, 1412 Airport Way, Fairbanks, AL 99701, USA;* [3]*Raptor Research and Technical Assistance Center, Boise State University, Boise, ID 83725, USA;* [4]*Aberdeen Proving Grounds, Aberdeen, MD 21010, USA;* [5]*Dept. of Zoology, Brigham Young University, Provo, UT 84602 USA*

Summary
The M13 repeat detects different levels of genetic variation in falcons. First, this minisatellite probe reveals typically highly variant restriction fragments that show no apparent unequal distribution between the sexes. Secondly, the M13 repeat detects sets of fragments that are only present in DNAs from female falcons. The level of polymorphism displayed by the sex-linked fragments is greatly reduced relative to most autosomal minisatellites. In addition, the size of these fragments (in kilobase pairs) is species-specific among Mauritius kestrels (*Falco punctatus*) and peregrines (*Falco peregrinus*). Variation observed at one of the sex-linked fragments in peregrines has proven to be useful in distinguishing a subset of the *tundrius* subspecies of this endangered raptor. This correlation has enabled a genetic test to be used to examine the representation of *tundrius* peregrines during mass migration.

Introduction

The peregrine falcon (*Falco peregrinus*) is found as a nesting species on all continents except Antarctica, and on all major islands except Iceland and New Zealand. North American and Greenlandic peregrines are divided into three subspecies based on morphological characteristics and nesting habitat (White and Boyce, 1988). *F.p. tundrius* nests in the northerly tundra regions of North America and is the only subspecies known to nest in Greenland. *F.p. anatum* nests in northern Mexico, throughout the western continental United States, and in the interior forested regions of Canada and Alaska where it intergrades with *tundrius*. The third North American subspecies, *F.p. pealei*, nests along the Pacific coastal regions of Alaska, Canada and Washington State. Unlike *pealei* and southern *anatum* populations, which are relatively sedentary, the *tundrius* and northern *anatum* breeding populations are highly migratory. Nestlings banded in Alaska (*tundrius* and *anatum*) and in

Greenland (*tundrius*) have been recovered during the boreal winter ("wintering") in South America (Ambrose and Riddle, 1988; Yates *et al.*, 1988).

Although the ecology of the peregrine has been studied extensively (see Cade, 1982 for a recent review), a lack of genetic investigations has hampered the ability to gain insights into several other important aspects of the biology of this species, including population genetics, evolution, biosystematics, and genome organization. In addition, a stably inherited set of genetic markers developed from surveys of known breeding populations would be of value to studies of migration and the possible existence of population-specific wintering sites.

As an approach to some of these issues, we are using molecular genetic techniques to study the nuclear DNA of the peregrine and other closely related species in the genus *Falco*. Specifically, a variety of different DNA sequence probes are being used to detect varying levels of polymorphism in the genome of the peregrine. Unique sequences and heterochromatic tandem repeat probes have been isolated by molecular cloning. These probes detect polymorphisms that arise primarily by base substitution and are generally useful in comparisons made between distinct populations (Longmire *et al.*, 1988). In addition, we are using minisatellite probes in experimental situations where very high levels of genetic resolution are required, such as in determinations of maternity and paternity. Minisatellites are clusters of short tandem repeat sequences. In humans, where minisatellites have been most extensively studied, these repeats are distributed along the length of the chromosomes, but show a higher density near the telomeres (Nakamura *et al.*, 1988). Variation results from differences in the number of basic repeating units within allelic clusters, and the level of polymorphism revealed by multilocus minisatellite probes is sufficient for individual-specific DNA fingerprinting (Jeffreys *et al.*, 1985a, b). The minisatellite probe that we have used most extensively is the tandem repeat found within the 282 base pair (bp) ClaI/HaeIII restriction fragment of bacteriophage M13 (Vassart *et al.*, 1987). The M13 repeat can be used directly to screen Southern blots containing restricted genomic DNAs, and to retrieve complementary clones from recombinant phage and cosmid libraries made from partially digested human and peregrine DNAs. A cloned minisatellite sequence (pV47-2) recovered in this manner from a human chromosome 16-specific library has proven useful in multilocus fingerprinting of humans and other species (Longmire *et al.*, 1990). M13-complementary clones isolated from peregrine libraries are being used at low stringency as multilocus probes, and at high stringency to detect individual hypervariable loci in peregrine DNA (work in progress).

We report here on the direct use of the M13 in studies of geographic variation in the genome of the peregrine. Besides detecting highly polymorphic autosomal loci, the phage repeat also reveals certain sets of restriction fragments only in DNAs from female falcons. Variation exhibited by the sex-linked fragments provides information about the genetic differentiation of peregrines at the population level, and makes a first step toward distinguishing migratory populations en route between their nesting and wintering grounds.

Materials and Methods

Blood Collection and DNA Isolation

As previously reported, we have developed a blood collection protocol that minimizes inconvenience and difficulties for field personnel (Longmire *et al.*, 1988). This is an important consideration since these falcon blood samples are often collected in remote regions (such as the Arctic). Between 0.1 and 0.5 ml of freshly drawn peripheral blood is added to a sterile 15 ml polypropylene tube containing 5 ml lysis buffer (0.1 M Tris-HCl pH 8.0, 0.1 M EDTA, 0.01 M NaCl, and 0.5% w/v SDS). Samples were stored in the field and shipped to the laboratory at ambient temperature. We have found that these samples can be stored for several years at room temperature without DNA degradation. For the isolation of DNA, proteinase K is added to 0.5 mg per ml, and the samples are incubated overnight at 37°C on a tube rotator. The tube rotation step is important for obtaining homogeneous lysates and significantly increases consistency of DNA yields between samples. The samples are extracted once with an equal volume of phenol that has been saturated with TE buffer (0.01 M Tris-HCl pH 8.0, 0.001 M EDTA). The extraction step is carried out for approximately 1 hour on a tube rotator at 37°C. Tube rotation has also been found to significantly enhance the efficiency of this extraction step. Following centrifugation for 10 min. 22°C at 800 xg, the aqueous phases were removed and dialyzed at 4°C for approximately 24 hours against three changes of TE buffer (or until the odor of phenol could no longer be detected). Final purified DNAs were quantitated by UV spectroscopy.

DNA Restrictions, Southern Blotting, and Hybridization

Five to ten micrograms of each DNA were digested with at least a 10-fold unit excess of restriction enzyme using buffer conditions recommended by the supplier (New England Bio Labs). Restricted DNAs were electrophoresed in 20 × 25 cm 0.8% agarose gels at 30–35 volts for approximately 30 hours and subsequently transferred onto Zetabind nylon membranes. Resulting blots were prehybridized for 2–4 hours at

42°C in 6 × SSC, 35% formamide, 0.005 M EDTA (pH 8.0), and 0.25% (w/v) powdered milk (Vassart *et al.*, 1987). Hybridization was carried out overnight at 42°C in the same solution containing 1×10^6 cpm per ml probe. The 282 bp HaeIII/ClaI fragment of M13mp8 was gel purified and labeled with ^{32}P to specific activities greater than 1×10^8 cpm per μg using the primer extension method described by Feinberg and Vogelstein (1983). Post-hybridization washes were as follows: 2 × 15 min. at 22°C in 2 × SSC, 0.1% SDS, and twice for 15 min. at 50°C in 1 × SSC, 0.1% SDS. Washed blots were autoradiographed at −70°C in cassettes containing intensifying screens.

Results and Discussion

M-13 Fingerprints and Detection of Sex-Linked Minisatellite Fragments in Falcons

We have used the M13 tandem repeat to examine DNAs from several species of falcons. Figure 1 shows the patterns obtained in HaeIII digests of DNAs from peregrines and Mauritius kestrels (*Falco punctatus*). The M13 probe detects considerable levels of polymorphism in the peregrine but reduced variation in Mauritius kestrels. Owing to habitat destruction, this kestrel was reduced to just two known breeding pairs prior to re-expansion by captive propagation and reintroduction (Temple, 1986). The decreased level of polymorphism seen in Mauritius kestrel DNAs is most likely attributable to the extreme bottleneck that this species has gone through (Longmire *et al.*, in prep.).

In addition to detecting HaeIII restriction fragments that display no apparent sex-linkage, the M13 repeat also hybridizes to certain sets of fragments that are present only in females. Since in birds females are the heterogametic sex, these fragments almost certainly map to the sex, or W, chromosome. In the Mauritius kestrel these fragments occur at 11.5 and 7.0 kb, and are non-polymorphic among the females that were sampled. In the peregrine, the sex-linked fragments are observed at 13.7, 7.4 and 6.1 kb. Whereas the 13.7 and 6.1 kb fragments were observed in all of the female peregrines that were examined, the 7.4 kb fragment is polymorphic and not present in all females.

Breeding Population Distribution of the 7.4 kb Sex-Linked Fragment in Peregrines

DNA fingerprinting using minisatellite probes has proven to be a powerful method for examining very close levels of relationship (i.e. maternity and paternity) within human and wildlife populations

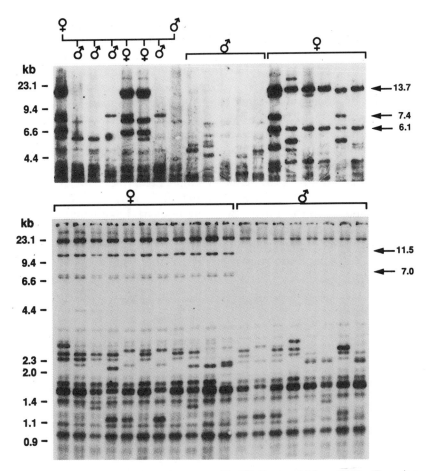

Figure 1. Hybridization of the M13 repeat to HaeIII digests of falcon DNAs. Ten micrograms of each DNA was digested with excess HaeIII, electrophoresed within 0.8% agarose gels, and the resulting blots were hybridized to the M13 repeat. Male and female family members and random individuals are as indicated. The top panel contains peregrine DNAs, and the lower panel shows results obtained with Mauritius kestrel DNAs. Arrows indicate sex-linked fragments in each species. Size standards, in kilobase (kb) pairs, were derived from HindIII digested bacteriophage lambda DNA, and from HaeIII digested ØX174 DNA.

(Jeffreys *et al.*, 1985a; Burke and Bruford, 1987). However, because the level of heterozygosity displayed by most minisatellite loci is very high (>90%), it is unlikely that these hypervariable alleles will become fixed within large breeding groups (Lynch, 1988). Hence, our efforts to develop population specific markers in the peregrine have centered mainly around probes that detect reduced levels of variation within geographically distinct groups (such as heterochromatic and unique sequence probes that detect base substitution-type polymorphisms).

222

Although the sex-linked fragments are detected in falcons by a minisatellite probe, the level of polymorphism observed at the 7.4 kb fragments is significantly reduced relative to autosomal-linked minisatellite fragments (the frequency of the 7.4 kb fragment among all female peregrines tested in this study was 0.66, whereas the mean occurrence among individuals of other polymorphic bands showing no apparent sex linkage was approximately 0.12). In essence, the 7.4 kb fragment is dimorphic in that it is either present or absent in the genomes of the female peregrines that are displayed in Fig. 1. Because of this reduced level of variation, we decided to examine the distribution of this fragment among different breeding populations. These included female peregrines from Argentina, the southwestern United States, Greenland, Ungava Bay and Rankin Inlet, Canada (data from these two were pooled), Lake Athabasca, and two river drainages in Alaska (the Tanana and the Colville).

Blots containing peregrine DNAs from the southwestern United States and Greenland are shown in Figs. 2 and 3 respectively, and the

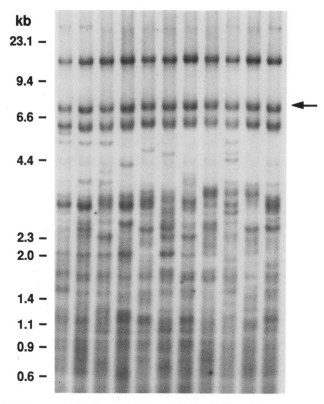

Figure 2. Hybridization of the M13 repeat to HaeIII digest of DNAs from female *anatum* peregrines nesting in the southwestern United States. Arrow indicates location of 7.4 kb sex-linked fragment. Size standards are as described in Fig. 1.

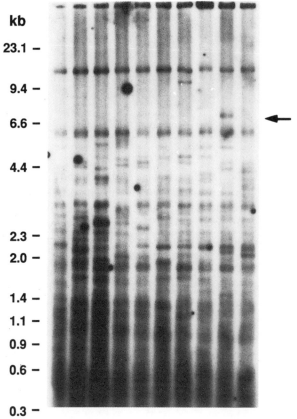

Figure 3. Hybridization of the M13 repeat to HaeIII digests of DNAs from female *tundrius* peregrines nesting in Greenland. Arrow indicates location of 7.4 kb sex-linked fragment. Size standards are as described in Fig. 1.

entire breeding population data set is summarized in Tab. 1. Interestingly, the absence of the 7.4 kb fragment ("null-7.4 kb type") was observed only among individuals of the *tundrius* subspecies. In addition, within *tundrius* the null-7.4 kb type shows a clinal distribution with a high frequency of representation among Greenland peregrines (0.90), and a steadily declining frequency westward in birds nesting in Ungava Bay/Rankin Inlet (0.50), and the Colville River of Alaska (0.29).

Use of the Null 7.4 kb Type of Monitor the Migration of Arctic Peregrines

The finding that the null-7.4 kb type was exhibited only in *tundrius* peregrines enabled a preliminary analysis to be made of the representation of these birds during seasonal migration. With this in mind, we examined female peregrines that were sampled during migration at

Table 1. Representation of the null-7.4 kb type in peregrines of defined origin

Breeding area	Subspecies	Freq. null-7.4 kb type
S.W. United States	anatum	0 ($n = 11$)
Lake Athabasca, Canada	anatum	0 ($n = 6$)
Tanana River, Alaska	anatum	0 ($n = 7$)
Greenland	tundrius	0.90 ($n = 10$)
Rankin/Ungava, Canada	tundrius	0.50 ($n = 6$)
Colville River, Alaska	tundrius	0.29 ($n = 7$)
Argentina	cassini	0 ($n = 5$)

Assateague Island, the Dry Tortugas, Padre Island, and at a wintering site at Peru. Blots containing DNAs from falcons sampled on Padre Island and in Peru are shown in Figs. 4 and 5 respectively, and all of the migration data are summarized in Tab. 2. It is interesting to note that although the frequency of the null-7.4 kb type is fairly high among the migrants sampled on Assateague, the Dry Tortugas, and Padre Island (0.50, 0.29, and 0.45 respectively), only one of the eleven female peregrines that were sampled in Peru shows this marker. This result suggests that a large portion of the peregrines migrating along the Eastern and Gulf coasts of the United States are wintering elsewhere in Central or South America. To address this issue more fully, an effort is currently

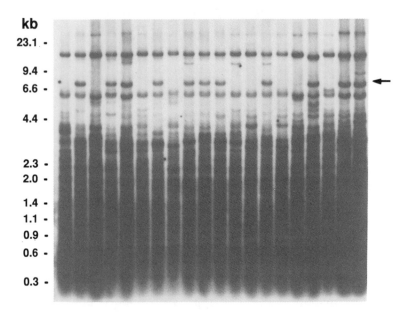

Figure 4. Hybridization of the M13 repeat to HaeIII digests of DNAs isolated from female migrant peregrines sampled on Padre Island, Texas. Arrow indicates locations of the 7.4 kb sex-linked fragment. Size standards are as described in Fig. 1.

Table 2. Representation of the null-7.4 kb marker in migrating and wintering peregrines

Location sampled	Freq. null-7.4 kb type
Assateague Island, Maryland (migrants)	0.50 ($n = 20$)
Dry Tortugas, Florida (migrants)	0.29 ($n = 7$)
Padre Island, Texas (migrants)	0.45 ($n = 20$)
Peru (wintering)	0.09 ($n = 11$)

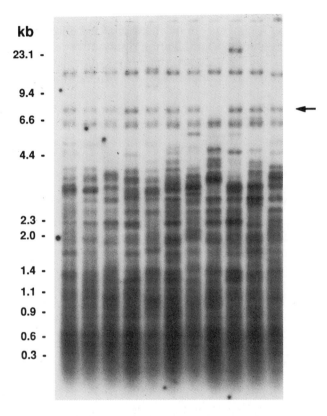

Figure 5. Hybridization of the M13 repeat to HaeIII digests of DNAs isolated from wintering female peregrines sampled in Peru. Arrow indicates location of the 7.4 kb sex-linked fragment. Size standards are as described in Fig. 1.

underway to obtain additional blood samples from other wintering sites.

General Discussion

The M13 repeat (Vassart *et al.*, 1987) detects different levels of genetic variation in falcon species. On the one hand, the probe reveals highly polymorphic restriction fragments that show no apparent unequal distribution between the sexes. These autosomal-linked hypervariable fragments will be useful in determinations of maternity and paternity within captive and natural populations. In addition, the M13 repeat hybridizes to certain sets of restriction fragments that are present only in females. These sex-linked fragments are monomorphic among Mauritius kestrels and only slightly variant in the peregrines that were examined. The reason that the sex-linked minisatellite fragments show drastically reduced levels of polymorphism is not known at this time. However, it may be that these fragments are localized in a region of the W chromosome where homologous recombination does not occur (i.e. – the non-pairing region). If this is the case, the nonpolymorphic nature of these sex-linked fragments may support Jeffreys' original model for unequal crossing over as a mechanism for generating minisatellite diversity (Jeffreys et al., 1985a).

The ability to detect sex-specific restriction fragments could be important in studies of several aspects of falcon biology. First, these female-specific fragments provide a genetic test for determinations of gender. Also, the sizes of the sex-linked fragments are species-specific, at least among Mauritius kestrels and peregrines. If this is true for other species of falcons, it becomes possible to identify the sex of an unknown sample, as well as the species of origin (if the sample is from a female). The species-specific nature of these sex-linked fragments may also be useful in systematic studies carried out within the genus *Falco* and the family *Falconidae*.

Secondly, the identification of sex-specific DNA sequences will be valuable in studies of genome organization. Owing to their unique size, these fragments can be selectively cloned and identified in recombinant DNA libraries. Using chromosome walking techniques the recovered clones could serve as initiation points for mapping purposes, and for obtaining additional sex-specific sequences. Given that minisatellites are often located very near to expressed genes (Bell et al., 1982; Proudfoot and Gil, 1982; Capone et al., 1983; Jeffreys et al., 1985a), chromosome walks of reasonable distances could possibly yield coding sequences that are expressed in a sex-specific manner. Studies of such coding regions could conceivably yield new information concerning the genetic basis of sex determination, especially in species where females are the heterogametic gender (Page et al., 1988).

Lastly, variation exhibited by the sex-linked minisatellite fragments has proven to be useful in peregrine population studies. Among the female peregrines that were examined, all members of *anatum* and *cassini* displayed three HaeIII restriction fragments that strongly hybridized to the M13 repeat. These fragments were observed at 13.7, 7.4, and 6.1 kb. Within *tundrius*, all females displayed the same 13.7 and 6.1 kb fragments that were found in *anatum* and *cassini*, but a lack of the 7.4 kb fragment was seen at a high frequency (0.90) in Greenlandic peregrines. The null-7.4 kb type steadily decreased in frequency among *tundrius* peregrines nesting further westward in Ungava Bay/Rankin Inlet (0.50), and on the Colville River of Alaska (0.29). It is interesting to speculate on the east-west clinal distribution of the null-7.4 kb marker within *tundrius*. Although we can suggest several possibilities, it is not feasible to determine if regional differences in the frequency of the null-7.4 kb type are a function of directional gene flow until we have examined peregrines from several other populations. These would include peregrines from Britain and Scandinavia (*F.p. peregrinus*), eastern Asia (*F.p. calidus*), and from the Siberian tundra (*F.p. japanensis*).

One speculative scenario is that the Greenland population was originally founded by a small number of peregrines largely of the null-7.4 kb type from northeastern Canada. Given that Greenland is rather isolated geographically, the marker has remained highly represented into the present time. The steady decrease of the null-7.4 kb marker further westward on the North American continent itself may be the result of genetic intermixing between *tundrius* and *anatum*. The peregrine is considered to be a highly philopatric species (Cade, 1982; Ambrose and Riddle, 1988). However, researchers in Alaska have recently identified a female peregrine nesting on the Colville River that was originally banded as a nestling approximately 450 km to the southeast on the Porcupine River (T. Swem and D. Mossop, unpubl. data). Since the peregrine population nesting on the Colville River is considered to be *tundrius*, and those on the Porcupine River are considered to be *anatum*, this occurrence documents genetic infiltration of *tundrius* by *anatum*. Thus, genetic data reported here taken together with field observations and morphometric data (White and Boyce, 1988) indicate that genetic intergradation might occur between *tundrius* and *anatum* peregrines where tundra and taiga habitats occur in close juxtaposition. If genetic exchange has occurred in both directions between *tundrius* and *anatum*, we would expect to eventually find the null-7.4 kb type present in peregrines considered to be *anatum*. With this in mind, we are currently collecting numerous blood samples from *anatum* peregrines nesting in close proximity to *tundrius* (such as on the Porcupine River).

Although the null-7.4 kb type was not found in all peregrines currently considered to be *tundrius*, all females that did display the marker were of the *tundrius* subspecies. Hence, although additional samples will have to be examined, current data strongly suggests that the null-7.4 kb type provides a valuable genetic marker for distinguishing a subset of *tundrius* peregrines. This has allowed the use of a genetic test to identify putative members of an avian subspecies during migration. Individuals bearing the null-7.4 kb type were found to be moderately represented among migrants sampled on Assateague Island, the Dry Tortugas, and on Padre Island (0.50, 0.29, and 0.45 respectively), and only slightly represented in the wintering population of peregrines present in Peru (0.09). This data may provide some initial evidence for the existence of population-specific wintering sites.

The result that a minisatellite probe proved to be informative in a study of population differentiation and migration was unexpected, as the value of minisatellites has previously been considered to be limited to investigations of very close genetic relationships (such as maternity and paternity). It should be noted that the general applicability of this approach to interpopulation studies in other species may be limited. The reason for such limitation is that the informative nature of the 7.4 kb fragment is attributable to a reduced level of polymorphism due most likely to sex-linkage, and sex-linked minisatellites may not be found in all species. Kashi *et al.* (1990) have recently described a cloned fragment of DNA containing a minisatellite and a $(TG)_n$ microsatellite that reveals monomorphic sex-specific patterns in cattle. Thus although unusual, the observance of sex-linked minisatellite fragments in falcons is not unique, and the occurrence of sex-specific minisatellites will only become known as more species are studied. A distinct disadvantage to the use of any sex-specific marker is that only members of the heterogametic gender can be analyzed.

Perhaps a more productive general approach to comparisons made between populations involves the use of DNA sequence probes that detect variation at the base substitution level. We have previously found that a cloned heterochromatic element is useful in differentiating peregrines from Argentina and Greenland (Longmire *et al.*, 1988). In addition, work currently in progress indicates that single copy cosmids retrieved from a library made from partially digested peregrine DNA will provide new markers that should increase the resolving power of this molecular approach to the study of peregrine populations. Markers for peregrine populations nesting in the taiga habitats of North America would be of benefit in order to more precisely distinguish boreal and arctic breeding falcons during migration and on their wintering grounds in Latin America.

Acknowledgements

The authors thank the many individuals who helped with this work. We thank C. Hildebrand and R. Moyzis for many helpful discussions, and L. Thompson and L. Martinez for manuscript preparation. P. Stacey kindly provided critical review of the manuscript. Excellent laboratory assistance was contributed by L. Clepper, C. Hardekopf, A. Lewis, E. Saxman, J. Shermer, A. Simmons, and J. Stark. Falcon blood samples were generously supplied by C. Anderson, S. Baker, O. Beingolea, P. Bente, D. Bird, D. Boyce, M. Bradley, D. Brimm, B. Burnham, F. Clunie, G. Court, S. Crowe, J. Enderson, K. Falk, R. Fyfe, P. Harrity, W. Heck, N. Hilgert, R. Hunter, A. Jenkins, C. Jones, E. Levine, W. Mattox, S. Moller, D. Morizot, D. Mossop, J. Parrish, K. Riddle, C. Sandford, C. Shank, C. Schultz, T. Swem, G. Vasina, J. Weaver, and M. Yates. In addition, blood samples were kindly provided by the Peregrine Fund Inc., the United States Fish and Wildlife Service, the Canadian Wildlife Service, the Greenland Peregrine Falcon Survey, the Padre Island Peregrine Falcon Survey, and by many North American falconers.

This work was conducted under the auspices of the U.S. Department of Energy with support from the Life Sciences Division of the Los Alamos National Laboratory and the U.S. Army.

References

Ambrose, R. E., and Riddle, K. E. (1988). Population dispersal, turnover, and migration of Alaska peregrines. In: Cade, T. J., Enderson, J. E., Thelander, C. G., and White, C. M. (eds), Peregrine Falcon Populations; their Management and Recovery. The Peregrine Fund Inc., Boise, Idaho, pp. 677–684.

Bell, G. I., Selby, M. J., and Rutter, W. J. (1982). The highly polymorphic region near the human insulin gene is composed of simple tandemly repeated sequences. *Nature* (London) 295: 31–35.

Burke, T. and Bruford, M. W. (1987). DNA fingerprinting in birds. *Nature* (London) 327: 149–152.

Cade, T. J. (1982). The Falcons of the World. Comstock Cornell Univ. Press, Ithaca, N.Y.

Capone, D. J., Chen, E. Y., Levinson, A. D., Seeburg, P. H., and Goeddel, D. V. (1983). Complete nucleotide sequence of the T24 human bladder carcinoma oncogene and its normal homologue. *Nature* (London) 302: 33–37.

Feinberg, A. P., and Vogelstein, B. (1983). A technique for radiolabeling DNA restriction fragments to high specific activity. *Anal. Biochem.* 132: 6–13.

Jeffreys, A. J., Wilson, V., and Thein, S. L. (1985a). Hypervariable minisatellite regions in human DNA. *Nature* (London) 314: 67–73.

Jeffreys, A. J., Brookfield, J. F. Y., and Semeonoff, R. (1985b). Positive identification of an immigration test-case using human DNA fingerprints. *Nature* (London) 317: 818–819.

Kashi, Y., Iraqi, F., Tikochinski, Y., Ruzitski, B., Nave, A., Beckmann, J. S., Freidmann, A., Soller, M., and Gruenbaum, Y. (1990). (TG)$_n$ uncovers sex-specific hybridization pattern in cattle. *Genomics* 7: 31–36.

Longmire, J. L., Lewis, A. K., Brown, N. C., Buckingham, J. M., Clark, L. M., Jones, M. D., Meincke, L. J., Meyne, J. M., Ratliff, R. L., Ray, F. A., Wagner, R. P., and Moyzis, R. K. (1988). Isolation and molecular characterization of a highly polymorphic centromeric tandem repeat in the family *Falconidae*. *Genomics* 2: 14–24.

Longmire, J. L., Kraemer, P. K., Brown, N. C., Hardekopf, L. C., and Deaven, L. L. (1990). A new multi-locus DNA fingerprinting probe: pV47-2. *Nucleic Acids Res.* 18: 1658.

Nakamura, Y., Carlson, M., Krapcho, K., Kanamori, M., and White, R. (1988). New approach for isolation of VNTR markers. *Am. J. Hum. Genet.* 43: 854–859.

Lynch, M. (1988). Estimation of relatedness by DNA fingerprinting. *Mol. Biol. Evol.* 5: 584–599.

Page, D. C., Mosher, R., Simpson, E. M., Fisher, E. M. C., Mardon, G. Pollack, J., McGillivray, B., de la Chapelle, A., and Brown, L. G. (1988). The sex-determining region of the human Y chromosome encodes a finger protein. *Cell* 51: 1091–1104.

Proudfoot, N. J., and Gil, A. (1982). The structure of the human zeta-globin gene and a closely linked, nearly identical pseudogene. *Cell* 31: 553–563.

Temple, S. A. (1986). Recovery of the endangered Mauritius kestrel from an extreme population bottleneck. *Auk* 103: 632–633.

Vassart, G., Georges, M., Monsieur, R., Brocas, H., Lequarre, A. S., and Christophe, D. (1987). A sequence in M13 phage detects hypervariable minisatellites in human and animal DNA. *Science* 235: 683–684.

White, C. M., and Boyce, A., Jr. (1988). An overview of peregrine falcon subspecies. In: Cade, T. J., Enderson, J. E., Thelander, C. G. and White, C. M. (eds), Peregrine Falcon Populations; their Management and Recovery. The Peregrine Fund Inc. Boise, Idaho, pp. 789–810.

Yates, M. A., Riddle, K. E., and Ward, F. P. (1988). Recoveries of peregrine falcons migrating through the eastern and central United States. In: Cade, T. J., Enderson J. E., Thelander, C. G., and White, C. M. (eds), Peregrine Falcon Populations; their Management and Recovery. The Peregrine Fund Inc. Boise, Idaho, pp 471–484.

DNA Fingerprinting: Approaches and Applications
ed. by T. Burke, G. Dolf, A. J. Jeffreys & R. Wolff
© 1991 Birkhäuser Verlag Basel/Switzerland

Hypervariable DNA Markers and their Applications in the Chicken

M. W. Bruford*, and T. Burke

Department of Zoology, University of Leicester, University Road, Leicester LE1 7RH, Great Britain
Present address: Molecular Genetics Unit, Institute of Zoology, Regent's Park, London NW1 4RY, Great Britain

Summary
In this paper, the current and anticipated applications of hypervariable molecular genetics in the chicken (*Gallus gallus*) are reviewed. Current areas of research using multilocus DNA fingerprinting include population genetics of inbred lines, identification of chimeras, genomic introgression studies and identification of alleles which co-segregate with quantitative traits. Data from research in our laboratory on the cloning and characterization of locus-specific probes in chickens are presented. Twenty-five highly polymorphic minisatellites isolated from two chicken genomic libraries have been tested for mutation, heterozygosity and line specificity. Locus-specific minisatellite probes form a potentially important component of a proposed genome map in chickens. They offer several distinct advantages over the use of multilocus probes, and minisatellites linked to quantitative trait loci could prove important in marker assisted selection (MAS) programmes.

Introduction

Current research in chicken hypervariable molecular genetics appears to be taking three broad courses: *1.* Identification and characterization of the genetic make-up of stocks/lines/breeds of chicken by means of multilocus fingerprinting. This is being applied to inbreeding estimation, identification of chimeras, measuring genome introgression in selection programmes and commercial protection. *2.* Investigating minisatellite alleles which respond to selection using the multilocus approach. *3.* Isolation and characterization of individual minisatellite loci for use as locus-specific probes both in mapping important genes and possibly the entire chicken genome.

In this introduction the approaches outlined above and the various methods being employed in their application will be discussed.

1. The Use of Multilocus DNA Fingerprints to Characterize Chicken Lines

Most lines, stocks or breeds of chickens being maintained today have been artificially selected for specific phenotypes and originate from a limited number of founder individuals. These lines are usually kept as closed flocks and are, as a consequence, often highly inbred. This inbreeding is detected in multilocus DNA fingerprints (Jeffreys et al., 1985a) as a higher than expected degree of similarity between individuals, and is usually quantified as a band-sharing coefficient (Jeffreys et. al., 1985b; Burke and Bruford, 1987). Gilbert et al. (1990) have shown the consequence of inbreeding to be a greatly reduced fingerprint variability in genetically isolated wild populations of the Californian Channel Island fox, and Jeffreys et al. (1987) observed similarly reduced levels of variability in laboratory strains of mice. These two studies have also shown that geographically or artificially isolated populations may have markedly different minisatellite profiles due to sampling or fixation of different sizes alleles at minisatellite loci. In chickens this effect might be accentuated as a consequence of linkage disequilibrium between minisatellite loci and genes under intense artificial selection.

Kuhnlein et al. (1989) demonstrated that DNA fingerprint patterns could be used to distinguish different chicken lines from each other and to calculate genetic distance between lines, provided that levels of intra-population variability were similar. The calculated distances correctly reflected the well documented breeding history of these lines. The analysis of fingerprint patterns has been further applied by Kuhnlein et al. (1990) to derive a means of estimating inbreeding coefficients in chicken lines from minisatellite band frequencies, which may be useful in estimating the likelihood of response to selection in lines with different inbreeding coefficients.

Multilocus DNA fingerprinting has also been applied to improving the efficiency of gene introgression in breeding programmes (Hillel et al., 1990). As minisatellites have been shown to be dispersed throughout the genomes of most vertebrate species (e.g. Jeffreys et al., 1986; Burke and Bruford, 1987; Burke et al., 1989; Bruford et al., 1990; and in preparation), DNA fingerprint patterns can be regarded as being representative of the whole genome in introgression studies. Thus if it is desired to transfer a trait into a recipient strain, the number of backcross generations required to return the population genome to that resembling the original strain, whilst still retaining the desired introgressed trait, can be reduced by selecting the individuals whose fingerprint pattern most resembles the recipient. Hillel et al. (1990) demonstrated this approach using ibex-goat hybrids. It could, however, be applied to other domestic species, including chickens.

The fact that lines of chickens can be characterized by their finger-print patterns has also been used in detecting somatic chimerism and germline transmission of donor cell types in transgenic experiments in chickens (Petitte *et al.*, 1990). Plymouth Barred Rock embryonic stem cells were injected into eggs from a highly inbred Dwarf White Leghorn line. The fingerprint of a phenotypically chimeric male was found to contain bands that were not present in the Dwarf White Leghorn line-specific fingerprint. These appeared to originate from the Plymouth Barred Rock line, and were transferred at a low level to the offspring of a cross between the chimera and Dwarf White Leghorn, demonstrating germline incorporation of the donor cells. This work holds promise for stable incorporation of desired genes into lines by transgenesis, and offers an alternative to the method employed by Hillel *et al.* (1990).

The application of DNA fingerprinting to line identification in chick-ens has potentially important implications in commercial strains. The ability to identify chickens as originating from a certain breeder's lines should assist in problems of commercial protection and confirming the origins of malperforming birds. However, with the large numbers of birds normally held in commercial stocks, the effective population size may be high enough to maintain high levels of heterozygosity and prevent the occurence of diagnostic alleles.

2. Identification of Alleles which Respond to Selection

Another area of research where multilocus DNA fingerprinting is being applied to chickens is in identifying bands within fingerprint patterns which increase in frequency when populations of birds are selected for specific traits. Kuhnlein and Zadworny (1990) have described alleles at a locus detected by the M13 hypervariable region that is associated in several different lines with resistance to Marek's disease. Plotzky *et al.* (1990) and Kuhnlein and Zadworny (1990) have also used the approach of utilising DNA mixes from individuals either within families (to analyse paternal genetic input) or populations which show measurably different phenotypes for the selected traits. The mixes are taken from groups of individuals present in each tail of the phenotype distribution, and their DNA fingerprints are compared. In analyses of families and of inbred lines, the differences are mainly seen in the intensities of particu-lar bands. In more outbred or divergently selected lines it is more likely that the simple presence or absence of a band will be observed. This technique is likely to be extensively used over the next few years, and might be particularly applicable to animals of low profligacy such as cattle (Plotzky *et al.*, 1990) where large paternal half-sib families are available. The eventual aim of this research is to isolate and clone minisatellite loci responding to selection and use them as probes in marker assisted selection (MAS) programmes.

3. Isolation and Characterization of Locus-specific Probes and their Applications

The approach described in section 2 is a potentially quick and effective means of establishing whether or not there are any minisatellite loci obviously linked to quantitative trait loci (QTL). However, given the size of the chicken genome (1.2×10^9 bp) and the large number of linkage groups (chromosome number $2n = 78$), compared with the relatively small number of loci detected on a genetic fingerprint (Bruford et al., 1990), it is unlikely that alleles which respond to selection will be found for a large number of these traits, even if many different multilocus systems were to be used.

Clearly, a more complete, though labour intensive, method of localizing QTLs would be to saturate the genome with locus-specific polymorphic markers. This approach has already been applied with great effect in the tomato in which it has allowed several quantitative traits to be resolved into discrete Mendelian factors (Tanksley et al., 1982; Paterson et al., 1988). In species where the genome already has extensive marker coverage, hypervariable sequences can also be used to extend the map further. Julier et al. (1990) have applied human derived hypervariable sequences as multilocus probes (at low stringency) in the mouse. These were tested against experimental crosses between inbred strains, and the segregating loci were analysed for linkage with other markers and assigned to specific chromosomal locations. An extension of this approach is being applied in cattle, where more than 130 polymorphic markers have already been identified and isolated. These are currently being used to increase the extent of the known linkage groups within the bovine map (Georges et al., 1990).

In chickens there is a paucity of well defined genomic markers available for linkage studies. Analysis of segregations in chicken pedigrees of hypervariable DNA fragments detected with the 33.6, 33.15, $3'\alpha$ HVR and M13 probes has demonstrated the presence of many unlinked independently segregating loci (Bruford et al., 1990; and in preparation). There are now many more multilocus fingerprint probes available, and through their application to chickens they may potentially be used to isolate very many informative markers which might contribute to a genome map. As they are highly polymorphic and likely to be informative in most crosses (see Fig. 3), they will provide an ideal base component from which to build a more detailed map. With this in mind we have recently started to clone and characterize locus-specific minisatellite probes in the chicken, and some of the data resulting from these experiments are presented here.

Materials and Methods

1. Construction and Analysis of Genomic Libraries

Chicken minisatellite sequences were isolated from two genomic DNA libraries cloned into charomid vectors (Saito and Stark, 1986) following the general method described in Armour *et al.* (1990) and Hanotte *et al.* (1990). Full details will be provided elsewhere (Bruford and Burke, in preparation).

Briefly, DNA samples from 20 unrelated individuals were pooled, and the pooled sample was digested to completion with NdeII. Two size fractions, 2–7 kb and 11–21 kb, were isolated by electroelution onto dialysis membranes and ligated into charomids 9–36 (36 kb long) and 9–30 (30 kb long) respectively (Saito and Stark, 1986). The ligation mixture was used to infect *E. coli* strain NM554 bacteria (recA⁻, mcrA⁻, mcrB⁻) (Raleigh *et al.* 1988), and the transformed bacteria were plated onto Luria Bertani agar plates in the presence of 50 μg/ml ampicillin (Sigma).

Library 1 (2–7 kb × charomid 9–36) yielded approximately 115,000 recombinants, and library 2 (11–21 kb × charomid 9–30) yielded approximately 35,000 recombinants. Approximately three haploid genome equivalents of insert DNA (4,300 colonies for library 1, 1700 colonies for library 2) (Hanotte *et al.*, 1990) were transferred into individual wells in microtitre plates containing Luria broth with 10% glycerol, and stored at −20°C. The libraries were screened by replication onto Amersham Hybond-N nylon membranes, which were then hybridized to the multilocus probes 33.6, 33.15 and M13.

2. Analysis of Minisatellite Clones

Recombinant DNA was extracted by alkaline lysis (Birnboim and Doly, 1979) and was digested with Sau3AI. The insert DNA was then collected by electroelution onto dialysis membranes. Probes were tested for allelic variability against MboI digested DNA samples: two commercial egg-laying birds, one red junglefowl and one North Holland blue. Once probes with high allelic variability had been identified, they were further characterized, as follows.

Mendelian inheritance, germline stability and linkage were tested using four families with a total of 106 parent/offspring combinations.

Heterozygosity was tested using DNA samples from 20 unrelated chickens of diverse origin (including one Sonnerat's junglefowl – *Gallus sonnerati*). Egg laying line specificity was tested using DNA samples of six individuals each from three different commercially available crosses. Broiler line specificity was tested using three individuals each from seven different commercially available crosses.

Results

1. Genome Library Screening

Library 1 (2–7 kb fraction) yielded 56 positive (intensely hybridizing) clones when probed with the Jeffreys multilocus probe 33.6. Hybridization with 33.15 revealed 72 positives, however in 15 cases both probes hybridized to the same clone.

Library 2 (11–21 kb fraction) yielded 9 positive clones when probed with 33.6, and 20 positives with 33.15. Both probes hybridized to the same clone in three cases. M13 HVR yielded 5 positive clones. However, of these, two were detected by 33.15 and one was detected by both 33.6 and 33.15.

2. Analysis of Positive Recombinants

A selection of positively hybridizing cloned insert DNAs were isolated from both libraries and 49 clones were hybridized to a panel of four unrelated chickens, as described. Figure 1 shows a selection of the results obtained from these experiments, ranging from the detection of highly polymorphic minisatellite bands to uninformative satellite patterns. Table 1 shows a summary of the characteristics of each clone tested. Heterozygosity values have been calculated, because of the small sample size (Wong *et al.*, 1986), from pair-wise comparisons of alleles in the four individuals.

Five clones revealed variable patterns of more than two bands in the individuals tested (see c*Gga*MS22 in Fig. 1). The similarity of these patterns almost invariably reflected the levels of relatedness between the individuals and they are therefore potentially highly valuable markers. An analysis of segregation among these probe loci is currently being carried out.

To date, 49 clones have been tested, among which four sequences were each isolated twice. Of these, 25 contain polymorphic sequences, three detect uninformative monomorphic loci, five detect multiple band patterns with informative variation at possibly more than one locus, and twelve detect one or more uninformative presumed satellite DNA

Kb

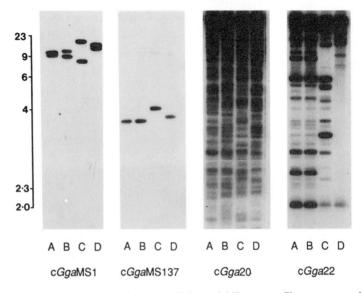

| A B C D | A B C D | A B C D | A B C D |
| cGgaMS1 | cGgaMS137 | cGga20 | cGga22 |

Figure 1. A selection of results from the allelic variability tests. Clones were probed as described against four unrelated chickens. A = ISA egg layer, B = ISA egg layer, C = red junglefowl, D = North Holland blue. Samples A and B originate from different lines, but are more closely related than other birds.

patterns. The presence of such a large proportion of clones which detect satellite DNA families is not surprising as we have found elsewhere (Bruford *et al.*, in preparation) that the Jeffreys probe 33.15 detects a presumed satellite sequence which has multiple internal sites for HinfI and AluI. Some of this sequence is likely to be present in the NdeII-libraries. In summary, informative minisatellite markers have been isolated in 51% of the clones tested.

Table 1. Characteristics of the different loci detected by 45 charomid probes

Library	Cloned fragment size	Specific "minisatellite" probes		Mean hetero-zygosity*	Complex pattern probes
		Monomorphic	Polymorphic		
1	2–7 kb	1	9	0.754	9
2	11–21 kb	2	16	0.809	8

*Includes monomorphic minisatellites.

3. Detailed Analysis of Polymorphic Minisatellite Locus Specific Probes

To date, eight of the polymorphic (and one monomorphic) minisatellite locus-specific clones described above have been tested more extensively as follows.

(a) Pedigree Analysis

Each probe was tested against four families containing a total of 53 offspring. The purpose of these experiments was to confirm Mendelian inheritance, detect mutation events and test for linkage among the cloned loci. The transmission of parental alleles was not significantly different from Mendelian expectation at any of the loci. No electrophoretically resolvable mutation event was detected in any family with any of the probes used, suggesting that the cloned loci are among those minisatellites having a relatively high germline stability.

For the eight probes tested, parental alleles were homozygous in 43% of cases. This figure is relatively high for minisatellite loci but is not surprising as families 1 and 2 comprised a cross between two closed flock lines, and families 3 and 4 resulted from a backcross between each of these lines and the F_1 generation. The high homozygosity also restricted the analysis of segregation of parental alleles. Nevertheless, there is, as expected, no evidence of linkage among any of the small number of loci tested to date.

We concluded that one probe, cGgaMS2, detected a locus on the Z chromosome (Bruford and Burke, in preparation). In avian species, females are the heterogametic sex and hence are hemizygous for this locus.

Figure 2 shows a typical pedigree of two parents and 13 offspring probed with a highly heterozygous autosomal probe – cGgaMS75.

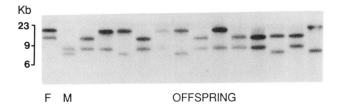

cGgaMS75

Figure 2. MboI digested samples of a family of two parents and thirteen offspring probed with cGgaMS75. Both parents are heterozygotes. This type of informative segration will be of particular value in linkage studies.

238

Kb

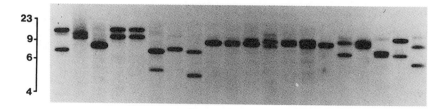

cGgaMS1

Figure 3. 20 unrelated individual chicken DNA samples digested with MboI and probed with cGgaMS1. 15 individuals (75%) are heterozygotes, and there are 15 detectably different alleles.

(b) Heterozygosity

A measure of heterozygosity was ascertained for each polymorphic locus as the proportion of a standard panel of 20 unrelated chickens of diverse origin that were heterozygotes. As most of the individuals originated from highly inbred populations, the value that would be obtained in an outbred population would be expected to be higher. Mean heterozygosity for the minisatellite probes tested was 73.6% (s.d. 15.4). Figure 3 shows a typical heterozygosity autoradiograph produced with cGgaMS1. In this case 15 out of 20 (75%) individuals are resolvable heterozygotes, with 15 detectably different alleles. It should be noted that a Sonnerat's junglefowl (Gallus sonnerati) sample was included in these gels and an apparently homologous locus of equal hybridizing strength was detected with six of the eight probes.

(c) Commercial Line Specificity

The probes were tested for their power to discriminate between different commercial egg-laying and broiler lines. These tests have proved largely negative to date and this is probably due to a combination of the variability shown by these probes, and the higher than anticipated level of intra-line genetic variation in the commercial populations. The egg layer tests were carried out against six individuals from each of three different commercial lines, and the highest number of individuals sharing an apparently unique line-specific allele with any of the probes tested was four (in 16% of the line tests). We are currently using further probes. Figure 4 shows a typical result of one of the tests using cGgaMS75 where two of the lines show a line characteristic allele in four of the six individuals. Broiler lines were tested using three individuals from each of seven different commercial crosses and because of the extremely low sample size, fragments were only considered line

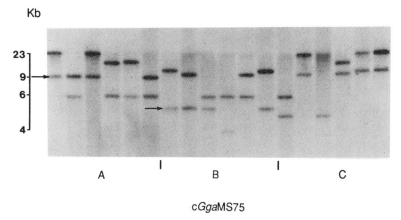

cGgaMS75

Figure 4. Egg-layer line specificity test. Samples of six chickens each from three lines probed with cGgaMS75. Lines A and B have unique alleles (arrowed) present in four individuals.

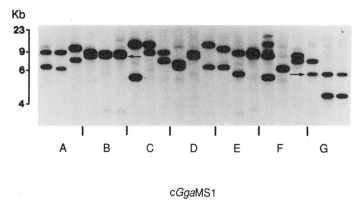

cGgaMS1

Figure 5. Broiler line tests. Three individuals each from seven lines probed with cGgaMS1. Lines B and G have unique sized alleles (arrowed) in all individuals.

characteristic if they were present in all three birds. Using this approach we were only able to assign line specificity in 3.5% of cases as intra- and inter-line variation was even greater than in the egg-laying lines.

Figure 5 illustrates the type of result typically obtained with the broiler lines (here using cGgaMS75). We are now applying the multi-band probes to this problem, as they appear to exhibit a lower level of variation than the locus-specific minisatellite sequences. However, the large effective population sizes characteristic of most commercial lines may still prove an impossible problem to overcome.

Conclusions

Hypervariable DNA analysis is likely to have a major impact on poultry breeding over the coming years. The current applications of multilocus DNA fingerprinting are likely to remain important – particularly in areas such as introgression studies, transgenics and inbreeding estimation where there is a practical advantage in sampling as much of the genome as simultaneously as possible.

We have demonstrated that the large scale isolation of minisatellite locus-specific probes is possible. There are now many different multilocus probes available with which to screen genomic libraries, and these should allow the isolation of many new sets of markers. An important recent related development is the discovery in mammalian genomes of polymorphic simple sequences or "microsatellite" loci (Weber and May, 1989; Litt and Luty, 1989; Tautz, 1989) which consist of very short, polymorphic repeat sequences such as $(CA)_n$ and $(GA)_n$. Microsatellite sequences are short enough for the efficient use of polymerase chain reaction DNA amplification. Once amplified, the variants at these loci can be conveniently resolved by agarose or acrylamide gel electrophoresis (Love et al., 1990). Additionally, these sequences appear to be very common in the genomes of higher vertebrates – they are probably at least an order of magnitude more common than minisatellites. Kashi et al. (1990) have recently demonstrated the presence of poly (TG) sequences in the chicken genome, opening the way for these loci to be exploited. These sequences are beginning to form an important part of genome maps in mice (Love et al., 1990) and cattle (Georges et al., 1990), and it seems likely that they will also make a significant contribution to the chicken mapping effort.

The chicken genome mapping project is now almost a reality, and a preliminary meeting of interested research groups took place at the 4th World Congress on Genetics Applied to Livestock Production in Edinburgh on 26th July 1990. Discussion focussed on the importance of worldwide collaboration on this project, bearing in mind that the expertise in (a) developing markers and (b) producing crosses between birds of diverse phenotype such that phenotypic and genotypic variance is maximized, rarely coexist within research groups. It is important that the mistakes made with other commercially important plants and animals (Roberts, 1990), where a lack of communication between groups has stifled progress, are not repeated in this project. The future for increasing knowledge of the largely empty chicken map is clearly dependent on cooperation, and with it the prospects look very exciting not only for chicken studies, but also for avian biology as a whole.

Acknowledgements
This work was supported by an SERC studentship and a grant from ICI Cellmark Diagnostics. We thank A. J. Jeffreys and ICI Cellmark for providing multilocus probes 33.6 and 33.15. We also thank M. B. Smiley of Institut Sélection Animale, France for providing samples and some financial support. A. J. Jeffreys, J. A. L. Armour and O. Hanotte provided invaluable help and advice during this study. The human minisatellite probes 33.6 and 33.15 are the subject of patent applications. Commercial enquires should be addressed to ICI Cellmark Diagnostics, 8 Blacklands Way, Abingdon, Oxon OX14 1DY, Great Britain.

References

Armour, J. A. L., Povey, S., Jeremiah, S., and Jeffreys, A. J. (1990) Systematic cloning of human minisatellites from ordered array charomid libraries. *Genomics.* (in press).

Birnboim, H. C., and Doly, J. (1979) A rapid alkaline extraction procedure for screening recombinant plasmid DNA. *Nucl. Acids Res.* 7: 1513–1523.

Bruford, M. W., Burke, T., Hanotte, O., and Smiley, M. B. (1990) Hypervariable markers in the chicken genome. *Proc. 4th World Congr. Genet. Appl. Livest. Prod.* 13: 139–142.

Burke, T., and Bruford, M. W. (1987) DNA fingerprinting in birds. *Nature* 327: 149–152.

Burke, T., Davies, N. B., Bruford, M. W., and Hatchwell, B. J. (1989) Parental care and mating behaviour of polyandrous dunnocks *Prunella modularis* related to paternity by DNA fingerprinting. *Nature* 338: 249–251.

Georges, M., Mishra, A., Sargeant, L., Steele, M., and Zhao, X. (1990) Progress towards a primary DNA marker map in cattle. *Proc. 4th World Congr. Genet. Appl. Livest. Prod.* 13: 107–111.

Gilbert, D. A., Lehman, N., O'Brien, S. J., and Wayne, R. K. (1990) Genetic fingerprinting reflects population differentiation in the Calfornia Channel Island fox. *Nature* 344: 764–767.

Hanotte, O., Burke, T., Armour, J. A. L., and Jeffreys, A. J. (1990) Hypervariable minisatellite DNA sequences in the Indian peafowl *Pavo cristatus*. *Genomics* (in press).

Hillel, J., Schaap, T., Haberfeld, A., Jeffreys, A. J., Plotzky, Y., Cahaner, A., and Lavi, U. (1990) DNA fingerprints applied to gene introgression in breeding programs. *Genetics* 124: 783–789.

Jeffreys, A. J., Wilson, V., and Thein, S. L. (1985a) Hypervariable "minisatellite" regions in human DNA. *Nature* 314: 67–73.

Jeffreys, A. J., Wilson, V., and Thein, S. L. (1985b) Individual-specific "fingerprints" of human DNA. *Nature* 316: 76–79.

Jeffreys, A. J., Wilson, V., Thein, S. L., Weatherall, D. J., and Ponder, B. A. J. (1986) DNA "fingerprints" and segregation analysis of multiple markers in human pedigrees. *Amer. J. Hum. Genet.* 39: 11–24.

Jeffreys, A. J., Wilson, V., Kelly, R., Taylor, B. A., and Bulfield, G. (1987) Mouse DNA "fingerprints": analysis of chromosome localization and germ-line stability of hypervariable loci in recombinant inbred strains. *Nucl. Acids Res.* 15: 2823–2836.

Julier, C., de Gouyon, B., Georges, M., Guénet, J.-L., Nakamura, Y., Avner, P., and Lathrop, G. M. (1990) Minisatellite linkage maps in the mouse by cross-hybridization with human probes containing tandem repeats. *Proc. Natl. Acad. Sci. USA* 87: 4584–4589.

Kashi, Y., Tikochinsky, Y., Genislav, E., Iraqi, F., Nave, A., Beckmann, J. S., Gruenbaum, Y., and Soller, M. (1990) Large restriction fragments containing poly-TG are highly polymorphic in a variety of vertebrates. *Nucl. Acids Res.* 18: 1129–1132.

Kuhnlein, U., Dawe, Y., Zadworny, D., and Gavora, J. S. (1989) DNA fingerprinting: a tool for determining genetic distances between strains of poultry. *Theor. Appl. Genet.* 77: 669–672.

Kuhnlein, U., Zadworny, D., Dawe, Y., Fairfull, R. W., and Gavora, J. S. (1990) Assessment of inbreeding by DNA fingerprinting: Development of a calibration curve using defined strains of chickens. *Genetics* 125: 161–165.

Kuhnlein, U., and Zadworny, D. (1990) Molecular aspects of poultry breeding. *Proc. 4th World Congr. Genet. Appl. Livest. Prod.* 16: 21–30.

Litt, M., and Luty, J. A. (1989) A hypervariable microsatellite revealed by *in vitro* amplification of a dinucleotide repeat within the cardiac muscle actin gene. *Am. J. Hum. Genet.* 44: 397–401.

Love, J. M., Knight, A. M., McAleer, M. A., and Todd, J. A. (1990) Towards construction of a high resolution map of the mouse genome using PCR-analysed microsatellites. *Nucl. Acids Res.* 18: 4123–4130.

Paterson, A. H., Lander, E. S., Hewitt, J. D., Peterson, S., Lincoln, S. E., and Tanksley, S. D. (1988) Resolution of quantitative traits into Mendelian factors by using a complete linkage map of restriction fragment length polymorphisms. *Nature* 335: 721–726.

Petitte, J. N., Clark, M. E., Lui, G., Verrinder Gibbins, A. M., and Etches, R. J. (1990) Production of somatic and germ-line chimeras in the chicken by transfer of early blastoderm cells. *Development* 108: 185–189.

Plotzky, Y., Cahaner, A., Haberfeld, A., Lavi, U., and Hillel, J. (1990) Analysis of association between DNA fingerprint bands and quantitative traits using DNA mixes. *Proc. 4th World Congr. Genet. Appl. Livest. Prod.* 13: 133–136.

Raleigh, E. A., Murray, N. E., Revel, H., Blumenthal, R. M., Westanay, D., Reith, A. D., Rigby, P. W. J., Elhai, J., and Hanahan, D. (1988) *McrA* and *mcrB* restriction phenotypes of some *E. coli* strains and implications for gene cloning. *Nucl. Acids Res.* 16: 1563–1575.

Roberts, L. (1990) An animal genome project? *Science* 248: 550–552.

Saito, I., and Stark, G. R. (1986) Charomids: cosmid vectors for efficient cloning and mapping of large or small restriction fragments. *Proc. Natl. Acad. Sci. USA* 83: 8664–8668.

Tanksley, S. D., Medina-Filho, H., and Rick, C. M. (1982) Use of naturally occurring enzyme variation to detect and map genes controlling quantitative traits in an interspecific backcross of tomato. *Heredity* 49: 11–25.

Tautz, D. (1989) Hypervariable of simple sequences as a general source for polymorphic DNA markers. *Nucl. Acids Res.* 17: 6463–6471.

Weber, J. L., and May, P. E. (1989) Abundant class of human DNA polymorphism which can be typed using the polymerase chain reaction. *Am. J. Hum. Genet.* 44: 388–396.

Wong, Z., Wilson, V., Jeffreys, A. J. and Thein, S. L. (1986) Cloning a selected fragment from a human DNA "fingerprint": isolation of an extremely polymorphic minisatellite. *Nucl. Acids Res.* 14: 4605–4616.

DNA Fingerprinting: Approaches and Applications
ed. by T. Burke, G. Dolf, A. J. Jeffreys & R. Wolff
© 1991 Birkhäuser Verlag Basel/Switzerland

Cloning of Hypervariable Minisatellite and Simple Sequence Microsatellite Repeats for DNA Fingerprinting of Important Aquacultural Species of Salmonids and Tilapia

P. Bentzen, A. S. Harris, and J. M. Wright*

Marine Gene Probe Laboratory and Department of Biology, Dalhousie University, Halifax, Nova Scotia, Canada B3H 4J1

Summary
DNA fingerprints reveal a class of highly polymorphic variable number tandem repeat (VNTR) loci that offer considerable potential as genetic markers in aquaculture and fisheries research. We have obtained DNA fingerprints from three commercially important fish species, Atlantic salmon (*Salmo salar*), rainbow trout (*Oncorhynchus mykiss*) and tilapia (*Oreochromis niloticus*) using the Jeffreys polycore probes, 33.15 and 33.6. The complexity of multilocus fingerprints, however, limits their utility in many contexts. To circumvent this problem we have cloned a number of highly polymorphic VNTRs for use as genetic markers: minisatellites from salmon and tilapia, and microsatellites from rainbow trout.

Both mini- and microsatellite loci appear to be abundant in fish genomes. Discrete classes of minisatellites, distinguishable by their hybridization to the 33.15 and 33.6 probes, occur in both salmonids and tilapia. Minisatellite arrays isolated from genomic libraries show a tendency, particularly in salmon, to occur in clusters. Cross-hybridization data suggest that polymorphic mini- and microsatellite loci are conserved in related salmonids.

We expect VNTR markers to find applications in quantititative genetic analyses, assessment of inbreeding, and identification of particular strains in aquaculture. In fisheries research VNTRs will offer logistic advantages over conventional genetic markers and may facilitate high resolution analyses of genotype/environment interactions.

Introduction

In recent years it has become apparent that important applications exist in disparate areas of aquaculture and fisheries research for genetic markers that are more polymorphic than those revealed by conventional techniques such as protein electrophoresis (Hallerman and Beckmann, 1988; Fields *et al.*, 1989; Harris *et al.*, 1991). Highly polymorphic genetic markers could be used to identify particular strains as well as aid breeding programs and quantitative genetic analyses in aquaculture. In fisheries research such markers would offer improved prospects for studies of population structure and migration, and could facilitate analyses of genotype/environment interactions that may be implicated

---

*Correspondence to J. M. Wright

in the extreme variations in survival of early life history stages observed in many fishes.

The advent of DNA fingerprinting (Jeffreys *et al.*, 1985a, b) has revealed the existence of an extensive class of genetic loci that are sufficiently polymorphic to serve as markers for these applications. DNA fingerprints highlight loci containing arrays of tandemly repeated short DNA sequences in which differences between alleles are generated by variation in the number of repeating units. Such loci are known as variable number tandem repeats (VNTRs) (Nakamura *et al.*, 1987). Highly polymorphic VNTRs appear to be ubiquitously distributed in eukaryotic genomes (Ryskov *et al.*, 1988; Tautz and Renz, 1984; Tautz, 1989).

A major class of VNTR loci comprises the minisatellites, in which the tandem arrays are composed of oligonucleotide repeats that are typically 9–65 base pairs (bp) in length and frequently GC-rich (Jarman and Wells, 1989). Variation in the number of tandem repeats can be extreme, with alleles at some loci differing in size by as much as 25 kilobase pairs (kbp) (Wong *et al.*, 1987). Some minisatellite loci comprise the most variable DNA sequences known, with heterozygosities approaching 100%, and mutation rates in excess of 2% per generation (Jeffreys *et al.*, 1988a; Gyllensten *et al.*, 1989; Kelly *et al.*, 1989).

Little is known about the genomic distribution of most minisatellites, and the data that are available suggest much as yet unresolved complexity. Segregation analyses of minisatellites visualized in multilocus fingerprints have revealed low levels of linkage in several mammalian and avian species, but not in inbred strains of mice, where levels of cosegregation of fingerprint bands are high (Jeffreys *et al.*, 1986; 1987; Jeffreys and Morton, 1987; Burke and Bruford, 1987; Burke *et al.*, 1989; Gyllensten *et al.*, 1989). In contrast to the low incidence of linkage observed in most fingerprints, sequence analysis of genomic clones has indicated that minisatellite arrays frequently occur in close association with each other as well as with other repetitive elements (Royle *et al.*, 1988; Armour *et al.*, 1989; Gyllensten *et al.*, 1989; Kashi *et al.*, 1990; Rogaev, 1990). Segregation, linkage, and *in situ* hybridization data for minisatellite loci revealed by locus-specific probes have shown them to be concentrated in pro-terminal regions in human DNA, but not in mouse DNA, where the loci studied have proven to be dispersed interstitially on chromosomes (Jeffreys *et al.*, 1987; Royle *et al.*, 1988). The genomic distribution and linkage relationships of minisatellite loci are important considerations for their use as genetic markers. Dispersed markers that segregate independently offer the maximum amount of information for studies of genome organization and the estimation of relatedness among individuals.

Microsatellites, also known as simple sequence repeats, comprise another class of loci that exhibit variation in tandem repeat number. In microsatellites the tandem arrays are formed from short nucleotide motifs of mono-, di-, tri- and tetranucleotide repeats such as $(AT)_n$ or $(GT)_n$ that frequently extend over less than 300 bp but may be much larger (Tautz, 1989). Although some authors apply the term "VNTR" exclusively to minisatellites (e.g., Love et al., 1990) we follow Litt and Luty (1989) in extending the use of the term to include microsatellites, since these loci satisfy the definition of "VNTR" created by Nakamura et al. (1987).

Some microsatellites exhibit a less coherent form referred to as cryptic simplicity, in which the repetitive motifs occur in scrambled arrangements in regions of coding and non-coding DNA showing biased nucleotide composition (Tautz et al., 1986). Microsatellites are dispersed at approximately 10 kbp intervals throughout the genome, embedded in unique DNA (Tautz, 1989). Microsatellite loci in the 100–300 bp size range, cloned from human, whale, Drosophila and mouse DNA, have been amplified by the polymerase chain reaction (PCR) using primers specific to unique flanking domains (Tautz, 1989; Litt and Luty, 1989; Weber and May, 1989; Love et al., 1990). These loci show high levels of polymorphism.

The technique of DNA fingerprinting allows numerous highly polymorphic VNTR loci to be detected simultaneously. Southern blots of genomic DNAs digested by restriction endonucleases such as HaeIII or HinfI are hybridized under conditions of low stringency to probes containing versions of the repeating units. For minisatellites, a variety of "core" sequences have been shown to reveal discrete classes of loci sharing similar repetitive motifs in their minisatellite arrays (Jeffreys et al., 1985a, b; Vassart et al., 1987; Georges et al., 1988; Fowler et al., 1988; Vergnaud, 1989). Probes based on various tri- and tetranucleotide simple repeats also reveal numerous polymorphic loci that can be resolved on Southern blots (Ali et al., 1986; Schafer et al., 1988). In each case, the result is a highly complex profile of variable and nonvariable bands, referred to as a DNA "fingerprint".

The power of the DNA fingerprinting technique as a means for revealing genetic variation has led to its widespread use in many organisms, including fishes (Georges et al., 1988; Lloyd et al., 1989; Fields et al., 1989; Harris et al., 1991; Turner et al., 1990; Nanda et al., 1990). However, the large number of loci detected in conventional DNA fingerprints severely limits their utility in many contexts where information is desired on the allelic states of particular loci. The complexity of the band profiles obtained means that in most cases allelic pairs of bands specific to individual loci cannot be identified. In addition, information on the state of some polymorphic loci is likely to be lost due to chance comigration of bands representing unrelated loci.

The inherent difficulties in the interpretation of multilocus fingerprints can be avoided by examining VNTR loci individually. This approach requires the cloning and characterization of specific VNTR markers. Two possibilities then exist: either the entire cloned VNTR or one or both domains of unique DNA that flank the VNTR array can be used to probe Southern blots under conditions of high stringency to reveal a single locus (Wong *et al.*, 1987; Armour *et al.*, 1990; Hanotte *et al.*, 1991). Alternatively, primers specific to the flanking domains of the VNTR array can be used to amplify the VNTR locus from genomic DNA samples by the polymerase chain reaction (PCR) (Jeffreys *et al.*, 1988b, 1990). The resulting allelic fragments can then be resolved by electrophoresis.

In an effort to bring VNTR markers to bear on important areas of aquaculture and fisheries research, we have cloned mini- and microsatellite loci from a number of commercially important fishes and aquatic invertebrates in our laboratory. Here we report on our efforts with three species of fish, the salmonids Atlantic salmon (*Salmo salar*) and rainbow trout (*Oncorhynchus mykiss*), and a species of tilapia (*Oreochromis niloticus*). Both Atlantic salmon and rainbow trout support important aquacultural, sport and commercial fishing industries. Tilapia are important in aquaculture in warm-climate areas of the world, and serve as a quick-breeding, logistically convenient model for quantitative genetic studies of fishes.

In addition to these practical concerns, both groups of fishes are of particular evolutionary interest. Tilapia, and the family Cichlidae to which they belong, have undergone extensive radiation (Kornfield, 1984). The salmonid fishes are ancient tetraploids still in the process of rediploidization (Allendorf and Thorgaard, 1984). Consequently, both groups of fishes offer excellent opportunities for study of the genomic organization and evolutionary dynamics of mini- and microsatellite loci in this important group of vertebrates.

Results

Multilocus Fingerprints

We have produced DNA fingerprints of tilapia and rainbow trout (Figs. 1 and 2) as well as Atlantic salmon (not shown) using two probes, 33.15 and 33.6, that contain tandemly repeated core motifs from human minisatellite arrays (Jeffreys *et al.*, 1985a, b). In each of the three species the two polycore probes yield band profiles that are distinct from each other, but similarly complex and polymorphic. In our hands, salmonid DNA consistently produces fingerprints with more background hybridization within DNA lanes than tilapia DNA, despite the use of identical methodology for the two types of fish. This background hybridization may stem from dispersed repetitive elements in salmonid

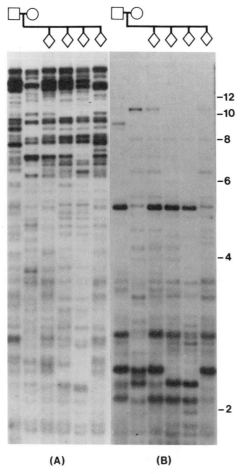

Figure 1. DNA fingerprints of tilapia. Southern blot of HinfI-digested genomic DNAs of two parents and four offspring. Replicate filters probed either with 33.6 (A) or 33.15 (B) polycore probes. Hybridization conditions: 16 hours at 58°C in 1% BSA, 7% SDS, 0.263 M sodium phosphate, pH 7.4, 1 mM EDTA (Westneat *et al.*, 1988). Wash conditions: 10 min at room temperature, then 20 min at 58°C, in 2 × SSC, 0.1% SDS (1 × SSC = 150 mM NaCl, 15 mM Na$_3$ Citrate).

DNA that have previously been postulated as an explanation for the observation that salmon DNA (when used as a carrier in Southern analyses) blocks the hybridization of minisatellite probes (Vassart *et al.*, 1987; Chimini *et al.*, 1989). Whatever its cause, the extensive background hybridization in salmonid fingerprints makes them less clear, and hence less informative, than tilapia fingerprints. This practical problem associated with multilocus DNA fingerprints derived from salmonids as well some other groups of fishes (unpubl. data) constitutes another reason to seek locus-specific probes in studies of VNTR polymorphism in fishes.

248

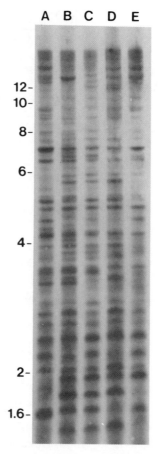

Figure 2. DNA fingerprints of rainbow trout. Southern blot of HaeIII-digested genomic DNAs of five full-sibs probed with 33.15 polycore probe. Hybridization and wash conditions as in Fig. 1.

Cloning of Minisatellites

We have cloned minisatellites from salmon and tilapia using a strategy similar to that employed for humans (Nakamura *et al.*, 1987; Wong *et al.*, 1987), birds (Gyllensten *et al.*, 1989) and cattle (Kashi *et al.*, 1990). DNA from one individual of each species was partially digested with MboI and cloned into the EMBL3 λ vector (Karn *et al.*, 1980) to create genomic libraries. Aliquots of each library were plated out, the resulting plaques blotted onto nylon membranes (Benton and Davis, 1977) and then hybridized to a 1:1 mixture of radioactively labelled 33.15 and 33.6 probes using hybridization conditions as specified (Fig. 1). About 1.6% of the plaques from the salmon library and 1% of those from the tilapia library hybridized to the probe mixture. Assuming a mean insert size of

15 kbp within each library, these results suggest that minisatellite arrays related in sequence to the 33.6 and 33.15 probes occur once every 1000 and 1500 kbp in salmon and tilapia DNA, respectively.

We selected 30 plaques from each library that hybridized strongly to the probe mixture on the assumption that these were likely to contain relatively long minisatellite arrays. There is evidence that the longest minisatellite arrays are also the most prone to mutation, and hence the most polymorphic (Jeffreys *et al.*, 1985a; Jarman *et al.*, 1986). We purified DNA from these clones, digested them with SalI (to release the

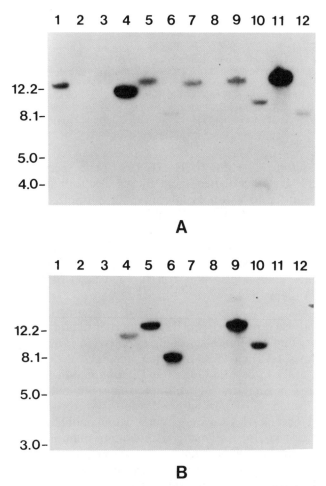

A

B

Figure 3. Differential hybridization of minisatellite clones to 33.15 and 33.6 polycore probes. Southern blot of salmon clones digested with SalI to release genomic inserts. A, probed with 33.15. B, probed with 33.6. Most clones hybridized preferentially to 33.15 (e.g. lanes 1,7,11), some to 33.6 (e.g. lane 6) and a minority with similar intensities to both probes (e.g. lanes 5,9). Note presence of two restriction fragments that hybridized preferentially to different probes in clone represented in lane 10. Hybridization and wash conditions as in Fig. 1.

250

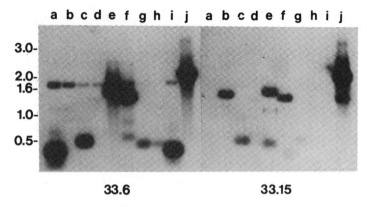

a b c d e f g h i j   a b c d e f g h i j

3.0-

2.0-
1.6-

1.0-

0.5-

33.6   **33.15**

Figure 4. Evidence that minisatellite arrays occur in clusters. Southern blot of tilapia clones digested with HinfI to isolate minisatellite arrays probed sequentially with 33.6 and 33.15 probes. The 1.7 kbp band evident in most clones probed with 33.6 derives from the EMBL vector. Most clones produced restriction fragments which hybridized preferentially to 33.6 (e.g. lanes a,g,h) or 33.15 (e.g. lane b) but some fragments hybridized strongly to both probes (e.g. lane c). Clones represented in lanes e,f,j, produced at least one fragment that hybridized to 33.6, and another that hybridized to 33.15. Hybridization and wash conditions as in Fig. 1.

genomic inserts) and HinfI or HaeIII (to isolate specific minisatellite arrays), and size fractionated them by gel electrophoresis. Following Southern transfer to nylon membranes, we hybridized the clones sequentially to the 33.15 and 33.6 probes. All but five salmon and three tilapia clones screened in this manner hybridized preferentially to one or the other polycore probe (Figs. 3 and 4). This result, along with the multilocus fingerprints (Figs. 1 and 2), indicates that the 33.15 and 33.6 probes identify two largely non-overlapping classes of minisatellites in salmon and tilapia.

When digested by HinfI and subjected to Southern analysis, three tilapia clones each revealed two fragments that hybridized to the polycore probes (Fig. 4). Similar screening of SalI- and HaeIII-digests of the salmon clones revealed 28 clones that produced 2–5 fragments that hybridized to 33.15 or 33.6 (Fig. 3 and data not shown). These results suggest three possibilities: (1) multiple minisatellite-bearing fragment may have ligated and jointly inserted into the vector; (2) one or more HinfI, SalI or HaeIII sites may occur within individual minisatellite arrays; (3) cloned DNA segments may contain multiple distinct arrays with restriction sites in the intervening non-repetitive DNA sequences. We favour the third possibility over the first two, because the large number of salmon clones involved and the presence in some clones of as many as five fragments that hybridized to the polycore probes make the first possibility unlikely, and because multiple minisatellite-bearing fragments were also observed when some of these clones were digested with other enzymes and screened with the polycore probes (data not shown), thus making the second possibility unlikely. In some cases, these closely linked arrays hybridized to different polycore probes. In five salmon and three tilapia clones at least one restriction fragment hybridized preferentially to the 33.15 and another to the 33.6 probe (Figs. 3 and 4). This result indicates that arrays containing different repetitive motifs occur in close proximity in the DNAs of salmon and tilapia.

We assayed the clones for their ability to detect polymorphism at specific minisatellite loci. Restriction fragments of insert DNA that hybridized to either 33.15 or 33.6 were purified from agarose gels by excising the appropriate bands from the gel and recovering the DNA using the glass milk procedure (Geneclean, BIO 101 Inc.; Vogelstein and Gillespie, 1979). The excised DNA fragments were labelled radioactively by the random priming method (Feinberg and Vogelstein, 1983, 1984) and hybridized to Southern blots bearing HinfI- or HaeIII-digested genomic DNA of several unrelated individuals. Six of the salmon clones and all of the tilapia clones tested in this manner yielded polymorphic band profiles of varying complexity.

The complexity of the band profiles obtained with clones from both species varied from one or two bands per lane (corresponding to the homo- or heterozygous condition, respectively, for a single locus) to as many as 24 variable bands (corresponding to numerous polymorphic loci). This variation between single and multilocus specificity was in part a function of the stringency of the hybridizations. For example, for salmon clone Ssa1, an increase in the stringency of hybridization (produced by a 5% increase in formamide concentration in the hybridization buffer) resulted in a shift from a complex multilocus fingerprint to a greatly simplified one in which a single locus predominates (Fig. 5). In other instances the same level of hybridization stringency revealed a single polymorphic locus with some clones, and complex multilocus fingerprints with others (Figs. 6 and 7). Variation in the specificity of clones under comparable conditions of stringency may reflect variation in the relative sizes of the minisatellite arrays and the unique flanking domains present in the clones. Clones bearing minisatellite alleles in which the amount of unique flanking DNA is high in proportion to the size of the repetitive array might tend to be more locus-specific as probes than clones in which the proportion of flanking DNA is low.

Preliminary screening with clones that acted as locus-specific probes revealed high levels of polymorphism. In salmon, three minisatellite loci each exhibited three alleles among four unrelated individuals when screened with locus-specific probes. A fourth salmon minisatellite locus exhibited four alleles in six unrelated individuals. In tilapia, two alleles of a minisatellite locus were detected in four individuals from the same aquacultural strain (data not shown).

For at least one salmon clone, Ssa1, hybridization at high stringency revealed two loci, a locus evident after short exposures of the autoradio-

252

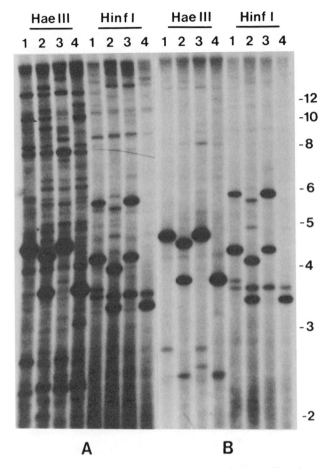

Figure 5. Effect of hybridization stringency on specificity of minisatellite clones as probes. Southern blot of HaeIII- and HinfI-digested genomic DNAs of four unrelated Atlantic salmon probed with 6 kbp SalI fragment of clone Ssa1. Hybridization conditions: A, 50% formamide; B, 55% formamide; both filters, 16 hours at 42°C in 5 × SSC. Wash conditions: 0.2 × SSC, 0.1% SDS at 65°C. "Strongly hybridizing" and "weakly hybridizing" loci revealed by clone Ssa1 (see text) are evident in HaeIII digests in the size ranges 3.5–5.0 and 2.0–3.0 kbp, respectively. Additional bands associated with the strongly hybridizing locus in HinfI digests (e.g. intense 6 kbp bands in individuals 1 and 3) likely result from partial methylation of a HinfI site flanking the primary locus minisatellite array.

graph, and an additional locus evident only after longer exposures (Fig. 5). The second locus could be an unrelated minisatellite exhibiting sequence similarity in its repetitive motif to that of the strongly hybridizing locus, or it could be a diverged paralogous relative of the first locus generated by the tetraploidization of the salmonid genome.

Analysis of salmon and tilapia families demonstrated that minisatellite loci revealed by clones from both species exhibit patterns of inheritance consistent with Mendelian expectations. This is most clear in

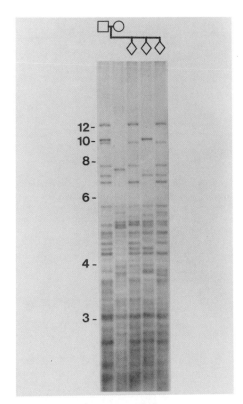

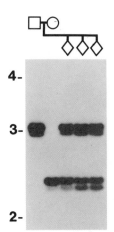

Figure 6. Mendelian inheritance of a minisatellite locus in tilapia. Southern blot of HinfI-digested DNA from blood of two parents and three offspring probed with tilapia clone Oni1. Parents are homozygous for different alleles; offspring are heterozygous. The lighter bands that appear below the intense bands in each lane are probably due to partial methylation of a HinfI site flanking the minisatellite array in each individual. HaeIII digests of the same DNA samples probed with Oni1 produced only a single band for each allele (data not shown). Hybridization conditions: 16 hours at 65°C in 55% formamide, 5 × SSC. Wash conditions: 65°C, 0.2 × SSC, 0.1% SDS.

Figure 7. Multilocus DNA fingerprints of tilapia produced using a tilapia clone as probe. Southern blot of HinfI-digested genomic DNAs of two parents and three offspring probed with clone Oni2. The multilocus band profiles in this figure appear to be a subset of those produced at lower stringency with the 33.6 probe. Hybridization and wash conditions as in Fig. 6.

single-locus autoradiographs (Figs. 6 and 8), but is also evident with multilocus fingerprints (Fig. 7). In all cases polymorphic bands segregate codominantly in approximately Mendelian ratios.

We tested one of our Atlantic salmon clones, Ssa1, as a probe on a Southern blot bearing DNAs of three species of Pacific salmon from the related genus, *Oncorhynchus*. Clone Ssa1 hybridized to the DNA of all three species under conditions of high stringency, revealing highly polymorphic band profiles for each species (Fig. 9). In a similar

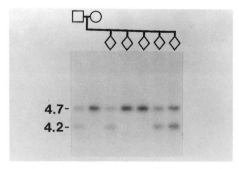

Figure 8. Mendelian inheritance of a minisatellite locus in salmon. Southern blot of HaeIII-digested genomic DNA of two parents and five offspring probed with clone Ssa1. Locus shown is the primary locus revealed by this probe (Fig. 5). Male parent and three offspring are heterozygous; female parent and remaining progeny are homozygous. Hybridization and wash conditions as in Fig. 5B.

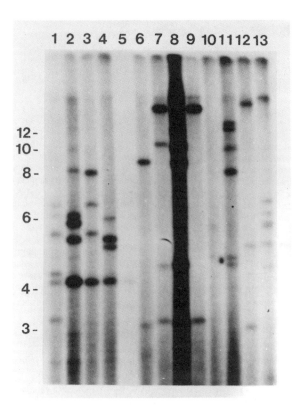

Figure 9. Hybridization of clone Ssa1 to DNA from related salmonids reveals polymorphic minisatellite loci. Southern blot of HaeIII-digested genomic DNAs probed with Ssa1. Lanes 1–5, chum salmon (*Oncorhynchus keta*); 6–9, chinook salmon (*O. tshawytscha*); 10–13, coho salmo (*O. kisutch*). Hybridization and wash conditions as in Fig. 5A.

experiment, tilapia clone Oni1 hybridized to polymorphic sequences in the DNAs of related species of *Oreochromis*, *Tilapia* and *Sarotherodon* (data not shown). These results suggest that specific polymorphic minisatellite loci are conserved among related genera of salmonids and tilapia.

Comparisons of HinfI- and HaeIII-generated band profiles revealed anomalies attributable to the HinfI enzyme. In single-locus profiles of salmon (Fig. 5) and tilapia (Fig. 6) digestion by HinfI consistently produced two bands per allele where HaeIII produced only one. In each instance, one or both bands produced by HinfI were clearly less intense than the corresponding HaeIII band. These apparently substoichiometric HinfI bands appeared despite the use of excess enzyme, and were generated reproducibly in repeat digests of the same samples. Taken in conjunction with the reported methylation sensitivity of HinfI (Kessler and Holtke, 1986), these results suggest that the additional HinfI bands are the product of partial methylation of a HinfI site flanking a minisatellite array. As a consequence, we no longer use HinfI to digest salmon or tilapia DNA, and recommend against its use on other fishes as well. It is important to note that the true nature of additional methylation-generated HinfI bands might go unrecognized in multilocus fingerprints.

Cloning of Microsatellites

We have cloned a number of microsatellites from rainbow trout using the approach of Tautz (1989). Among the microsatellites we have isolated from trout are simple sequence repeats consisting of $(AT)_n$ and $(GA)_n$. The most interesting, however, is a tri-nucleotide purine, purine, pyrimidine (PuPuPy) repeat, composed predominately of GAC and GAT nucleotide triplets. Hybridization of the PuPuPy sequence to a Southern blot bearing restriction endonuclease digests of genomic DNA from closely related salmonids indicates that the PuPuPy sequence is dispersed in the genomes of these fishes (Fig. 10).

Amplification of rainbow trout genomic DNA using PCR and primers complementary to unique (non-repeat) sequences flanking the PuPuPy repeat produced at least four fragments resolved by agarose gel electrophoresis per individual (Fig. 11). It is unlikely that the multiple fragments amplified are due to artifacts of the PCR process as only a single fragment was amplified from the cloned PuPuPy sequence. It appears, therefore, that there are at least two loci with closely similar flanking domains in the rainbow trout genome. Such loci might stem from the ancestral tetraploidization of the salmonid genome.

Variation in the sizes of the amplified fragments indicates that the PuPuPy loci are highly polymorphic. Since the band profiles vary from

256

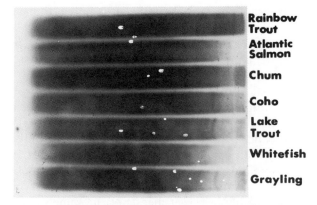

Figure 10. A dispersed microsatellite in the genomes of salmonid fishes. Southern blot of AluI-digested genomic DNAs of salmonids probed with the rainbow trout PuPuPy sequence. Hybridization conditions: 16 hours at 42°C in 5 × SSPE, 42% formamide, 1% SDS, 1 × Denhardt's, 100 μg/ml of tRNA. Wash conditions: 0.5 × SSC, 0.1% SDS at 50°C. Effective separation range of the gel was 0.2–23 kbp.

individual to individual, they represent simple DNA fingerprints. The degree of band-sharing is high among closely related individuals, suggesting that the bands are inherited allelically in Mendelian fashion.

We tested the primers derived from the rainbow trout PuPuPy flanking domains on several other salmonid species. In each case a number of fragments of varying size were produced by PCR amplification (Fig. 11). This suggests that the PuPuPy loci are conserved and remain polymorphic in other salmonid species.

Discussion

We have cloned several highly polymorphic mini- and microsatellite loci from the salmonids, Atlantic salmon and rainbow trout, and the tilapia, *O. niloticus*. Our studies provide preliminary evidence that the genomic organization of these loci in fishes is similar to that observed in previously studied vertebrates. Both mini- and microsatellite loci appear abundant in fish genomes. Minisatellites occur in discrete classes distinguishable by their hybridization to the 33.15 and 33.6 polycore probes and show a tendency, particularly in salmon, to occur in clusters. Our results suggest that some mini- and microsatellite loci in salmonids occur in multiple versions that may stem from the tetraploidization of the salmonid genome. Further studies, including subcloning and sequencing of the polymorphic VNTR loci, and detailed analyses of their structure, distribution and evolutionary dynamics, are in progress.

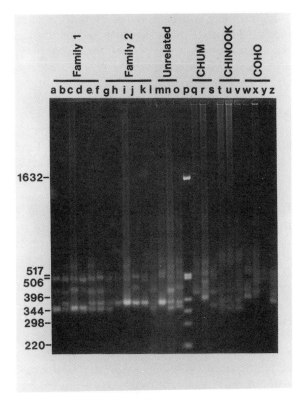

Figure 11. Simple DNA fingerprints generated by PCR amplification of the PuPuPy microsatellite in Pacific salmonids. Primers complementary to the unique flanking sequences of the cloned microsatellite from rainbow trout were used to amplify by PCR the microsatellite from 40 ng of genomic DNA from rainbow trout (two full-sib families, lanes a–l, and unrelated individuals, lanes m–o), chum salmon (unrelated individuals, lanes q–s), chinook salmon (unrelated individuals, lanes t–v) and coho salmon (unrelated individuals in lanes w–y). Lane z shows the PCR product from the cloned microsatellite DNA. Lane p contains molecular weight markers (indicated at left in base pairs). PCR conditions: denaturation of genomic DNA at 95°C for 1 min; primer annealing at 50°C for 2 min; and polymerase extension at 72°C for 1 min. PCR products were analyzed by 2.4% agarose gel electrophoresis.

Application of VNTR Markers to Aquaculture and Fisheries

We anticipate a number of applications of VNTR markers to aquaculture and fisheries (Hallerman and Beckmann, 1988; Fields *et al.*, 1989; Harris *et al.*, 1991). For instance, analyses of VNTR polymorphism can be used to simultaneously simplify and improve the execution of genetic studies in aquaculture. In conventional designs, the progeny of different crosses must be segregated in separate rearing containers because of the impracticality of physically tagging newly emerged fish fry. Such segregation confounds the genetic effects under study with potentially significant environmental influences on phenotype (Uraiwan and Doyle, 1986;

Tave, 1986). This problem can be avoided by rearing families together and then using VNTR markers to identify the progeny of particular crosses. Unlike conventional electrophoretic markers previously advocated for this purpose (Moav *et al.*, 1976), VNTR loci are likely to be variable enough to allow the familial relationships of progeny to be accurately determined without need for preselection of parental stock for suitable genetic markers.

VNTR markers, analyzed separately with locus-specific probes or jointly in multilocus fingerprints, can be used to estimate levels of inbreeding in aquacultural broodstock (Kuhnlein *et al.*, 1990). VNTR markers may also provide the means to identify particular strains of aquacultural stock, both to protect proprietary interests and to monitor the effects of escaped or deliberately released stocks on wild populations. The latter issue is a matter of particular concern in the case of salmonid fishes, for which aquacultural production of exotic or domesticated strains poses the threat of genetic contamination of nearby wild stocks (e.g., Taggart and Ferguson, 1986; Anon, 1989).

In fisheries research analysis of VNTR polymorphism offers a potentially powerful alternative to established methods of genetic analysis, such as protein electrophoresis and restriction endonuclease analysis of mitochondrial DNA (mtDNA) (Ryman and Utter, 1987) in studies of the structure, migratory habits and degree of mixing of populations. For some species, including Atlantic salmon, the utility of conventional protein and mtDNA markers in these contexts is limited by low levels of genetic variation (e.g., Davidson *et al.*, 1989; Waldman *et al.*, 1988; Smith *et al.*, 1989; Graves *et al.*, 1984). This limitation is unlikely to apply to highly polymorphic VNTR markers.

Analysis of VNTR polymorphism also presents important logistic advantages over previous methods of genetic analysis. Conventional isozyme and mtDNA analyses require fresh or carefully frozen tissue, a difficult requirement when samples come from remote locations, as is often the case in fisheries and studies of artisanal aquacultural operations. In contrast, the small amounts of fish blood or other tissue required for Southern analysis of VNTR polymorphism can readily be stored and transported without need for refrigeration (unpubl. data). In addition, unlike tissue requirements for isozyme techniques, extraction of blood samples for VNTR analysis need not involve the sacrifice of the individuals being studied, a critical advantage in the case of breeding stock.

Analysis of VNTR polymorphism via PCR amplification of mini- and microsatellite loci offers additional logistic advantages. The relatively minute amount of tissue required further minimizes the invasiveness of studies on living individuals, and extends the potential scope of fisheries studies to early life history stages such as pelagic larvae and eggs. The latter application would not only permit the inclusion of early life history stages of fishes in conventional population genetic analyses, but would also make possible a new and potentially important line of experimental research. Many commercially important fishes exhibit tremendous variation in survival during their early life history stages, and genotype/environment interactions may be important in this variation (e.g., Mork and Sundnes, 1985a, b). The ability to use PCR amplification of VNTR markers to infer the parentage of larvae would make possible, for the first time, experiments in which genetic influences on survivorship are assessed by rearing fish larvae from several crosses together in experimental mesocosms under varying conditions.

Finally, PCR may permit the extension of population genetic studies into the past. PCR amplification of VNTR loci from dried or alcohol preserved specimens could be used to compare historical allele frequences with those of the present in a manner similar to that employed in recent PCR-facilitated studies of mtDNA from museum specimens (e.g. Thomas *et al.*, 1990). This approach would allow retrospective analysis of impacts on fish populations of events such as fish stocking programs and habitat alterations.

Acknowledgements
Technical assistance was provided by D. Denti and K. Prinowski. Pacific salmon samples were generously provided by R. Withler. This work was supported by a Natural Sciences and Engineering Research Council of Canada (NSERC) grant to J.M.W. and funding from the Nova Scotia government Department of Industry, Trade and Technology to the Marine Gene Laboratory. A.S.H. and P.B. were supported by NSERC postgraduate and postdoctoral fellowships, respectively.

References

Ali, S., Muller C. R., and Epplen, J. T. (1986) DNA fingerprinting by oligonucleotide probes specific for simple repeats. *Hum. Genet.* 74: 239–243.

Allendorf, F. W., and Thorgaard, G. H. (1984) Tetraploidy and the evolution of salmonid fishes. In: Turner, B. J. (ed.) Evolutionary Genetics of Fishes, Plenum, New York, pp. 1–54.

Anon. (1989) Report of Dublin meeting on genetic threats to wild stocks from salmon aquaculture. North Atlantic Salmon Conservation Organization (NASCO) paper CNL(89)19. Edinburgh, Scotland, 33 pp.

Armour, J. A. L., Wong, Z., Wilson, V., Royle, N. J., and Jeffreys, A. J. (1989) Sequences flanking the repeat arrays of human minisatellites: association with tandem and dispersed repeat elements. *Nucleic Acids Res.* 17: 4925–4935.

Armour, J. A. L., Povey, S., Jeremiah, S., and Jeffreys, A. J. (1990) Systematic cloning of human minisatellites from ordered array charomid libraries. *Genomics* 8: 501–512.

Benton, W. D., and Davis, R. W. (1977) Screening λ gt recombinant clones by hybridization to single plaques *in situ. Science* 196: 180.

Burke, T., and Bruford, M. W. (1987) DNA fingerprinting in birds. *Nature* 327: 149–152.

Burke, T. Davies, N. B., Bruford M. W., and Hatchwell, B. J. (1989) Parental care and mating behaviour of polyandrous dunnocks *Prunella modularis* related to paternity by DNA fingerprinting. *Nature* 338: 249–251.

Chimini, G., Mattei, M.-G., Passage, E., Nguyen, C., Boretto, J., Mattei, J.-F., Jordan, B. R. (1989) *In situ* hybridization and pulsed-field gel analysis define two major minisatellite loci: 1q23 for minisatellite 33.6 and 7q35-q36 for minisatellite 33.15. *Genomics* 5: 316–324.

260

Davidson, W. S., Birt, T. P., and Green, J. M. (1989) A review of genetic variation in Atlantic salmon, *Salmo salar* L., and its importance for stock identification, enhancement programmes and aquaculture. *J. Fish. Biol.* 34: 547–560.

Feinberg, A. P., and Vogelstein, B. (1983) A technique for radiolabelling DNA restriction endonuclease fragments to high specific activity. *Anal. Biochem.* 132: 6–13.

Feinberg, A. P., and Vogelstein, B. (1984) Addendum. *Anal. Biochem.* 137: 266–267.

Fields, R. W., Johnson, K. R., and Thorgaard, G. H. (1989) DNA fingerprints in rainbow trout detected by hybridization with DNA of bacteriophage M13. *Trans. Am. Fish. Soc.* 118: 78–81.

Fowler, S. J., Gill, P., Werrett, D. J., and Higgs, D. R. (1988) Individual specific DNA fingerprints from a hypervariable region probe: alpha-globin 3'HVR. *Hum. Genet.* 79: 142–146.

Georges, M., Lequarre, A.-S., Castelli, M., Hanset, R., and Vassart, G. (1988) DNA fingerprinting in domestic animals using four different minisatellite probes. *Cytogenet. Cell. Genet.* 47: 127–131.

Graves, J. E., Ferris, S. D., and Dizon, A. E. (1984) Close genetic similarity of Atlantic and Pacific skipjack tuna (*Katsuonis pelamis*) demonstrated with restriction endonuclease analysis of mitochondrial DNA. *Mar. Biol.* 79: 315–319.

Gyllensten, U. B., Jakobsson, S., Temrin, H., and Wilson, A.C. (1989) Nucleotide sequence and genomic organization of bird minisatellites. *Nucleic Acids Res.* 17: 2203–2215.

Hallerman, E. M., and Beckmann, J. S. (1988) DNA-level polymorphism as a tool in fisheries science. *Can. J. Fish. Aquat. Sci.* 45: 1075–1087.

Hanotte, O., Burke, T., Armour, J. A., and Jeffreys, A. J. (1991) Hypervariable minisatellite DNA sequences in the Indian peafowl *Pavo cristatus*. *Genomics* 9: 587–597.

Harris, A. S., Bieger, S., Doyle, R. W., and Wright, J. M. (1991) DNA fingerprinting of tilapia. *Oreochromis niloticus*, and its application to aquaculture genetics. *Aquaculture* 92: 157–163.

Jarman, A. P., Nicholls, R. D., Weatherall, D. J., Clegg, J. B., and Higgs, D. R. (1986) Molecular characterization of a hypervariable region downstream of the human α-globin gene cluster. *EMBO J.* 5: 1857–1863.

Jarman, A. P., and Wells, R. A. (1989) Hypervariable minisatellites: recombinators or innocent bystanders? *Trends Genet.* 5: 367–371.

Jeffreys, A. J., Wilson, V., and Thein, S. L. (1985a) Hypervariable "minisatellite" regions in human DNA. *Nature* 314: 67–73.

Jeffreys, A. J., Wilson, V., and Thein, S. L. (1985b) Individual-specific "fingerprints" of human DNA. *Nature* 316: 76–79.

Jeffreys, A. J., Wilson, V., Thein, S. L., Weatherall, D. J., and Ponder, B. A. J. (1986) DNA "fingerprints" and segregation analysis of multiple markers in human pedigrees. *Am. J. Hum. Genet.* 39: 11–24.

Jeffreys, A. J., and Morton, D. B. (1987) DNA fingerprints of dogs and cats. *Animal Genet.* 18: 1–15.

Jeffreys, A. J., Wilson, V., Kelly, R., Taylor, B. A., and Bulfield, G. (1987) Mouse DNA 'fingerprints': analysis of chromosome localization and germ-line stability of hypervariable loci in recombinant inbred strains. *Nucleic Acids Res.* 15: 2823–2837.

Jeffreys, A. J., Royle, N. J., Wilson, V., and Wong, Z. (1988a) Spontaneous mutation rates to new length alleles at tandem-repetitive hypervariable loci in human DNA. *Nature* 332: 278–281.

Jeffreys, A. J., Wilson, V., Neumann R., and Keyte, J. (1988b) Amplification of human minisatellites by the polymerase chain reaction: towards DNA fingerprinting of single cells. *Nucleic Acids Res.* 16: 10953–10971.

Jeffreys, A. J., Neumann, R., and Wilson, V. (1990) Repeat unit sequence variation in minisatellites: a novel source of DNA polymorphism for studying variation and mutation by single molecule analysis. *Cell* 60: 473–485.

Karn, J. M., Brenner, S., Barnett, L., and Casareni, G. (1980) Novel bacteriophage λ cloning vector. *Proc. Natl. Acad. Sci. USA* 77: 5172.

Kashi, Y., Iraqi, F., Tikochinski, Y., Ruzitsky, B., Nave, A., Beckmann, J. S., Friedmann, A., Soller, M., and Gruenbaum, Y. (1990) $(TG)_n$ uncovers a sex-specific hybridization pattern in cattle. *Genomics* 7: 31–36.

Kelly, R., Bulfield, G., Collick, A., Gibbs, M., and Jeffreys, A. J. (1989) Characterization of a highly unstable mouse minisatellite locus: evidence for somatic mutation during early development. *Genomics* 5: 844–856.

Kessler, C., and Holtke, H.-J. (1986) Specificity of restriction endonucleases and methylases – a review (Edition 2). *Gene* 47: 1–153.

Kornfield, I. (1984) Descriptive genetics of cichlid fishes. In: Turner, B. J. (ed.) Evolutionary Genetics of Fishes, Plenum, New York, pp. 591–616.

Kuhnlein, U., Zadworny, D., Dawe, Y., Fairfull, R. W., and Gavora, J. S. (1990) Assessment of inbreeding by DNA fingerprinting: development of a calibration curve using defined strains of chickens. *Genetics* 125: 161–165.

Litt, M., and Luty, J.A. (1989) A hypervariable microsatellite revealed by *in vitro* amplification of a dinucleotide repeat within the cardiac muscle actin gene. *Am. J. Hum. Genet.* 44: 397–401.

Lloyd, M. A., Fields, M. J., and Thorgaard, G. H. (1989) BKm minisatellite sequences are not sex associated but reveal DNA fingerprint polymorphisms in rainbow trout. *Genome* 32: 865–868.

Love, J. M., Knight, A. M., McAleer, M. A., and Todd, J. A. (1990) Towards construction of a high resolution map of the mouse genome using PCR-analysed microsatellites. *Nucleic Acids Res.* 18: 4123–4130.

Moav, R., Brody, T., Wohlfarth, G., and Hulata, G. (1976) Applications of electrophoretic genetic markers to fish breeding. I. Advantages and methods. *Aquaculture* 9: 217–228.

Mork, J., and Sundnes, G. (1985a) Haemoglobin polymorphism in Atlantic cod (*Gadus morhua*): allele frequency variation between year classes in a Norwegian fjord stock. *Helgolander Meersunters* 39: 55–62.

Mork. J., and Sundnes, G. (1985b) O-group cod (*Gadus morhua*) in captivity: differential survival of certain genotypes. *Helgolander Meeresunters* 39: 63–70.

Nakamura, Y., Leppert, M., O'Connell, P., Wolff, R., Holm, T., Culver, M., Martin, C., Fujimoti, E. Hoff, M., Kumlin, E., and White, R. (1987) Variable number of tandem repeat (VNTR) markers for human gene mapping. *Science* 235: 1616–1622.

Nanda, I., Feichtinger, W., Schmid, M., Schroder, J. H., Zischler, H., and Epplen, J. T. (1990) Simple repetitive sequences are associated with differentiation of the sex chromosomes in the guppy fish. *J. Mol. Evol.* 30: 456–462.

Rogaev, E.I. (1990) Simple human DNA-repeats associated with genomic hypervariability, flanking the genomic retroposons and similar to retroviral sites. *Nucleic Acids Res.* 18: 1879–1885.

Royle, N. J., Clarkson, R. E., Wong, Z., and Jeffreys, A. J. (1988) Clustering of hypervariable minisatellites in the proterminal regions of human autosomes. *Genomics* 3: 352–360.

Ryman, N., and Utter, F. (eds.) (1987) Population Genetics and Fishery Management, University of Washington, Seattle.

Ryskov, A. P., Jincharadze, A. G., Prosnyak, M. I., Ivanov, P. L., and Limborska, S. A. (1988) M13 phage DNA as a universal marker for DNA fingerprinting of animals, plants and microorganisms. *FEBS Letters* 233: 388–392.

Schafer, R., Zischler, H., and Epplen, J. T. (1988) (CAC)$_5$, a very informative oligonucleotide for DNA fingerprinting. *Nucleic Acids Res.* 16: 5196.

Smith, P. J., Birley, A. J., Jamieson, A., and Bishop, C. A. (1989) Mitochondrial DNA in the Atlantic cod, *Gadus morhua*: lack of genetic divergence between eastern and western populations. *J. Fish. Biol.* 34: 369–377.

Taggart, J. B., and Ferguson, A. (1986) Electrophoretic evaluation of a supplemental stocking programme for brown trout, *Salmo trutta* L. *Aquaculture and Fisheries Management* 17: 155–162.

Tautz, D. (1989) Hypervariability of simple sequences as a general source for polymorphic DNA markers. *Nucleic Acids Res.* 17: 6463–6471.

Tautz, D., and Renz, M. (1984) Simple sequences are ubiquitous repetitive components of eukaryotic genomes. *Nucleic Acids Res.* 12: 4127–4138.

Tautz, D., Trick, M., and Dover, G. A. (1986) Cryptic simplicity in DNA is a major source of DNA variation. *Nature* 332: 652–658.

Tave, D. (1986) Genetics for fish hatchery managers. AVI, Westport, Connecticut.

Thomas, W. K., Pääbo, S., Villablanca, F. X., and Wilson, A. C. (1990) Spatial and temporal continuity of kangaroo rat populations shown by sequencing mitochrondrial DNA from museum specimens. *J. Mol. Evol.* 31: 101–112.

Turner, B. J., Elder, J. F. Jr., Laughlin, T. F., and Davis, W. P. (1990) Genetic variation in clonal vertebrates detected by simple-sequence DNA fingerprinting. *Proc. Natl. Acad. Sci. USA* 87: 5653–5657.

Uraiwan, S., and Doyle, R.W. (1986) Replicate variance and the choice of selection procedure for tilapia (*Oreochromis niloticus*) stock improvement in Thailand. *Aquaculture* 57: 27–35.

Vassart, G., Georges, M., Monsieur, R., Brocas, H., Lequarre, A.-S., and Christophe, D. (1987) A sequence of M13 phage detects hypervariable minisatellites in human and animal DNA. *Science* 235: 683–684.

Vergnaud, G. (1989) Polymers of random short oligonucleotides detect polymorphic loci in the human genome. *Nucleic Acids Res.* 17: 7623–7630.

Vogelstein, B., and Gillespie, D. (1979) Preparative and analytical purification of DNA from agarose. *Proc. Natl. Acad. Sci. USA* 76: 615–619.

Waldman, J. R., Grossfield, J., and Wirgin, I. (1988) Review of stock discrimination techniques for striped bass. *N. Am. J. Fish. Management* 8: 410–425.

Weber, J. L., and May, P. E. (1989) Abundant class of human DNA polymorphisms which can be typed using the polymerase chain reaction. *Am. J. Hum. Genet.* 44: 388–396.

Westneat, D. F., Noon, W. A., Reeve, H. K., and Aquadro, C. F. (1988) Improved hybridization conditions for DNA 'fingerprints' probed with M13. *Nucleic Acids Res.* 16: 4161.

Wong, Z., Wilson, V., Patel, I., Povey, S., and Jeffreys, A. J. (1987) Characterization of a panel of highly variable minisatellites cloned from human DNA. *Ann. Hum. Genet.* 51: 269–288.

DNA Fingerprinting: Approaches and Applications
ed. by T. Burke, G. Dolf, A. J. Jeffreys & R. Wolff
© 1991 Birkhäuser Verlag Basel/Switzerland

Genetic Factors Accountable for Line-specific DNA Fingerprint Bands in Quail

J. Hillel[a], O. Gal[a], T. Schaap[b], A. Haberfeld[a], Y. Plotsky[a],
H. Marks[c], P. B. Siegel[d], E. A. Dunnington[d], and A. Cahaner[a]

[a]Dept. of Genetics, Faculty of Agriculture, The Hebrew University, Rehovot 76100, Israel;
[b]Dept. of Human Genetics, Hadassah – Hebrew University Medical Center, Jerusalem,
Israel; [c]USDA-ARS, University of Georgia, Athens, GA 30602, USA; [d]Dept. of Poultry
Science, Virginia Polytechnic Institute and State University, Blacksburg, VA 24061, USA

Summary
DNA fingerprints, prepared from mixes of DNA of individuals sampled from lines of
Japanese quail selected for high or low 4-week body weight, were used to evaluate the relative
contribution of several evolutionary forces to genetic diversity among populations. Comparisons between lines – two replicates of each selection direction and a control unselected line –
were used to determine the frequency of line-specific DNA fingerprint bands produced by each
of three major evolutionary forces: 1) mutation; 2) genetic drift; 3) selection. The latter force
is expected to generate line-specific bands only if there is linkage disequilibrium between DNA
fingerprint loci and quantitative loci (QTLs) controlling body weight. Using probes 33.6 and
R18.1, an average of 48.4 DNA fingerprint bands in each line were analyzed. On average, 27.8
bands were found to be line-specific among the 96.8 (2 × 48.4) bands analyzed in an average
comparison between pairs of lines. Based on the frequencies of line-specific bands in each
particular comparison, it was calculated that 21% of the line-specific bands were due to
mutation, 11% due to a single genetic drift event, 11% due to selection, 21% due to the
combined effects of genetic drift and selection, 22% due to double independent events of
genetic drift, and 14% due to undefined factors.
    Although evidence was found for a high frequency of genetic changes attributable to genetic
drift, and a higher than expected frequency of linkage disequilibrium, the emphasis of this
report is on the methodology suggested rather than on the particular results.

Introduction

Since their discovery in 1985 (Jeffreys *et al.*, 1985a) DNA fingerprints
have been used intensively as molecular genetic markers in a wide range
of applications. Although the main emphasis has been on the identification of individuals, parentage testing and sibship verification (Jeffreys *et
al.*, 1985a, b, 1986; Jeffreys and Morton, 1987; Gill *et al.*, 1986; Hillel *et
al.*, 1989), other uses that have also been reported include linkage
analysis (Plotsky *et al.*, 1990; Thein *et al.*, 1988; Jeffreys *et al.*, 1987),
gene transfer (Hillel *et al.*, 1990) and evolutionary studies (Reeve *et al.*,
1990; Gilbert *et al.*, 1990). The most common use of DNA fingerprint
information in evolutionary studies is for the estimation of relatedness
between populations and lines, as reflected in the degree of similarity

between the DNA fingerprint patterns of the analyzed groups (Kuhn-lein et al., 1989, 1990).

DNA Mixes Approach

Population analysis of DNA fingerprint patterns is extremely difficult when individuals are being evaluated because of the volume of laboratory experiments required, difficulties in comparing information from different gels, and the complexity of drawing a common DNA fingerprint pattern representing the entire analyzed population. Mixing DNA from individuals representing a line has been used by Dunnington et al. (1990) to overcome these difficulties. Mixes of DNA can be used efficiently for linkage analysis between DNA fingerprint loci and quantitative trait loci (QTLs) (Plotsky et al., 1990), for line identification (Dunnington et al., 1990), and for measuring relatedness between populations of common origin (Dunnington et al., 1991). Two DNA mixes from two lines of White Leghorn chickens divergently selected for antibody response to sheep red blood cells (Siegel and Gross, 1980; Martin et al., 1990) are presented in Fig. 1 to demonstrate that in DNA fingerprints of DNA mixes all fragments other than those from extremely polymorphic loci are present in each mix.

Also, the relative intensity of the same band in different samples or populations represents differences in band frequencies. Highly mutable and highly polymorphic bands are not useful for evolutionary studies when the analyzed populations are separated by many generations. Therefore, DNA mixes may provide advantages in evolutionary studies by diluting the contribution of the evolutionarily rapidly changing loci.

Line-Specific Bands

In several experiments, DNA fingerprints of DNA mixes prepared from individuals of divergently selected lines originating from a common base population were found to be clearly different. In White Rock chickens selected over 31 generations for high or low 8 week body weight (Dunnington and Siegel, 1985), approximately 50% of the bands were line-specific (Dunnington et al., 1990). Although genetic variability within the selected lines was appreciable, as reflected by 54% band sharing between individuals within lines, this pattern of line-specificity was observable even when samples as small as 5 chickens per line were used in the DNA mixes (Fig. 2). These differences in band frequency apparently reflect differences in the entire genome as

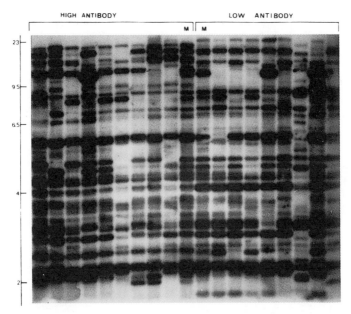

Figure 1. DNA fingerprints of individuals and of their DNA mixes (M). DNA fingerprints of nine White Leghorn chickens from a line selected for high antibody response to sheep red blood cells (SRBC) and the DNA mix (M) are shown in the left part of the figure. Similarly, DNA fingerprints of eight birds from a line selected for low antibody response are shown with the DNA fingerprint of their DNA mix in the right part of the figure. DNA fingerprints of the DNA mixes represent faithfully the frequency of a given band when the two lines are compared.

chickens from these selected lines differ six-fold in body weight and in many other measured traits (Dunnington *et al.*, 1986). In such selection experiments, the lines are not only subjected to divergent selection, but gene flow is also blocked, making them separately subject to mutation and genetic drift. Furthermore, any attempt to interpret these band frequencies as the result of selection should assume a linkage disequilibrium state in the original base population, between the line-specific bands and alleles of genes contributing quantitatively to the selected traits. To date, there is a paucity of reports that suggest such disequilibrium states occur at high frequency. There is also a lack of simple and reliable ways to evaluate the relative importance of these three genetic factors (mutation, genetic drift and linkage disequilibrium) in the production of line-specific bands in selected lines, such as those described above. Had identifying the evolutionary factors accountable for each of the line-specific bands been feasible, a better understanding of these important evolutionary forces would have been achieved.

266

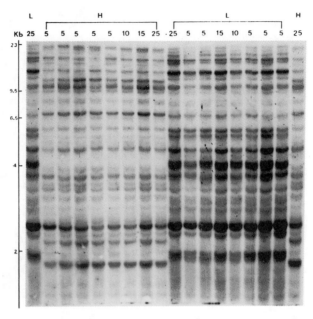

Figure 2. Line-specificity of DNA fingerprint bands. DNA fingerprints of DNA mixes from White Rock chickens selected over 31 generations for high (H) or low (L) 8-week body weight. The number of individuals in each mix is indicated at the top of each lane. Notice the similarity between the various mixes of different samples within each line and the large number of DNA fingerprint bands found in all mixes of one line and absent in the mixes of the other line. The lines are switched in the extreme left-hand and right-hand lanes to aid comparison.

Replicate Selected Lines with a Control Unselected line

Divergent selection experiments focused on the sampling of individuals from single replicates of the last generation alone do not allow the detailed analysis outlined above. In contrast, one could draw the optimal situation in which such analysis would be much easier. The ideal structure should include the following components:

(a) DNA from individuals of the original common base population;
(b) Two replicated lines from each of the two opposite directions of selection;
(c) An unselected control population;
(d) Individual DNA samples from the control and each of the selected populations at each of n generations of the selection experiment.

This structure should be accompanied by a careful genetic and sampling plan to avoid involving more factors such as founder effect and inbreeding. In the following section, an experiment for Japanese quail is presented, in which only components (b) and (c) of the above list are included. In this experiment, there were 20 generations of divergent mass selection for 4-week body weight (Darden and Marks, 1988). Two replicates of each selection direction and an unselected control population were included. Each of the 5 lines was reproduced by 30 pair-matings per generation. These 30 males and 30 females were selected (or taken at random in the control line) from 180–200 quail per line that were evaluated in each generation.

The Quail Experiment

Blood samples were obtained from 20 individuals from each of four lines and 11 individuals from the fifth line (one of the low weight lines).

Table 1. Possible explanations of the actual frequency of specific distributions of presence (1) and absence (0) of DNA fingerprint bands, independently detected by probes R18.1 and 33.6, among an unselected control population (C) of quail, and replicated lines selected divergently for high (H) or low (L) body weight

Presence (1) of bands					Band frequency			
H1	H2	L1	L2	C	R18.1	33.6	Total	Interpretation
1	0	0	0	0	2	0	2	
0	1	0	0	0	1	4	5	
0	0	1	0	0	2	5	7	Mutation
0	0	0	1	0	0	0	0	
0	0	0	0	1	0	1	1	
0	1	1	1	1	2	0	2	
1	0	1	1	1	2	0	2	
1	1	0	1	1	0	0	0	Genetic drift
1	1	1	0	1	2	2	4	(GD)
1	1	1	1	0	0	0	0	
1	1	0	0	0	1	1	2	
0	0	1	1	0	0	0	0	Linkage
0	0	1	1	1	2	0	2	disequilibrium
1	1	0	0	1	0	0	0	(LD)
1	0	0	0	1	1	0	1	
0	1	0	0	1	2	3	5	LD + GD
0	0	1	0	1	0	2	2	
0	0	0	1	1	1	0	1	
0	1	0	1	1	0	0	0	
0	1	1	0	1	0	5	5	
0	1	1	1	0	0	2	2	
1	0	0	1	1	1	0	1	2 × GD
1	0	1	0	1	1	0	1	
1	0	1	1	0	1	0	1	
1	1	0	1	0	0	0	0	
1	1	1	0	0	0	1	1	
1	0	1	0	0	1	0	1	
1	0	0	1	0	2	0	2	Comigration or
0	1	1	0	0	1	1	2	unidentified
0	1	0	1	0	2	0	2	factors

268

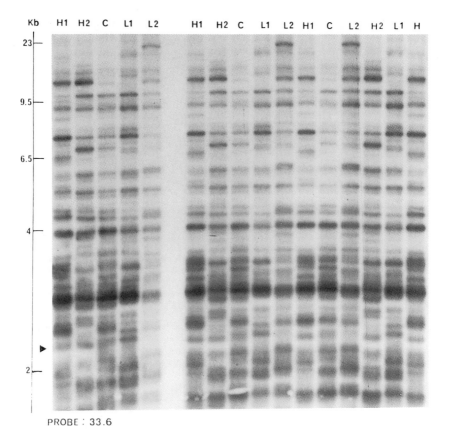

PROBE : 33.6

Figure 3. DNA fingerprints of quail lines. DNA fingerprints of DNA mixes of quail lines, selected for high (H) and low (L) 4-week body weight over 20 generations. H1 and H2 are two replicates of line H, L1 and L2 are two replicates of lines L and C is the unselected control line. The five lines are located three times with different neighbours to allow reliable comparisons of band presence and absence in the five lines.

The interpretation and frequency for each combination of presence and absence of DNA fingerprint bands in each of the five lines are presented in Tab. 1 for bands detected by Jeffrey's probe 33.6 and by probe R18.1 (Haberfeld *et al.*, 1991). DNA fingerprints of the mixes are shown in Figs. 3 and 4 for Probe 33.6. The rationale of the interpretation was as follows:

a. Evolutionary stable bands: Bands of similar intensities in all the lines were interpreted as monomorphic, i.e., stable through the evolution process. Analyses of individuals indicate that some of these bands are polymorphic within lines while others are monomorphic (data not shown).

b. Mutation: Bands present in only one of the five lines were considered likely to have been generated by a mutation in the particular line.

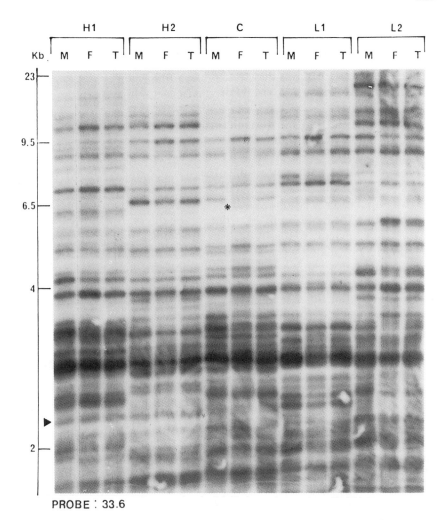

PROBE : 33.6

Figure 4. DNA fingerprints of males (M), females (F) and total (T) quail selected lines. DNA fingerprints of DNA mixes representing each replicate of the line selected for high (H1 and H2) and low (L1 and L2) 4-week body weight. C is the unselected control line. The high similarity between the DNA fingerprint patterns of the three mixes within each line demonstrates the repeatability and hence the reliability of the basic results.

c. Genetic drift as sole factor: Absence of a band in only one line and presence in the other four lines was interpreted as loss due to genetic drift resulting from sampling error in previous generations.

d. Linkage disequilibrium as sole factor: When a band was observed in both replicates of a given direction of the selection and absent in the two replicates of the opposite direction, it was interpreted as resulting from selection, based on linkage disequilibrium between this band and alleles of a QTL controlling body weight or any genetically correlated quantitative trait in the base population. Although this interpretation was made irrespective of the control line, presence of this band in the control line would indicate the existence of linkage disequilibrium in the original base population more definitively.

e. Combination of linkage disequilibrium and genetic drift: These two factors could occur simultaneously and independently upon a given band in different lines. In such a case, the band should be present in the control line and in one of the other selected lines. This interpretation assumes that a state of linkage disequilibrium in the original base population was followed by selection and later by genetic drift accounting for band removal from one of the two replicates.

f. Two independent events of genetic drift: Such a theoretically rare state was postulated when a band was present in any three lines and absent in the remaining two lines. Whereas this interpretation is not the only one, it is probably the simplest.

g. Comigration or other unidentified factors: Four cases were observed in which we could draw no simple interpretation. These cases are characterized by absence of a band in the control line and presence in only one replicate of each of the two selection directions. A possible interpretation, but not very likely to occur, is the comigration of two different bands to the same position on the gel.

Integrative Analysis

Based on the data shown in Tab. 1, one can look for line-specific bands between any pair of the five lines. These ten comparisons (Tab. 2) show that, on average, 28.5% of bands per line are line-specific while the remaining 71.5% of the bands are common to the lines compared. From the rationale given above, analysis of each line-specific band could be made to determine possible causes for the differences between lines. Since difference between replicates cannot result from the combination of selection and linkage disequilibrium, these two comparisons were excluded from the integrative analysis shown in Tab. 3. The results of this analysis indicate the high relative weight of genetic drift as a factor generating genetic differences between populations, as being reflected in line-specific DNA fingerprint bands, either alone in one or two independent events or combined with selection and linkage disequilibrium. The mean rate of mutation to a detectably different allele length at minisatellite loci has been estimated to range from 0.003 (Jeffreys et al., 1988) to 0.0007 (Georges et al., 1990). The cumulative effect through 20 generations in the quail experiment is considered to account for 21% of the line-specific bands (Tab. 3).

Table 2. Summary of line-specific and common DNA fingerprint bands in the quail lines

Lines comparison	Number of bands			Percentage	
	Specific (S)	Common (C)	Total = S + 2 × C	Specific	Common
H1 L1	30	34	98	31%	69%
H1 L2	21	30	81	26%	74%
H2 L1	30	40	110	27%	73%
H2 L2	33	30	93	33%	67%
H1 C	27	33	93	29%	71%
H2 C	25	40	105	24%	76%
L1 C	23	42	107	21%	79%
L2 C	26	32	90	29%	71%
H1 H2	34	31	96	35%	65%
L1 L2	29	33	95	30%	70%
Average	27.8	34.5	96.8	28.5%	71.5%

The relatively high frequency of apparent linkage disequilibrium, either as the only factor or combined with genetic drift, is surprising. Biological verification of this finding can be achieved through the isolation and cloning of the bands concerned and their examination in segregating families resulting from crosses between the selected lines. If this interpretation is valid, the use of multilocus probes to identify bands linked to economically important QTLs will be a promising approach.

Table 3. Relative contribution of genetic factors causing line-specific DNA fingerprint bands detected by probes 33.6 and R18.1, summaraized from Tab. 1

Lines comparison	Number of spec. bands	Relative (%) contribution of factors[a]					
		Mut.	GD	LD	GD + LD	2 × GD	Com.
H1 L1	30	30	7	13	10	27	13
H1 L2	21	10	29	19	9	19	14
L2 L1	30	40	7	13	23	7	10
H2 L2	33	15	18	12	18	24	13
Sub aver.	29	24	15	14	15	19	13
H1 C	27	11	7	15	30	26	11
H2 C	25	24	8	16	16	20	16
L1 C	23	34	0	0	30	22	14
L2 C	26	4	15	0	31	35	15
Sub aver.	25	18	7	7	26	26	14
Average	27	21	11	11	21	22	14

[a]Mut. – mutation; GD – genetic drift; LD – linkage disequilibrium; Com. – comigration (or unidentified factor)

272

In summary, it must be remembered that the proposed approach is based on only one experiment, and there is no intent to determine conclusively the relative importance of significant evolutionary forces or to support definitively one or another evolutionary theory. Nevertheless, cumulative DNA fingerprint information obtained from experiments of a similar kind to that presented here may project light on controversial evolutionary theories.

References

Darden, J. R., and Marks, H. L. (1988) Divergent selection for growth in Japanese quail under split and complete nutritional environments. 1. Genetic and correlated responses to selection. *Poultry Sci.* 67: 519–529.

Dunnington, E. A., and Siegel, P. B. (1985) Long-term selection for 8-week body weight in chickens – Direct and correlated response. *Theor. Appl. Genet.* 71: 305–313.

Dunnington, E. A., Siegel, P. B., Cherry, J. A., Jones, D. E., and Zelenka, D. J. (1986) Physiological traits in adult female chickens after selection and relaxation of selection for 8-week body weight. *J. Anim. Breeding Genet.* 103: 51–58.

Dunnington, E. A., Gal, O., Plotsky, Y., Haberfeld, A., Kirk, T., Goldberg, A., Lavi, U., Cahaner, A., Siegel, P. B., and Hillel, J. (1990) DNA fingerprints of chickens selected for high and low body weight for 31 generations. *Anim. Genet.* 21: 221–231.

Dunnington, E. A., Gal, O., Siegel, P. B., Haberfeld, A., Cahaner, A., Lavi, U., Plotsky, Y., and Hillel, J. (1991) DNA fingerprint comparisons between selected populations of chickens. *Poultry Sci.* (in press).

Georges, M., Lathrop, M., Hilbert, P., Marcotte, A., Schwers, A., Swillens, S., Vassart, G., and Hanset, R. (1990) On the use of DNA fingerprints for linkage studies in cattle. *Genomics* 6: 461–474.

Gilbert, D. A., Lehman, N., O'Brien, S. J., and Wayne, R. K. (1990) Genetic fingerprinting reflects population differentiation in the California Channel Island fox. *Nature* 344: 764–767.

Gill, B., Jeffreys, A. J., and Werrett, D. J. (1985) Forensic application of DNA 'fingerprints'. *Nature* 318: 577–579.

Haberfeld, A., Yoffe, O., Plotsky, Y., and Hillel, J. (1991) DNA fingerprints of farm animals generated by microsatellite and minisatellite DNA probes. *Anim. Genet.* 22: 125–131.

Hillel, J., Plotsky, Y., Haberfeld, A., Lavi, U., Cahaner, A., and Jeffreys, A. J. (1989) DNA fingerprints of poultry. *Anim. Genet.* 20: 25–35.

Hillel, J., Schaap, T., Haberfeld, A., Jeffreys, A. J., Plotzky, Y., Cahaner, A., and Lavi, U. (1990) DNA fingerprints applied to gene introgression in breeding programs. *Genetics* 124: 783–789.

Jeffreys, A. J., Wilson, V., and Thein, S. L. (1985a) Hypervariable 'minisatellite' regions in human DNA. *Nature* 314: 67–73.

Jeffreys, A. J., and Morton, D. B. (1987) DNA fingerprints of dogs and cats. *Anim. Genet.* 18: 1–15.

Jeffreys, A. J., Brookfield, J. F. Y., and Semeonoff, R. (1985b) Positive identification of an immigration test-case using human DNA fingerprints. *Nature* 317: 818–819.

Jeffreys, A. J., Wilson, V., Thein, S. L., Weatherall, D. J., and Ponder, B. A. J. (1986) DNA 'fingerprints' and segregation analysis of multiple markers in human pedigrees. *Amer. J. Hum. Genet.* 39: 11–24.

Jeffreys, A. J., Royle, N. J., Wilson, V., and Wong, Z. (1988) Spontaneous mutation rates to new length alleles at tandem-repetitive hypervariable loci in human DNA. *Nature* 332: 278–281.

Kuhnlein, U., Dawe, Y., Zadworny, D., and Gavora, J. S. (1989) DNA fingerprinting: A tool for determining genetic distances between strains of poultry. *Theor. Appl. Genet.* 77: 669–672.

Kuhnlein, U., Zadworny, D., Dawe, Y., Fairfull, R. W., and Gavora, J. S. (1990) Development of a calibration curve using defined strains of chickens. *Genetics* 125: 161–165.

Martin, A., Dunnington, E. A., Gross, W. B., Briles, W. E., Briles, R. W., and Siegel, P. B. (1990) Production traits and alloantigen systems in lines of chickens selected for high or low antibody response to sheep erythrocytes. *Poultry Sci.* 69: 871–878.

Plotsky, Y., Cahaner, A., Haberfeld, A., Lavi, U., and Hillel, J. (1990) Analysis of genetic association between DNA fingerprint bands and quantitative traits using DNA mixes. Proc. 4th World Congr. Genet. Appl. Livestock Prod., Edinburgh, Vol. XIII pp. 133–136.

Reeve, H. K., Westneat, D. F., Noon, W. A., Sherman, P. W., and Aquadro, C. F. (1990) DNA fingerprinting reveals high levels of inbreeding in colonies of the eusocial naked mole-rat. *Proc. Natl. Acad. Sci. USA* 87: 2496–2500.

Royle, N. J., Clarkson, R. E., Wong, Z., and Jeffreys, A. J. (1988). Clustering of hypervariable minisatellites in the proterminal region of human autosomes. *Genomics* 3: 352–360.

Siegel, P. B., and Gross, W. B. (1980) Production and persistence of antibodies in chickens to sheep erythrocytes 1. Directional selection. *Poultry Sci,* 59: 1–5.

Thein, S. L., Jeffreys, A. J., Gooi, H. C., Cotter, F., Flint, J., O'Connor, N. T. J., and Wainscoat, J. S. (1988) Detection of somatic changes in human DNA by DNA fingerprint analysis. *Brit. J. Cancer* 55: 353–356.

DNA Fingerprinting: Approaches and Applications
ed. by T. Burke, G. Dolf, A. J. Jeffreys & R. Wolff

Identification of Markers Associated with Quantitative Trait Loci in Chickens by DNA Fingerprinting

U. Kuhnlein[a], D. Zadworny[a], J. S. Gavora[b], and R. W. Fairfull[b]

[a]*Department of Animal Science, Macdonald Campus of McGill University, Ste. Anne de Bellevue, Quebec, Canada H9X 1C0*; [b]*Animal Research Centre, Agriculture Canada, Ottawa, Ontario, Canada K1A 0C6*

Summary
Three approaches for identifying VNTR alleles associated with quantitative traits in chickens are described. One approach is based on the comparison of well-defined selected and non-selected control strains. The second approach is based on analyzing chickens within a breeding population ranked according to specific traits and the third approach involves segregation analysis. In this latter approach a large number of offspring of a single male segregating for a quantitative trait are produced and tested for trait association of the male DNA fingerprinting bands. In all cases pooled DNA samples of birds, rather than individual samples, are analyzed and band intensity is assumed to reflect the relative frequency of an allele. Examples from the literature and from our laboratory indicate that these methods permit the identification of DNA fingerprinting bands associated with quantitative traits. After developing locus-specific probes for these bands it should ultimately be possible to detect and map quantitative trait loci.

Introduction

Linkage analysis in large families between phenotypes and chromosomal markers has proven to be extremely useful in locating alleles responsible for human genetic disorders. In these cases success was dependent on being able to distinguish clearly the phenotypes associated with particular alleles. For quantitative traits, which by definition are determined by many different genes, the situation is more complex. Not only can the same phenotype arise from allelic variations in different genes, but environmental influences are often large and can obscure the assignment of a genotype based on phenotypic expression. It is therefore necessary to find new ways for mapping of genetic loci which determine quantitative traits.

The availability of a large number of well-defined differentially selected strains makes the chicken a good system in which to develop strategies for mapping quantitative trait loci (QTL). Several laboratories, including ours, have initiated such programs. However, only a few papers and reports pertaining to this subject have been published. The following overview on approaches aimed at identifying VNTR markers linked to QTLs in chickens by DNA fingerprinting therefore contains several previously unpublished results from our laboratory.

1. Marker Identification Based on Comparing Selected and Control Strains

A sensitive approach for identifying VNTR markers linked to QTLs is based on the analysis of DNA fingerprinting patterns of selected and control strains derived from the same genetic base. Many such strains are available, ranging from strains selected for egg production traits (rate of egg laying, growth rate, age at sexual maturity, egg shell strength, etc.) to strains selected for Marek's disease (MD) resistance and elevated humoral immune response.

Assuming that A and B are two alleles of a gene occurring at frequencies v_A and v_B, respectively, and if this gene is linked to a VNTR locus with alleles $M_1 \ldots M_n$, occurring at frequencies $m_k (k = 1 \ldots n)$, one would expect that a change in the frequency of allele A affects the frequency of the VNTR allele M_k according to the equation

$$\Delta m_k = (m_k^A/v_A - m_k^B/v_B) \cdot \Delta v_A$$

where and m_k^A and m_k^B are the frequencies of the allele M_k linked to allele A and B, respectively. Hence, neglecting randomization by recombination, the frequencies of VNTR alleles (i.e. bands in a DNA fingerprint) linked to alleles of QTLs which respond to selection are expected to be affected by selection, unless $m_k^A/v_A = m_k^B/v_B$.

An example where differential selection resulted in clear changes of the DNA fingerprinting pattern is shown in Fig. 1. Two Leghorn strains S and K were analyzed which had been derived in 1935 at Cornell from a common genetic base. Strain S had been selected for Marek's disease (MD) susceptibility, while strain K had been selected for MD resistance as well as egg production and related traits. The strains S and K represent extremes in terms of MD resistance and susceptibility, respectively (Grunder et al., 1972).

The most obvious difference between the two strains was the presence of a very intense hybridization signal in strain K. Less exposed autoradiographs revealed the presence of one or two bands per bird, which had molecular weights of approximately 20 kb and produced signals which were several-fold more intense than any other bands present in either strain K or strain S (see Fig. 3). They presumably consist of long repeats of a relatively simple sequence, since they were not susceptible to single and double-digestion by a series of 4-cutter enzymes (Kuhnlein et al., unpublished).

276

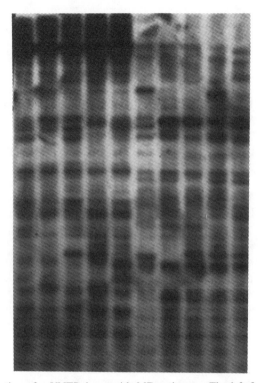

Figure 1. Co-selection of a VNTR locus with MD resistance. The left five lanes are DNA fingerprints of chickens from the MD resistant strain *K* and the right five lanes are DNA fingerprints of chickens from the MD susceptible strain *S*. The two strains had been derived at Cornell University in 1935–36 from a common base population, but founder birds of strain *K* included a few commercial birds introduced in 1936 and 1940. Thereafter, the strains were reproduced as closed breeding populations. Until 1966, strain *S* was selected for MD-suscep-tibility and strain *K* for MD-resistance as well as egg production and related traits (Hutt and Cole, 1957; Grunder *et al.*, 1972; Gavora *et al.*, 1979; Kuhnlein *et al.*, 1989a). All DNA fingerprints shown in this communication were carried out as described by Kuhnlein *et al.* (1989b, 1990) except that the agarose concentration in gels was 0.7% rather than 1%. DNA was digested with HaeIII and the DNA fingerprinting probe used was M13 DNA (Vassart *et al.*, 1987). The migration distance shown is from the origin of the gel to 1 kb.

Analysis of two additional sets of strains derived from unrelated founder populations (as judged from breeding history and the types of endogenous viral genes present in the respective breeding populations) confirmed that selection for MD-resistance was indeed associated with an increased frequency of two high molecular weight bands with a strong affinity for the particular hybridization probe (Kuhnlein and Zadworny, 1989; 1990).

Other markers whose frequencies respond to selection can be iden-tified by examining DNA fingerprints of individual chickens. However, such comparisons are difficult and labour intensive and it is more expedient to pool DNA samples from several chickens of each strain and compare the fingerprints of these DNA pools. In such DNA fingerprints, changes in the frequency of a band are expected to be reflected by a difference in band intensity.

The method of analyzing pools of DNA for the comparison of lines has been pioneered independently by Dunnington *et al.* (1990) and M. W. Bruford (pers. comm.). Dunnington *et al.* (1990) analyzed two lines of White Plymouth Rock chickens selected for 31 generations for high and low body weight, respectively. From a total of 54 bands scored, 48% were specific to just one of the lines and 52% were common to both lines.

The appearance of such line-specific bands might not simply reflect co-selection of the respective marker alleles with alleles at QTLs, but founder effects, genetic drift and mutation to new alleles might also contribute. It is therefore important to compare duplicated selected lines derived from a large base population to exclude such random effects. Recently Hillel *et al.* (1990) analyzed four strains of quail derived from a relatively large base population. Two lines were selected for high body weight and two for low body weight. The occurrence of bands in both lines selected in one direction, but in neither line selected in the other direction, indicates that at least some of the VNTR alleles might be linked to alleles at QTLs.

An example of this type of analysis that has been obtained in our laboratory is shown in Fig. 2. Three strains were derived in 1969 from a common genetic base which had been established from four commercial Leghorn strains. One strain was kept non-selected, whereas two strains were selected for egg production and related traits. Pools of 15 chickens were DNA fingerprinted, revealing about 12 clear bands per strain. The selected strains differed from the control strain by three bands.

The bands which responded to selection in the two independently selected strains were the same. This indicates that the band differences between selected and control strains were not due to mutation to new alleles, genetic drift or founder effects in the selected strains. Drift in the control strain as an alternative explanation is also unlikely, since throughout the breeding program its effective population size ($N_e = 457$) was larger than that of the selected strains ($N_e = 197$). Further, reproducible effects of selection on allelic frequencies in this set of strains were also observed with other marker genes (Kuhnlein *et al.*, 1989). Based on the analysis of individual chickens, band differences were due to changes in band frequency rather than the absence of a band in an entire strain.

Since DNA fingerprint bands might represent alleles of the same or of linked VNTR loci, it is not possible to draw conclusions about the number of QTL alleles which respond to selection. For this purpose probes which are specific for the VNTR alleles which respond to selection have to be prepared.

278

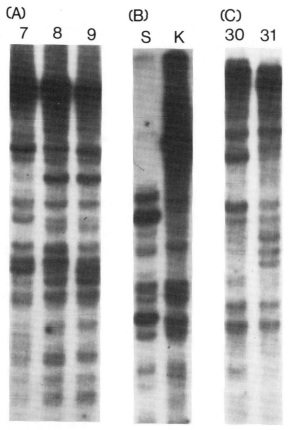

Figure 2. Fingerprints of pooled DNA samples of selected and control strains. DNA of 15 chickens of each strain was pooled, digested with MspI and probed with M13. Panel A: Strain 7 was formed from four commercial Leghorn stocks in 1958 and has been maintained without selection by random mating. Strains 8 and 9 were derived from strain 7 in 1969 and were selected for egg production and related traits (Gowe and Fairfull, 1984). Panel B: Strain S and K are White Leghorn strains derived from a common genetic base and selected for Marek's disease resistance and susceptibility, respectively. In addition, strain K was selected for egg production traits (see Fig. 1). Panel C: Strain 30 was established in 1983 as a reference population from several commercial meat-type stocks (Chambers et al., 1984) and since kept non-selected. Strain 31 was derived from strain 30 and selected for growth rate and feed efficiency.

2. Identification of Markers by Correlation Analysis with Quantitative Traits

From the results discussed above, a comparative analysis of differentially selected strains is sensitive and will presumably result in the identification of many VNTR alleles linked to QTLs. However, this approach is limited. Often only strains selected for composite traits are available and the analysis of a well-defined trait requires the development of specific strains with relatively large population sizes. Further, the quantitative contribution of a QTL allele to a trait remains unknown.

A complementary approach which is more informative is to identify marker genes associated with QTL alleles by correlation analysis. In this approach chickens in a breeding population are ranked according to a specific trait. The DNA of chickens found at the extreme ends of the phenotypic distribution is then pooled and subjected to DNA fingerprinting.

In the most simple case of two QTL alleles A and B, one having a dominant and additive effect, we expect that the two homozygotes will occur at the two extremes of the trait distribution. If v_A and v_B are the frequencies of alleles A and B in the breeding population and m_k^A and m_k^B are the frequencies of the VNTR allele M_k linked to allele A and B, respectively, the difference in frequency (Δm_k) of the VNTR allele M_k between the two homozygous populations would be

$$\Delta m_k = (m_k^A/v_A - m_k^B/v_B)$$

Figure 3 shows examples of this type of analysis in strains S and K. A total of 120 hens in each strain were ranked according to either body weight, rate of egg production or age at sexual maturity. Pools of DNA from 15 chickens from the tail ends of the distribution were analyzed by DNA fingerprinting and compared with each other and a random pool of DNA.

In strain S the only difference was observed in DNA pools from chickens from the two tail ends of the distribution of the rate of egg production. In both pools a 20 kb band was observed which was not present in the random pool or any of the other pools representing the other two traits. This band is apparently identical to a band present in 60% of the chickens of strain K. Strain K had been derived from the same genetic base as strain S, but had been selected for MD resistance, as well as egg production and related traits (see Fig. 1). Hence, a band whose frequency was shown to respond to selection for egg production traits and/or MD resistance was found to be associated with the rate of egg production.

The most obvious difference among the pools of strain K was a band which was only present in the DNA pool of chickens with the lowest body weight. This band is common to birds of strain S which, as an indirect consequence of selection, have a significantly lower body weight than the birds of strain K. Thus, a correlation between this marker allele and body weight is consistent with the selection results.

Additional differences in strain K were observed in the relative intensities of the two strongly hybridizing 20 kb bands. These differed

280

A

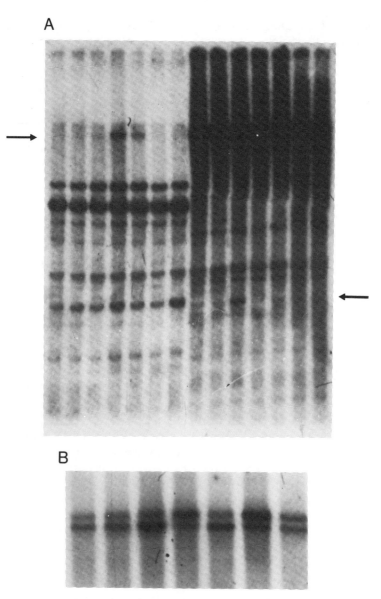

B

Figure 3. Fingerprints of pooled DNA samples from chickens ranked according to trait. Panel A: A total of 95 chickens of strain S and 110 of strain K (see Fig. 1) were ranked according to traits and 15 birds representing the extremes were identified. DNA pools were established for the extremes of each trait and compared with a random pool for DNA fingerprinting. Digestion was with MspI and M13 was used as a probe. Lanes 1–7 are pools from strain S and lanes 8–14 from strain K. The DNA pools were: random (lanes 1 & 8); high body weight at housing (lanes 2 & 9); low body weight at housing (lanes 3 & 10); high egg production (lanes 4 & 11); low egg production (lanes 5 & 12); late sexual maturity (lanes 6 & 13); early sexual maturity (lanes 7 & 14). The arrows indicate bands which differ among the pools of strain S and K, respectively. Panel B: Resolution of the high intensity bands (20 kb) present in the pools of strain K. Electrophoresis was carried out for 48 h, rather than the standard 17 h.

most obviously between the pools representing extremes of egg production and extremes of age at sexual maturity. It should be noted that the same two bands had been associated with MD resistance, and may thus be markers associated with several QTLs or a single QTL with pleiotropic effects.

3. Identification of Markers Linked to QTLs by Segregation Analysis

The resolution of the correlation described above is dependent on the presence of a few major alleles (or clusters of alleles) which determine the distribution of chickens according to a trait. If this is not the case, the pools at extremes of the trait distribution will not lead to a recognizable accumulation of a marker allele.

A more powerful strategy based on segregation analysis has been described by Plotsky *et al.* (1990). It consists of producing a large number of offspring of a single male which segregate for a quantitative trait. The offspring are then ranked according to the trait and pools of DNA from chickens are analyzed by DNA fingerprinting. In this case, the paternal bands will be seen against a background of random maternal bands and can be tested for association with a trait. Using this approach, Plotsky *et al.* (1990) were able to identify two DNA fingerprinting bands associated with fatness and leanness in broilers.

Segregation analysis is probably the most powerful of the three methods described above. However, it cannot simply make use of existing strains, but requires the setting up of a large breeding program.

4. Conclusion

The three approaches described above, when used in conjunction with many different minisatellites and microsatellites as DNA fingerprinting probes, should reveal a series of bands associated with different traits. Subsequently, locus-specific probes for these bands will have to be prepared to determine which DNA fingerprinting bands are allelic and/or represent loci which are linked to each other. This will permit the estimation of the number of single or clustered QTLs and provide the location of these QTLs within a genomic map of the chicken chromosomes.

Acknowledgements
This work was supported by a grant of the Natural Sciences and Engineering Council of Canada to U. Kuhnlein. We thank L. Volkov and the staff of the Animal Research Centre of Agriculture Canada for technical assistance.

References

Chambers, J. R., Bernon, D. E., and Gavora, J. S. (1984) Synthesis and parameters of new populations of meat-type chickens. *Theor. Appl. Genet.* 69: 23–30.

Dunnington, E. A., Gal, O., Plotsky, Y., Haberfeld, A., Kirk, T., Goldberg, A., Lavi, U., Cahaner, A., Siegel, P. B., and Hillel, J. (1990) DNA fingerprints of chickens selected for high and low body weight for 31 generations. *Anim. Genet.* 21: 247–257.

Gavora, J. S., Emsley, A., and Cole, R. K. (1979) Inbreeding in 35 generations of development of Cornell S strain of Leghorns. *Poultry Sci.* 58: 1133–1136.

Gowe, R. S. and Fairfull, R. W. (1980) Performance of six long term multi-trait selected Leghorn strains and three control strains, and a strain cross evaluation of the selected strains. Proc. 1980 South Pacific Poult. Sci. Conv., Auckland, N.Z. pp. 141–162.

Grunder, A. A., Jeffers, T. K., Spencer, J. L., Robertson, A., and Speckmann, G. W. (1972) Resistance of strains of chickens to Marek's disease. *Can. J. Anim. Sci.* 52: 1–10.

Hillel, J., Haberfeld, A., Gal, O., Plotsky, Y., and Cahaner, A. (1990) Genetic factors accountable for line-specific DNA fingerprinting bands in quail. In: Burke, T., Dolf, G., Jeffreys, A. J., and Wolff, R. (eds). DNA Fingerprinting: Approaches and Applications. Birkhäuser, Basel. pp 263–273. (This volume)

Hutt, F. B., and Cole, R. K. (1957) Control of leukosis in the fowl. *J. Am. Vet. Med. Assoc.* 131: 491–495.

Kuhnlein, U., Gavora, J. S., Spencer, J. L., Bernon, D. E., and Sabour, M. (1989a) Incidence of endogenous viral genes in two strains of White Leghorn chickens selected for egg production and susceptibility or resistance to Marek's disease. *Theor. Appl. Gen.* 77: 26–32.

Kuhnlein, U., Dawe, Y., Zadworny D., and Gavora J. S. (1989b) DNA fingerprinting: a tool for determining genetic distances between strains of poultry. *Theor. Appl. Gen.* 77: 669–672.

Kuhnlein, U., Sabour, M., Gavora, J. S., Fairfull, R. W., and Bernon, B. W. (1989c) Influence of selection for egg production and Marek's disease resistance on the incidence of endogenous viral genes in White Leghorns. *Poultry Sci.* 68: 1161–1167.

Kuhnlein, U., and Zadworny, D. (1989) DNA fingerprinting in chickens applied to assess genetic variability, strain relationships and to identify alleles which respond to selection. Proc. 38th Ann. National Poultry Breeders Roundtable. St. Louis. pp. 147–169.

Kuhnlein, U., and Zadworny, D. (1990) Molecular aspects of poultry breeding. Proc. 4th World Congr. on Genet. Applied to Livestock Prod. XVI: 21–30.

Plotsky, Y., Cahaner, A., Haberfeld, A., Lavi, U., and Hillel, J. (1990) Analysis of genetic association between DNA fingerprint bands and quantitative traits using DNA mixes. Proc. 4th World Congr. on Genet. Applied to Livestock Prod. XIII: 133–136.

Vassart, G., Georges, M., Monsieur, R., Brocas, H., Lequarre, A. S., and Christophe, D. (1987) A sequence in M13 phage detects hypervariable minisatellites in human and animal DNA. *Science* 235: 683–684.

Two-dimensional DNA-Fingerprinting in Animals

C. P. Schelling[a], E. Clavadetscher[b], E. Schärer[a], P. E. Thomann[a],
C. C. Kuenzle[b], and U. Hübscher[b]

[a]*Dept. of Laboratory Animal Science, University of Zürich, Winterthurerstr. 190,
CH-8057 Zürich, Switzerland;* [b]*Dept. of Pharmacology and Biochemistry,
University of Zürich, Winterthurerstr. 190, CH-8057 Zürich, Switzerland*

Summary
DNA-fingerprinting has become, during the last five years, an important method of genetic
analysis in medicine, veterinary medicine and other disciplines. The power of this technique,
especially for genetic linkage analysis, may be enhanced in humans by using the two
dimensional DNA-fingerprinting method. Here we show that this procedure can successfully
be applied to different animal species, e.g. pig, dog and mouse. Optimal conditions, however,
have to be determined for each species tested. With the use of marker systems as well as
computer programs it will be possible to evaluate complex two-dimensional spot patterns in
a short time and with high reliability.

Introduction

Tandemly repeated sequences within the genome of eukaryotes have
become of interest to geneticists as well as to researchers in related
disciplines. Some of these so-called minisatellites (Wyman and White,
1980; Bell *et al.*, 1982; Knott *et al.*, 1986) can show variation in the
number of repeat units and hence are called VNTRs (Variable Number
of Tandem Repeats). Multiallelism and high heterozygosities make
these sequences valuable genetic markers (Nakamura *et al.*, 1987). In
1985, Jeffreys and coworkers developed a method for detecting several
minisatellite loci simultaneously (Jeffreys *et al.*, 1985a). The resulting
genetic DNA-fingerprints can be used for individual-specific identification (Jeffreys *et al.*, 1985b; Gill *et al.*, 1985). Similar hypervariable
sequences occur in animals (Jeffreys and Morton, 1987; Georges *et al.*,
1988) and in plants (Dallas, 1988).

We originally introduced the one-dimensional DNA-fingerprinting
method for genetic quality control of mouse inbred strains (Signer,
1989; Thomann *et al.*, 1990). Since it has been shown that DNA-fingerprint patterns can be used for genetic linkage analysis in humans
(Jeffreys *et al.*, 1986) as well as in cattle (Georges *et al.*, 1990), we also
started to analyse genomic DNA from pigs and dogs using human
minisatellite probes. This allowed us to identify individuals and determine relationships within families. However, we detected fewer polymorphic bands in the DNA-fingerprint patterns of dogs and pigs as

compared to published reports (Jeffreys and Morton, 1987; Georges *et al.*, 1988).

The low resolution of agarose gels has prevented the more widespread use of DNA-fingerprinting with multilocus-probes in genetic linkage analysis. While the high molecular weight bands are resolved, short alleles can remain undetected, since bands in the lower molecular size range are usually poorly resolved or tend to run out of the gel during extended times of electrophoresis. Thus, important information can be lost, due to the failure to detect all alleles from all loci.

It is well known that elevated temperatures can be used to dissociate double stranded DNA molecules. In 1979, Fischer and Lerman demonstrated the analytical power of two-dimensional separation of DNA fragments including denaturant gradient gel electrophoresis when analysing restriction fragment patterns derived from the genome of different *Escherichia coli* strains. In the first dimension, the restriction fragments are separated according to size, and in the second dimension, which includes a gradient of denaturants, the fragments are separated according to their specific composition and base sequence. This is because double stranded DNA molecules exposed to increasing amounts of denaturants at an elevated constant temperature begin to dissociate in their so-called low-melting domains. This dissociation is highly sequence-dependent and as a result different fragments will melt at different concentrations of denaturants (for review see Lerman *et al.*, 1984).

These partially melted DNA molecules virtually stop migrating and give rise to defined spots. Only low-melting domains, which can be determined from the sequence using computer programs (Lerman *et al.*, 1984) are amenable to this technique. This method has been shown to detect single base-pair substitutions in DNA fragments (Fischer and Lerman, 1983; Uitterlinden and Vijg, 1990) and is therefore a valuable tool for analysing altered DNA sequences.

Two-dimensional separation of DNA fragments, including gel electrophoresis, has been used to improve resolution of DNA restriction fragments for DNA-fingerprinting using human multilocus-probes (Uitterlinden *et al.*, 1989). With this two-dimensional DNA-fingerprinting method over 600 spots could be detected per individual. About 150 spot polymorphisms were observed between parents and 70% of these were transmitted to their offspring. In addition, it could be shown that minisatellite alleles from the same locus, which strongly differ in total length but share a common adjacent 5′ or 3′ low-melting domain will migrate in the second dimension to almost the same isotherm (see Fig. 1), i.e. a zone of defined denaturant concentration in a gradient (Uitterlinden and Vijg, 1989; 1990).

Due to the low information content of one-dimensional DNA-fingerprint patterns in pigs and dogs we decided to introduce the two-dimensional DNA-fingerprinting method for these animals. In this minireview

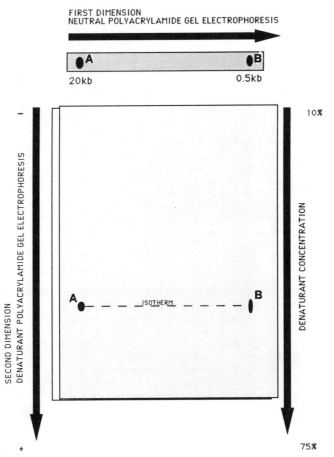

FIRST DIMENSION
NEUTRAL POLYACRYLAMIDE GEL ELECTROPHORESIS

A   B

20kb   0.5kb

SECOND DIMENSION
DENATURANT POLYACRYLAMIDE GEL ELECTROPHORESIS

DENATURANT CONCENTRATION

10%

A — — — ISOTHERM — — — B

75%

Figure 1. Schematic representation of the two-dimensional DNA-fingerprinting method. Minisatellite alleles A and B, which are different in their length, share a common low-melting domain in their 5′ or 3′ adjacent sequence and therefore migrate in the second dimension to a common isotherm.

we show two-dimensional DNA-fingerprint patterns of pig, dog and mouse and discuss the necessity to standardize the method for each species.

The Method

The two-dimensional DNA-fingerprinting method is based on the separation of DNA restriction fragments according to their size in the first dimension and according to their melting characteristics in the second dimension (Fig. 1). The procedure as applied to pigs, dogs and mice is described below.

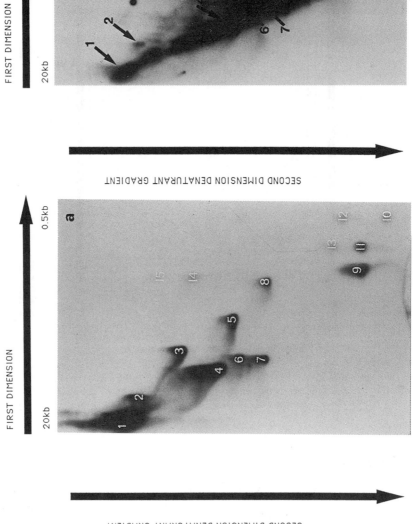

High molecular weight DNA was isolated and digested with the optimal restriction enzyme (see examples for dog, pig and mouse below). 5–10 μg of restriction fragments were separated according to their size in the first dimension in a neutral 6% polyacrylamide gel in 1 × TAE (Sambrook *et al.*, 1989) at 250 V for 3 hours at 50°C. After staining (0.1 μg ethidium bromide/ml TAE) and destaining the gel, the region between 20 kb and 0.5 kb was cut out of the gel, and directly put onto a 6% polyacrylamide gel containing a linear (10–75%) concentration gradient of denaturant (100% denaturant = 7.0 M urea/40% formamide). Electrophoresis was carried out at 225 V for 12 hours at 60 ± 0.5°C. After electrophoresis the gel was stained with ethidium bromide, destained as above and photographed. The gel was irradiated with 302 nm UV-light for 4 minutes to fragment the separated DNA. The restriction fragments were transferred onto nylon membranes (Hybond N plus, Amersham) by semi-dry blotting (PolyBlot, ABN SBD-100) for 65 minutes at 1 mA/cm² in 1 × TBE (Sambrook *et al.*, 1989). After the transfer, the nylon membrane was soaked for 10 minutes between well wetted (0.4 M NaOH, 0.6 M NaCl) Whatman papers. The membrane was then dried between two fresh Whatman papers and soaked again for 5 minutes in 100 mM Na phosphate (pH 7.2). The dried membrane was finally baked for 1 hour at 80°C.

Prehybridization and hybridization were carried out in glass bottles. The prehybridization solution (0.5 M Na phosphate, 1 mM EDTA, 7% (w/v) SDS) was added at least 30 minutes before hybridization, the temperature for both prehybridization and for hybridization being 65°C. Then the labelled probes were added to the bottles. After hybridization (60–90 minutes) two wash steps of 30 minutes each at 65°C were performed with 100 ml of a solution containing 2 × SSC, 0.1% (w/v) SDS. A third wash step was carried out at room temperature with 500 ml of the same solution. Finally, the membrane was dried and exposed to X-ray film.

Examples of Two-dimensional DNA-Fingerprint Patterns of Dog, Pig and Mouse

First, we had to standardize the conditions for the second dimension. The optimal temperature was 60°C ± 0.5°C for all animals tested (see

Figure 2. A marker system for two-dimensional DNA-fingerprints. 100 ng HaeIII and 100 ng BglII digested lambda-DNA were included and separated together with genomic restriction fragments (10 μg genomic pig DNA digested with HinfI). The figure shows autoradiograms of the same nylon membrane, first hybridized with ³²P-labelled total lambda-DNA (a), and then, after stripping of the membrane, hybridized with ³²P-labelled Jeffreys core probe 33.15 (b). Fifteen well-separated lambda-spots are visible and can easily be transposed as reference points to the two-dimensional DNA-fingerprint (see arrows).

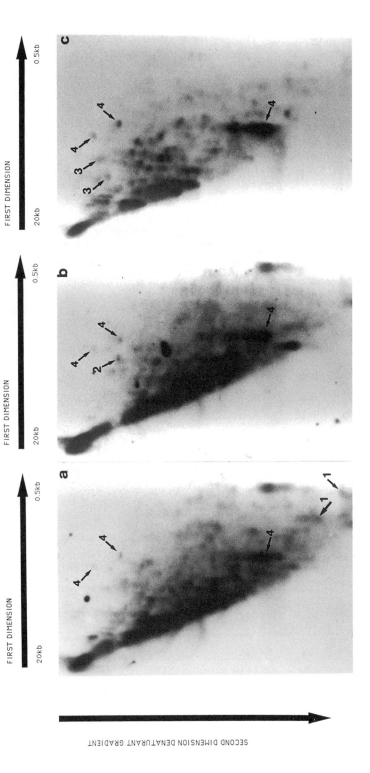

FIRST DIMENSION

20kb   0.5kb

SECOND DIMENSION DENATURANT GRADIENT

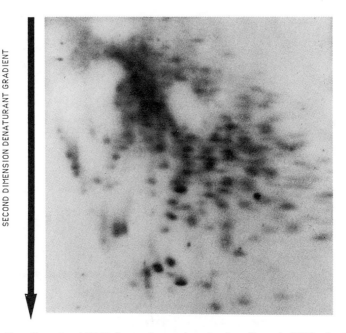

Figure 4. Two-dimensional DNA-fingerprint pattern of a dog. Genomic DNA of a Bernese Mountain Dog was digested with HinfI and separated by two-dimensional denaturant gradient gel electrophoresis. Minisatellites were detected with $(CAC)_n$ DNA probe, yielding about 100 well-separated spots.

also Uitterlinden *et al.*, 1989). In addition, we included an internal marker system (E. Mullaart, pers. comm.). Two samples of bacteriophage lambda DNA were separately digested with HaeIII and BglII, mixed and added to the genomic DNA restriction fragments (10 μg of pig DNA digested with HinfI). The mixture was then separated by two-dimensional denaturant gradient gel electrophoresis and transferred to a nylon membrane. When probed with ^{32}P-labelled lambda DNA, 15–20 well-separated spots of lambda DNA were visible (Fig. 2a). Figure 2b shows an autoradiogram of the same experiment, but probed with a ^{32}P-labelled Jeffreys' core probe 33.15. The localization of the

Figure 3. Different two-dimensional DNA-fingerprint patterns of a pig. Following the experiment shown in Fig. 2b (probe 33.15) the same membrane was successively hybridized with Jeffreys' core probe 33.6 (b) and with $(CAC)_n$ DNA probe (c), respectively. For better comparison an identical photograph of the hybridization with Jeffreys' core probe 33.15 (Fig. 2b) is added (a). All three probes detected different subsets of minisatellites (some examples are indicated by arrows: 1 for 33.15, 2 for 33.6 and 3 for $(CAC)_n$). Some minisatellites are detected by all three probes (arrow 4).

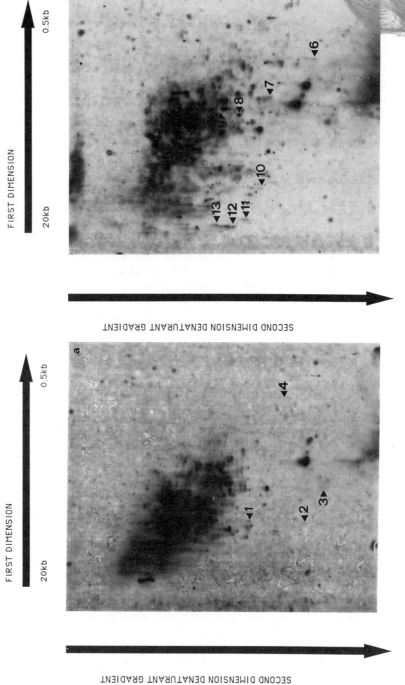

marker spots can easily be transposed to the two-dimensional DNA-fingerprint pattern, where they serve as reference points. In addition, known monomorphic spots in the DNA-fingerprint pattern can also be used as reference points. These marker systems are the basis for a computer assisted evaluation of two-dimensional DNA-fingerprint patterns, e.g. the GeIIabII, 2DGel Analysis System.

Figure 3 shows two-dimensional DNA-fingerprint patterns of a pig. The same membrane as in Fig. 2 was tested with different DNA probes. Following the experiment described for Fig. 2, the nylon membrane was successively hybridized to Jeffreys' core probe 33.6 (Fig. 3b) and to a $(CAC)_n$ probe (Fig. 3c). As a control the membrane was exposed to X-ray film for 48 hours after each removal of the radioactive material with 0.4 M NaOH at 42°C for 30 minutes. All three DNA probes detected different subsets of minisatellites, some of which are marked by arrows: arrow 1 marks minisatellites specific for 33.15, arrow 2 for 33.6 and arrow 3 for $(CAC)_n$, respectively; arrow 4 shows minisatellites detected by all 3 probes.

For the dog, the best two-dimensional DNA-fingerprint patterns were obtained using HinfI for digestion of the genomic DNA in combination with $(CAC)_n$ as the probe (Fig. 4). Pedigrees of Bernese Mountain Dogs were analyzed and the familial relationship could be determined within more than 80 spots. As a comparison, with the conventional one-dimensional DNA-fingerprinting method only 12–14 bands could be resolved. In addition two Bernese Mountain Dogs (full sibs), which had only two different bands in their one-dimensional DNA-fingerprint patterns, showed at least 10 different spots in their two-dimensional DNA-fingerprint patterns (Schelling *et al.*, to be published elsewhere).

A third example of the analytical power of this method is demonstrated in Fig. 5. Two very closely related mouse inbred strains, C57BL/6J/Zur (Fig. 5a) and C57BL/10SnJ/Zur (Fig. 5b), showed differences in their two-dimensional DNA-fingerprint patterns in more than 13 spots. The one-dimensional DNA-fingerprint pattern of the same strains had revealed only 6–7 differences among the various bands (Signer, 1989).

Concluding Remarks

The recently described two-dimensional DNA-fingerprinting method is applicable to various animal species. However, as exemplified for pig,

Figure 5. Two-dimensional DNA-fingerprint patterns of the two closely related mouse inbred strains C57BL/6J and C57BL/10SnJ. Genomic DNA was digested with HaeIII and minisatellites were detected by Jeffreys' core probe 33.6. The two-dimensional fingerprint patterns of C57BL/6J (a) and C57BL/10SnJ (b) can easily be distinguished. Spots indicated by arrows 1–4 are specific for C57BL/6J and spots indicated by arrows 5–13 are specific for C57BL/10SnJ.

dog and mouse, the method requires optimization for each species. In agreement with results obtained in humans (Uitterlinden *et al.*, 1989), the information content (number of minisatellites analysable) is increased. At the moment we estimate the real gain of additional useful information to be 2–5 fold. Inclusion of markers such as digested bacteriophage lambda DNA gives prominent spots that can be used to analyse two-dimensional DNA-fingerprint patterns from different experiments by using computer programs. The information content of the patterns may be further increased by including species-specific multilocus probes and locus-specific probes. Thus, the two-dimensional DNA-fingerprinting method may be valuable for veterinary medicine, e.g. to identify disease genes in any given animal species (see also Uitterlinden *et al.*, 1991).

Acknowledgements
We thank A. Uitterlinden for his help in introducing this method and A. J. Jeffreys for the human minisatellite probes. This work has been supported by the Bundesamt für Veterinär-wesen, by the UFA-Winterthur, by the Stiftung für Wissenschaftliche Forschung der Universität Zürich and the Canton of Zürich.

References

Bell, G. I., Selby, M. J., and Rutter, W. J. (1982) The highly polymorphic region near the human insulin gene is composed of simple tandemly repeating sequences. *Nature* 295: 31–35.

Dallas, J. F. (1988) Detection of DNA fingerprints of cultivated rice by hybridization with a human minisatellite DNA probe. *Proc. Natl. Acad. Sci. USA* 85: 6831–6835.

Fischer, S. G., and Lerman, L. S. (1979) Length-independent separation of DNA restriction fragments in two-dimensional gel electrophoresis. *Cell* 16: 191–200.

Fischer, S. G., and Lerman, L. S. (1983) DNA fragments differing by single base-pair substitutions are separated in denaturant gradient gels: correspondence with melting theory. *Proc. Natl. Acad. Sci. USA* 80: 1579–1583.

Georges, M., Lequarre, A. S., Castelli, M., Hanset, R., and Vassart, G. (1988) DNA fingerprinting in domestic animals using four different minisatellite probes. *Cytogenet. Cell. Genet.* 47: 127–131.

Georges, M., Lathrop, M., Hilbert, P., Marcotte, A., Schwers, A., Swillens, S., Vassart, G., and Hanset, R. (1990) On the use of DNA fingerprints for linkage studies in cattle. *Genomics* 6: 461–474.

Gill, P., Jeffreys, A. J., and Werrett, D. J. (1985) Forensic applications of DNA fingerprints. *Nature* 318: 577–579.

Jeffreys, A. J., Wilson, V., and Thein, S. L. (1985a) Hypervariable minisatellite regions in human DNA. *Nature* 314: 67–73.

Jeffreys, A. J., Brookfield, J. Y. F., and Semeonoff, R. (1985b) Positive identification of an immigration test-case using human DNA fingerprints. *Nature* 317: 818–819.

Jeffreys, A. J., Wilson, V., Thein, S. L., Weatherall, D. J., and Ponder, B. A. J. (1986) DNA fingerprints and segregation analysis of multiple markers in human pedigrees. *Am. J. Hum. Genet.* 39: 11–24.

Jeffreys, A. J., and Morton, D. B. (1987) DNA fingerprints of dogs and cats. *Anim. Genet.* 18: 1–15.

Lerman, L. S., Fischer, S. G., Hurley, I., Silverstain, K., and Lumelsky, N. (1984) Sequence-determined DNA separations. *Ann. Rev. Biophys. Bioeng.* 13: 399–423.

Knott, T. J., Wallis, S. C., Pease, R. J., Powell, L. M., and Scott, J. (1986) A hypervariable region to the human apolipoprotein B gene. *Nucleic Acids Res.* 14: 9215–9216.

Nakamura, Y., Leppert, M., O'Connell, P., Wolff, R., Holm, T., Culver, M., Martin, C., Fujimoto, E., Hoff, M., Kumlin, E., and White, R. (1987) Variable Number of Tandem Repeat (VNTR) markers for human gene mapping. *Science* 235: 1616–1622.

Sambrook, J., Fritsch, E. F., and Maniatis, T. (1989) Molecular Cloning (second edition), Cold Spring Harbour Laboratory Press.

Signer, E. N. (1989) Thesis: DNA – Fingerprints zur Ueberwachung der Reinerbigkeit von Mäuse-Inzuchstämmen. Universität Zürich.

Thomann, P. E., Signer, E. N., Schelling, C. P., and Hübscher, U. (1990) DNA-fingerprinting for quality assurance of inbred mice. Proceedings of the 4th FELASA Symposium, Lyon, France (in press).

Uitterlinden, A. G., Slagboom, P. E., Knook, D. L., and Vijg, J. (1989) Two-dimensional DNA fingerprinting of human individuals. *Proc. Natl. Acad. Sci. USA* 86: 2742–2746.

Uitterlinden, A. G., and Vijg, J. (1989) Two-dimensional DNA typing. *Trends in Biotechnology* 7: 336–341.

Uitterlinden, A. G., and Vijg, J. (1990) Denaturing gradient gel electrophoretic analysis of the human c Ha-ras 1 proto-oncogene. *Applied and Theoretical Electrophoresis* 1: 175–179.

Uitterlinden, A. G., and Vijg, J. (1990) Denaturing gradient gel electrophoretic analysis of human minisatellites. Electrophoresis (in press).

Uitterlinden, A. G., Meulenbelt, I., Mullaart, E., Slagboom, P. E., and Vijg, J. (1991) Genome scanning by two-dimensional DNA typing: a rapid method to map genetic traits in humans, animals and plants. *Electrophoresis* (in press).

Wyman, A. R., and White, R. (1980) A highly polymorphic locus in human DNA. *Proc. Natl. Acad. Sci. USA* 77: 6754–6758.

DNA Fingerprinting: Approaches and Applications
ed. by T. Burke, G. Dolf, A. J. Jeffreys & R. Wolff
© 1991 Birkhäuser Verlag Basel/Switzerland

Applications of DNA Fingerprinting in Plant Breeding

H. Nybom

Balsgård – Department of Horticultural Plant Breeding, Swedish University of Agricultural Sciences, Fjälkestadsvägen 123-1, S-29194 Kristianstad, Sweden

Summary
Several DNA minisatellite probes have yielded fragment profiles that appear very useful for plant breeding work. These fragment profiles show no variation when vegetatively propagated material is analyzed. Similarly, specimens obtained through selfing in inbreeding species exhibit identical profiles. In contrast, genetic recombination in cross-pollinating species results in highly variable, usually individual-specific fragment profiles. Thus different cultivars can be distinguished, as also can genotypes of wild species in natural populations. These fragment profiles can also be utilized in parentage analysis, as has already been conducted in rice and apples, thereby enabling us to elucidate the origin of insufficiently documented cultivars. Moreover, estimates of genetic variation based on similarity indices calculated from fragment profiles show a close association with known levels of genetic relatedness.

Introduction

DNA fingerprints have been obtained by hybridization of a suitable probe to nuclear minisatellites in many species representing various metazoan phyla, e.g. mammals, birds and fish (Burke, 1989; Turner *et al.*, 1990). Some of the probes used in these studies have also produced informative DNA fragment profiles in organisms as far removed as fungi (Braithwaite and Manners, 1989), protozoa (Rogstad *et al.*, 1989) and bacteria (Huey and Hall, 1989). In contrast, relatively few investigations have been carried out on plant material. The plants successfully studied using minisatellite analysis do, however, include gymnosperms as well as a diverse group of angiosperms (Tab. 1; see also review in Weising and Kahl, 1990). Thus, useful minisatellite DNA sequences appear to occur commonly in the plant kingdom, suggesting that DNA fingerprinting will eventually become a much appreciated tool in many areas of botanical research.

Plant material is often considered more difficult to work with than that from animals. In the present paper some commonly occurring problems will be addressed in the Methodology section, and protocols that have proven useful in plant studies will be referred to. So far, most plants investigated have been representatives of commercially important species. The main emphasis of these studies has consequently been

Table 1. Plant species in which minisatellite DNA analysis has been performed by use of the M13 repeat probe, the human-derived 33.6 and 33.15 probes, or artificial repeat probes

Plant species	Reference
Acer negundo	Nybom and Rogstad, 1990
Arabidopsis thaliana	Zimmerman *et al.*, 1989
Asimina triloba	Rogstad *et al.*, 1988a
Brassica napus	Weising *et al.*, 1990
Cicer arietinum	Weising *et al.*, 1989; Weising *et al.*, 1990
Citrus sinensis	Ryskov *et al.*, 1988
Glycine max	Ryskov *et al.*, 1988
Gossypium hirsutum	Ryskov *et al.*, 1988
Hordeum spontaneum	Weising *et al.*, 1990
Hordeum vulgare	Ryskov *et al.*, 1988; Weising *et al.*, 1989
Lens culinaris	Weising *et al.*, 1990
Linum bienne, L. usitatissimum	Zimmerman *et al.*, 1989
Lycopersicon esculentum	Rogstad *et al.*, 1988a
Malus × *domestica, M.* spp.	Nybom 1990a, b; Nybom *et al.*, 1990; Nybom and Schaal 1990a
Medicago sativa	Zimmerman *et al.*, 1989
Musa acuminata	Weising *et al.*, 1990
Oryza glaberrima, O. sativa	Dallas, 1988
Pinus torreyana	Rogstad *et al.*, 1988a
Polyalthia glauca	Rogstad *et al.*, 1988a
Poncirus trifoliata	Ryskov *et al.*, 1988
Populus deltoides, P. tremuloides	Rogstad *et al.*, 1988a, b
Prunus serotina	Nybom *et al.*, 1990
Rubus allegheniensis, R. flagellaris, R. idaeus, R. occidentalis, R. pensilvanicus, R. ursinus	Nybom and Hall 1991; Nybom *et al.*, 1989; Nybom *et al.*, 1990; Nybom and Schaal 1990b
Solanum tuberosum	Weising *et al.*, 1990
Vitis vinifera	Striem *et al.*, 1990

related to the needs of plant breeding, such as genotype identification, parentage analysis, and the estimation of genetic relatedness. The results of some of these studies will be discussed in the Applications section.

Methodology

DNA Extraction, Electrophoresis, Blotting and Hybridization

A major prerequisite for DNA fingerprint work is the availability of high quality DNA. For example, samples of DNA that have been successfully hybridized with rDNA probes may still not be clean enough to permit hybridization with minisatellite probes (B. Schaal, pers. comm). Fortunately, however, many plant species yield DNA of sufficient quality even with the rather quick and easy to carry out CTAB (hexa-decyltrimethylammonium bromide)-protocol described in Saghai-Maroof *et al.* (1984; for modifications see Nybom and Schaal, 1990b). A critical part of this procedure is the initial CTAB extraction step; many plants yield cleaner DNA when this step is carried out at room temperature or even on ice instead of at 60°C. The duration of this step is also important, and should for especially difficult materials not exceed 5 min (K. Wolff, pers. comm). Also, one must not despair if the isopropanol precipitation does not result in clearly discernible strands of DNA. On the contrary, I, at least, have usually obtained cleaner DNA when there was just a faint, cloudy precipitate or even none at all. In these cases the DNA is secured by spinning the tubes, pouring off the supernatant, and then carefully dissolving the DNA adhering to the tube walls with TE. Where this simple CTAB protocol does not result in DNA of sufficient quality, one may have to resort to more expensive and time-consuming methods involving a subsequent caesium chloride gradient centrifugation.

Samples of 3–10 μg DNA are digested with one, or sometimes two, restriction enzymes (usually AluI, DraI, HaeIII, HindIII, HinfI, RsaI or TaqI) at 5–10 units per μg. The DNA is then electrophoresed in 0.7–1.2% agarose gels. My best results were obtained by running a submersed, water-cooled gel in TPE buffer (0.09 M Tris, 0.0013% phosphoric acid, 2 mM EDTA, pH 8.0) at 1–2 V/cm. The resulting gels are usually Southern blotted onto a nylon or nitrocellulose filter. Hybridizations can, however, also be performed on the dried gels directly (Weising et al., 1989). Though several non-radioactive labelling methods have been described recently, all plant investigations published so far have utilized ^{32}P, usually by random-priming (Dallas, 1988; Rogstad et al., 1988a; Ryskov et al., 1988, Striem et al. 1990).

A number of hybridization protocols have been published. Those involving conspecific non-minisatellite probes usually contain formamide, SSC, Denhardt's solution and salmon sperm DNA. However, vertebrate carrier DNA must be avoided when the more general minisatellite probes are used, since it would otherwise act as a competitor (Vassart et al., 1987). Such protocols, in addition to formamide and SSC, thus contain only dried nonfat milk (Dallas 1988; Rogstad et al., 1988a, b), Denhardt's solution (Ryskov et al., 1988), or both (Zimmerman et al., 1989). However, I have had the best results with the protocol of Westneat et al. (1988): 7% SDS (can be substituted with sodium sarkosyl), 1 mM EDTA (pH 8.0), 0.263 M Na phosphate, pH 7.2 and 1% bovine serum albumin (fraction V). The first two washes are carried out for 15 min in 2 × SSC, 0.1% SDS at room temperature followed by a 15–45 min wash in the same solution at 60°C. Autoradiography is then performed at −80°C for 2–10 days with intensifying screens.

Choice of Probes

Variable levels of genetic variation have been reported in RFLP investigations utilizing congeneric probes. Gebhardt et al., (1989) were thus able to distinguish a large number of potato genotypes (*Solanum tubero-sum* subsp. *tuberosum*) utilizing probes from a potato cDNA library, four base cutter restriction enzymes, and denaturing polyacrylamide gels resulting in high resolution of small-sized DNA fragments. Fragment profile variation was also obtained in a study of wheat (*Triticum aestivum*) utilizing a conspecific probe with a minisatellite-like sequence (Martienssen and Baulcombe, 1989). However, in most cases the levels of variation encountered with these conspecific probes are relatively low, as reported, for example, by Nagamine *et al.* (1989) in a study of beets (*Beta vulgaris* and *B. nana*).

Apparently individual-specific fingerprints have more often been obtained by employing a tandem repeat sequence from the protein III gene of the M13 bacteriophage vector, originally reported by Vassart *et al.* (1987) to produce fingerprints in man and some other mammals. A number of different plant species have since been fingerprinted utilizing various M13 probes, consisting of either the entire M13 genome (Ryskov *et al.*, 1988) or shorter fragments containing the protein III gene minisatellite (Rogstad *et al.*, 1988a; Nybom *et al.*, 1989; Zimmerman *et al.*, 1989; Nybom, 1990a, b; Nybom and Schaal 1990a, b; Nybom and Rogstad, 1990; Nybom *et al.*, 1990; Nybom and Hall, 1991).

Another very useful set of minisatellite probes is the series derived from the human genome by Jeffreys and coworkers (Jeffreys *et al.*, 1985a). Thus the 33.15 probe has been used for cottonwood (*Populus deltoides*) and quaking aspen (*P. tremuloides*) (Rogstad *et al.*, 1988b), whereas the 33.15 and the 33.6 probes have been utilized for some species and cultivars of rice (*Oryza sativa* and *O. glaberrima*) (Dallas, 1988).

The hybridization patterns obtained using 33.6 and 33.15 are very different from one another (Dallas, 1988), as are also those derived from 33.6 and M13, whereas 33.15 and M13 appear to yield very similar patterns (unpubl. data). Thus the discriminatory power of DNA fingerprinting may be further enhanced by the subsequent hybridization of the same filters to several different minisatellite probes. Artificial probes consisting of simple repetitive sequences, such as $(GACA)_4$ and $(GATA)_4$, have also yielded variable DNA fragment patterns in several plant species (Weising *et al.*, 1989; 1990). These findings greatly increase the number of potentially useful probes.

Evaluation and Statistics

Hybridization of suitably digested DNA to a minisatellite probe usually results in the detection of a large number of fragments. These fragments tend to become progressively more crowded with diminishing size. Thus the larger ones, 8–10 kb, are relatively easy to score, whereas fragments smaller than 2–3 kb are almost impossible to evaluate unambiguously.

The number of useful fragments detected usually ranges between 10 and 30.

Fragment patterns are commonly evaluated by scoring the individual fragments as either present or absent for each individual sample. Computer-aided scanning of autoradiographs may become a valuable tool (Nybom, 1990c), especially if this information can be processed with computer software that allows comparisons of large numbers of fragment profiles (Gill et al., 1990).

A similarity index, D, can be calculated for each possible pairwise comparison between samples: $D_{AB} = 2N_{AB}/(N_A + N_B)$ where N_A and N_B are the numbers of fragments scored in A and B, respectively, and N_{AB} is the number of shared fragments, as described by Wetton et al., (1987). Assuming that the fragments are in linkage equilibrium, the probability that two different samples will exhibit identical fragment profiles can then be calculated as the mean D_{AB} raised to the mean number of fragments scored per individual.

Assuming that all fragments (i) are monomorphic within cultivars, (ii) occur independently of one another, and (iii) have equal frequencies within the material studied, Dallas (1988) instead calculates a considerably lower probability, $P = [x^2 + (1-x)^2]^{n/x}$. However, he proceeds to show that probably none of these assumptions hold true. Thus great care should be taken when evaluating probability estimates until more is known about the genetic properties of these DNA fragment profiles in each species under study.

Applications

Identification of Genotypes

The ability to properly identify different genotypes, whether these constitute commercial varieties or grow in natural populations of a potentially valuable species, is an important prerequisite for plant breeding. Isozyme analysis has frequently proven useful to a certain degree, but is often not sensitive enough to differentiate between all genotypes investigated (Bournival and Korban, 1987; Cousineau and Donnelly, 1989). In some cases, cultivars are more easily distinguished by RFLP analysis, as shown in, for example, maize (Zea mays) (Helentjaris et al., 1985; Evola et al., 1986), oilseed rape (Brassica napus) (Figdore et al., 1988), lettuce (Lactuca sativa) (Landry et al., 1987), soybean (Glycine max) (Apuya et al., 1988), rice (McCouch et al., 1988), potato (Gebhardt et al., 1989) and carnation (Dianthus caryophyllus) (Woodson, 1989). However, these studies utilize probes developed from random collections of single-copy DNA or cDNA sequences of each species or genus. In most cases the levels of genetic variation thus detected are not as high as with the minisatellite probes, and consequently a number of probe/enzyme combinations must be performed in order to distinguish all the cultivars being investigated.

However, by hybridization of DNA samples to minisatellite probes, such as the M13 repeat probe and the human-derived 33.6 and 33.15, we have an extremely sensitive tool for the detection of genetic variation. In an initial investigation, two varieties of barley (*Hordeum vulgare*) were analyzed and found to exhibit different fragment profiles after digestion with BspRI + HindIII, and hybridization to the M13 probe (Ryskov *et al.*, 1988).

DNA fragment profiles are generally not influenced by environmental conditions. Identical fragment profiles were thus found in material sampled from two widely separated branches of a large cottonwood tree, as also in material from a mature tree and its sucker plant, from DNA digested with HaeIII and hybridized to the M13 probe (Rogstad *et al.*, 1988a). A North American pawpaw (*Asimina triloba*) treelet and its sucker also yielded identical fragment profiles, as did two plants of an inbred tomato variety (*Lycopersicon esculentum*). Three trees of the narrowly endemic and genetically probably rather depauperate torrey pine (*Pinus torreyana*) resulted in two identical and one slightly different fragment profiles. On the other hand, clearly different fragment profiles were encountered in a pair of sibling seedlings from the cross-pollinating tropical tree species *Polyalthia glauca*. Similarly, six different trees of the dioecious and therefore obligately outcrossing cottonwood could all be distinguished by their fragment profiles, utilizing either the M13 or the 33.15 probe (Rogstad *et al.*, 1988a, b).

Dallas (1988) obtained unique fragment profiles in each of seven *O. sativa* cultivars and in two *O. glaberrima* cultivars, following digestion of the DNA with DraI and hybridization to the 33.6 probe. Since rice is perpetuated by selfing, levels of intra-cultivar variation were, on the other hand, expected to be very low. This was substantiated by finding identical fragment profiles in 14 seedlings of an *O. sativa* cultivar. The effect of somatic cell manipulation was investigated by comparing control plants of one cultivar with progeny produced by selfing plants regenerated from protoplasts. One very faint fragment was found in all regenerated plants but not in the control. The other fragments were instead identical in all plants investigated, suggesting that the regeneration process had no major effect on the hybridization pattern. However, the existence of minor differences, which would probably remain undetected, cannot be ruled out.

Inter-cultivar variation has also been reported in grapes (Striem *et al.* 1990) as well as in alfalfa (*Medicago sativa*) and flax (*Linum usitatissimum*), following hybridization to the M13 probe (Zimmerman *et al.*, 1989). On the other hand, no variation was found between five flax lines that were derived from the same inbred parent

300

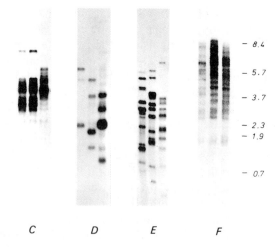

Figure 1. DNA samples hybridized to a (GATA)₄ probe (Weising *et al.*, 1990). C, three cultivars of *Solanum tuberosum*, digested with AluI; D, three cultivars of *Brassica napus* var. *oleifera*, digested with AluI; E, three accessions of *Cicer arietinum*, digested with TaqI; F, three accessions of *Lens culinaris*, digested with AluI. Molecular weight markers (in kb) are indicated to the right. [Reproduced through the courtesy of *Fingerprint News*].

but had been propagated as independent inbred lines for six generations.

High levels of genetic variation were found in three cultivars of oilseed rape and in three different accessions of chickpea (*Cicer arietinum*) by hybridization of the dried gels directly with the artificial (GATA)₄ minisatellite probe (Weising *et al.*, 1990) (Fig. 1). Some variation was also noted in samples of wild banana (*Musa acuminata*), potato and lentil (*Lens culinaris*). Hybridizing the gels instead with the (GACA)₄ probe detected overall lower levels of genetic variation. With this latter probe, however, considerable variation was found in barley (Weising *et al.*, 1989).

Three genera in the economically important angiosperm family Rosaceae were investigated by Nybom *et al.* (1990). Four trees of the wild cherry species *Prunus serotina* were analysed with four different restriction enzymes, each resulting in apparently individual-specific fragment profiles. Pairwise comparisons were performed and a mean D-value found of 0.10–0.34, depending on which restriction enzyme had been utilized. Assuming that fragments were independently inherited, the probability that different specimens have identical fragment profiles in this species was calculated to be between $3.0 \times 10^{-5} - 3.4 \times 10^{-3}$, also depending on which restriction enzyme had been utilized. Inter- and intraspecific DNA fragment profile variation was also found among some wild bramble (*Rubus*) plants, with an average D-value of 0.25 for comparisons among different fragment patterns following digestion with

HaeIII, and of 0.31 with HinfI. The probability of chance identity was similarly found as 5.5×10^{-6} and 7.4×10^{-8}, respectively. Finally, three trees each of the apple varieties Golden Delicious, Jonathan, Red Delicious and Rome Beauty (*Malus × domestica*) were analyzed (Fig. 2). These cultivars could all be distinguished with each one of five different restriction enzymes. The mean D-value for pairwise comparisons among cultivars varied between 0.27 and 0.48, depending on the restriction enzyme used. The probability of identical DNA fragment profiles in different cultivars was estimated to be between 4.2×10^{-5} – 0.031. Within-cultivar variation was found only in Rome Beauty and was most likely due to misidentification of the plant material.

In a study of *Rubus* cultivars, DNA samples from eight different blackberries and two raspberries were digested with HinfI and hybridized to the M13 probe (Nybom *et al.*, 1989) (Fig. 3). All cultivars proved to have unique fragment profiles with an average D-value of 0.55, even though some of the blackberry cultivars were very closely related. Two canes were sampled from each cultivar but no within-cultivar variation was found. The probability of finding identical DNA fragment patterns in different cultivars was estimated, as above, to be 2.4×10^{-3}. Similar results were obtained in another set of blackberry and raspberry cultivars, with an average D-value after digestion with HinfI of 0.46 and an estimated probability of chance identity of

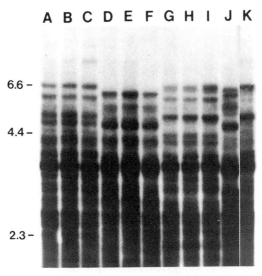

Figure 2. DNA samples from two to three trees each of four apple cultivars digested with RsaI and hybridized to the M13 repeat probe (Nybom *et al.*, 1990). A, B, C, Golden Delicious; D, E, F, Jonathan; G, H, I, Red Delicious; J, K, Rome Beauty (at least one of these is probably from misidentified plant material). Size markers were derived from λ DNA cut with HindIII. [Reproduced through the courtesy of *Theoretical and Applied Genetics*.].

302

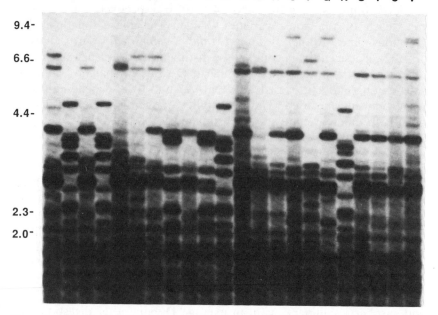

A B C D E F G H I J K L M N O P Q R S T U V

9.4-
6.6-
4.4-
2.3-
2.0-

Figure 3. DNA samples of some *Rubus* cultivars digested with HinfI and hybridized with the M13 repeat probe (Nybom *et al.*, 1989). Of each cultivar two, or in one case, four canes were sampled; A, G, Shawnee; B, D, K, R, Lowden Sweet; C, U, Comanche; E, M, Rosborough; F, P, Brazos; H, J, Titan; I, V, Cherokee; L, T, Cheyenne; N, S, Choctaw; O, Q, Darrow. Of these Titan is a red raspberry (*R. idaeus*), Lowden Sweet a purple raspberry (*R. idaeus* × *R. occidentalis*) and the remainder are blackberries (*Rubus* subgen. *Rubus*). Size markers were derived from λ DNA cut with HindIII. [Reproduced through the courtesy of *Acta Horticulturae*].

1.3×10^{-3} (Nybom and Hall, 1991). Additional samples from some of these specimens were instead digested with HaeIII which resulted in overall lower D-values. These were, however, well correlated with those obtained from digestion with HinfI. The main reason for the difference in D-values seems to be that digestion with HaeIII results in a higher proportion of large-sized and more easily evaluated fragments. Thus the choice of restriction enzyme is definitely an important factor in minisatellite analysis.

Many plant species have a clonal population structure due to extensive vegetative propagation. In some cases such clones can be further extended by the production of apomictically formed seeds (i.e. seeds formed without prior fertilization). Proper identification of different clones in nature, of course, greatly facilitates the collecting of genotypes to be used in future breeding work. Minisatellite analysis was thus applied to two natural *Rubus* populations in order to investigate their population structure (Nybom and Schaal, 1990b). Twenty canes from each species were collected along a 600 m stretch of road. The DNA was digested with HinfI and hybridized to the M13 probe. Five different DNA fragment profiles were found in the apomictic *R. pensilvanicus* (highbush blackberry) whereas the sexual *R. occidentalis* (black raspberry) instead was represented by 15 different fragment profiles. Moreover, identical fragment profiles were found in the latter only in canes separated by a maximum of 4 m. In the first-mentioned species, identical fragment profiles were found in canes upto 500 m apart, presumably resulting from dispersal of apomictically formed seeds.

Individual-specific DNA fragment profiles were found in all 21 investigated trees of box elder (*Acer negundo*) following digestion of the DNA with HinfI and hybridization to the M13 probe (Nybom and Rogstad, 1990). The average D-value was 0.55, and the probability of finding identical fragment profiles was 0.012. On the other hand, closely occurring trees of quaking aspen in several cases exhibited identical DNA fragment profiles (Rogstad, Nybom & Schaal, unpubl. data). Both species are sexual and outcross, but the latter propagates vegetatively and may thus form large clones.

Somatic mutations affecting, for example, plant structure and productivity give rise to varieties known as "sports". Some of these constitute an improvement of the original genotype, and have thus been registered and patented as varieties of their own. However, most of these sports deviate from the original variety only in minor characteristics and may thus be very difficult to tell apart. Minisatellite analysis has been successfully employed in human cancer research to detect somatic mutations (Hayward *et al.*, 1990). Thus a similar approach was attempted in a study of 15 different sports of Red Delicious apples (Nybom, 1990a). Unfortunately, no qualitative differences in fragment profiles were found. Some minor differences in relative fragment intensity were noted, which perhaps can be explained by the chimaeral nature of many of these Red Delicious mutations. The homogeneity of the Red Delicious sports can be contrasted to the results obtained in apple seedlings. Here, unique fragment profiles were encountered in every one of 55 and 64 apple seedlings, respectively, that had resulted from sexual recombination (Nybom, 1990b; Nybom and Schaal, 1990a) (Fig. 4).

In summary, the variability encountered in DNA fragment profiles following hybridization to minisatellite probes seems to be closely associated with the expected amount of genetic variation in the investigated samples. Thus sexual recombination in cross-pollinating species usually yields highly variable fragment profiles, whereas selfing in inbreeding species, as well as vegetative propagation, instead results in identical fragment profiles. Moreover, minisatellite analysis could also successfully distinguish those *Rubus* cultivars (Nybom *et al.*, 1989) that exhibit identical isozyme patterns when investigated with a large number of enzymes (Cousineau and Donnelly, 1989).

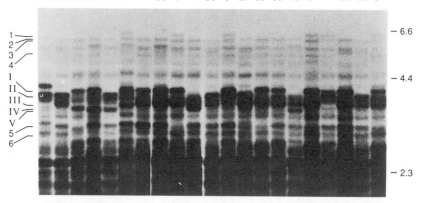

Figure 4. DNA samples, digested with TaqI and hybridized with the M13 repeat probe, of apple cultivars GD = Golden Delicious, JO = Jonathan, RD = Red Delicious, and of 16 seedlings that have JO as seed parent (Nybom and Schaal, 1990a). Fragments utilized in this study are denoted to the left, size markers derived by digesting λ DNA with HindIII, to the right. Fragments I–V are believed to constitute a multiallelic locus. [Reproduced through the courtesy of *Theoretical and Applied Genetics*.]

Parentage Analysis

Parentage analysis has been carried out using DNA fingerprinting in humans as well as in many animal species (reviewed in Burke, 1989). The first corresponding study in plants was carried out in the F_2 progeny of a mass crossing between two rice cultivars (Dallas, 1988). Fragments found in only one of the two grandparents were identified, and each F_2 offspring was then scored for presence or absence of these fragments. All fragments found in the offspring could also be found in one or both grandparental genotypes except in one case.

A study was also made of the inheritance of the scored fragments in 14 offspring plants through comparison with their grandparents (Dallas, 1988). The numbers of segregating fragments were analysed using separate comparisons for each fragment. However, due to low numbers, composite analyses are statistically more powerful in this material. Thus a chi-square test on the 5 fragments inherited from one grandparent, assuming a 3:1 segregation, yields $\chi^2 = 0.0$, and on the 6 fragments inherited from the other grandparent yields $\chi^2 = 3.10$, df $= 1$, $0.1 > p > 0.05$. In conclusion, the F_2 absence/presence ratios for these fragments do not differ significantly from 3:1, but a deviation may nevertheless be suspected for the second parent.

The occurrence of linkage between fragments was also investigated in this material (Dallas, 1988). Data for all possible non-identical pairs of fragments were thus combined and separated according to whether the fragment pair originated from the same grandparent or from different grandparents. Visual comparison of the numbers obtained showed a slight excess of grandparental classes and a corresponding deficiency of recombined classes, compared with the 9:3:3:1 ratios expected on the basis of independent segregation. However, these numbers lend themselves very nicely to chi-square analyses; those cases in which the two members of a fragment pair were present in different grandparents yield $\chi^2 = 8.84$, df = 3, $0.05 > p > 0.025$, and those cases where the two members instead were present in the same grandparent yield $\chi^2 = 12.9$, df = 3, p < 0.005. Thus at least some linked loci appear to have been included in this study.

Paternity of apple seedlings derived from open pollination in Golden Delicious and Jonathan was studied in an orchard where these two cultivars as well as Red Delicious were grown (Nybom and Schaal, 1990a). DNA was extracted from the collected seedlings and digested with TaqI, and in some cases also with RsaI. The resulting filters were hybridized with the M13 repeat probe. A total of 11 informative fragments were scored in the TaqI-digested gels, i.e. fragments occurring in one or two but not in all three of the parental genotypes (Fig. 4). Digestion with RsaI resulted in more large-sized fragments, with a total of 17 fragments being used for the evaluation. Of the 30 Golden Delicious seedlings analysed, 5 appeared to have been sired by Jonathan and 24 by Red Delicious, whereas the remainder showed some fragments not encountered in any of the putative parents and thus would have arisen by pollination from an outside source or by mutation. Of the 34 Jonathan seedlings, 9 appeared to have been sired by Golden Delicious, 16 by Red Delicious, 5 by either Golden Delicious or Red Delicious, and 2 by selfing, whereas another two exhibited fragments not encountered elsewhere. The analysis was facilitated by the finding of a seemingly multiallelic locus, with fragment I occurring in Red Delicious and Golden Delicious, fragments II and III in Jonathan, fragment IV in Golden Delicious and fragment V in Red Delicious. The fragments are similar in size, ranging between 3.2 and 3.9 kb, and hybridize very strongly to the M13 probe. Two of these fragments, or rarely one, occur in all offspring plants. Those cases where only one fragment was found could be explained either by homozygosity or by pollen from an outside source. Moreover, these fragments occur with similar intensity in the autoradiographs, as expected for alleles at the same locus since these would have the same minisatellite repeat sequence and thus also the same degree of similarity to the probe. However, the combination of fragments in the individual seedlings appears to deviate somewhat from a 1:1:1:1 distribution, with a deficiency in seedlings homozygous for fragment I (in the one case that could be tested statistically, $0.10 > p > 0.05$; Nybom and Schaal, 1990a). This deviation could result from the locus being linked to other genes for which homozygosity is selectively disadvantageous.

In another study, seedlings that derived from some ornamental apple trees were analysed (Nybom, 1990b). There were so many putative pollen parents that they could not in most cases be run on the same gel as the seedlings. This greatly complicated the study since only a few fragments could be matched unambiguously across gels. Thus paternity was assigned to only 21 of the 55 seedlings investigated.

The ability to trace the ancestry of commercially important but insufficiently documented cultivars may become an important tool in future plant breeding. Likewise, it should be possible to find fingerprint markers for some traits that are economically important but difficult to evaluate properly, as already reported in, for example, cattle (Georges et al., 1990).

Genetic Relatedness

Many factors may theoretically interfere with the usefulness of minisatellite analysis for the estimation of relatedness between pairs of individuals, especially for more distant relationships, as discussed by Lynch (1988). These are comigration of non-allelic markers, linkage disequilibria between marker loci, inability to observe and evaluate smaller fragments, possible linkage of marker loci with other loci under selection, and high and variable mutation rates.

Preliminary results obtained in plant studies appear to involve some of these problems concerning, for example, linkage disequilibria (Dallas, 1988), problems with evaluation of smaller fragments (Nybom and Rogstad, 1990), and linkage of marker loci with other loci under selection (Nybom and Schaal, 1990a). Results are also affected by which probe (Weising et al., 1989) as well as by which restriction enzyme has been utilized (Nybom, 1990b; Nybom et al., 1990; Nybom and Hall, 1991). Nevertheless, minisatellite analysis may provide a useful tool in estimating genetic relatedness among groups of individuals. Promising results have thus been reported in studies of genetic relatedness in various mammals (Gilbert et al., 1990; Hillel et al., 1990; Reeve et al., 1990) and birds (Wetton et al., 1987; Hillel et al., 1989; Kuhnlein et al., 1989, 1990).

A few such studies have also been conducted on plant material. In most cases the documented variation is lower than in animals, probably due to higher levels of selfing and inbreeding in plants. Comparison of some Rubus cultivar fragment profiles, however, resulted in D-values that appear to be closely associated with the expected level of genetic relatedness (Nybom et al., 1989). Thus comparisons among siblings yielded D-values of 0.63–0.90 and comparisons between parents and offspring 0.60–0.89, whereas comparisons among unrelated cultivars instead yielded 0.07–0.44.

Another study of *Rubus* species and cultivars was carried out in order to clarify the relationships of some old and insufficiently known cultivars (Nybom and Hall, 1991). The average D-value obtained from comparing all genotypes with each other was 0.46. Considerably higher values were, however, encountered in some cases. Thus a comparison between the presumably closely related Boysen and Young cultivars yielded $D = 0.62$, and comparisons between these and their putative parent, Austin Thornless, yielded $D = 0.67$ and $D = 0.82$, respectively. The blackberry cultivar Santiam has supposedly arisen from a cross between *R. ursinus* and Logan, which here is corroborated by the comparison between Santiam and Logan, yielding $D = 0.75$. Comparisons among parents and offspring in red raspberry yielded an average $D = 0.70$, and among half sibs $D = 0.87$.

A study of ornamental apple trees and their offspring included material from five different genotypes on which fruits were collected after open pollination. From each maternal parent, 6–12 seedling offspring were then obtained and subsequently analyzed by digesting the DNA with TaqI and, in some cases, subsequently also with RsaI, and hybridizing the resulting filters with the M13 probe (Nybom, 1990b). A close association was found between known levels of genetic relatedness and fragment profile similarities. Thus comparisons between a maternal genotype and its offspring, based on TaqI-digested samples, yielded an overall average $D = 0.71$, whereas comparisons among siblings and/or half-sibs yielded $D = 0.63$, and, for comparisons among unrelated individuals $D = 0.41$. These data were then analyzed in relation to expected probabilities of fragment sharing in obligately outcrossing organisms. Following Jeffreys *et al.*, (1985b), the mean population frequency, q, of resolvable alleles was calculated from $x = q^2 + 2q(1 - q) = 2q - q^2$, yielding $q = 0.18$ for the TaqI-digested samples.

The expected probability that a tree and its offspring share fragments was then calculated as: $(1 + q - q^2)/(2 - q)$ (J. F. Y. Brookfield in Burke and Bruford, 1987) yielding a probability of 0.63. This expected probability was somewhat higher than the D-value averages actually obtained for the five maternal genotypes and their sets of offspring (the offspring of one of the genotypes were divided onto two gels, and thus analysed separately): 0.55, 0.66, 0.85, 0.68, 0.72 and 0.72. This is especially pronounced in the genotype B4 ($D = 0.85$), which is considerably more similar to its offspring than predicted. However, no less than six fragments were shared by this genotype and all of its offspring, suggesting that B4 is more homozygous than any of the other original genotypes. These instead have only one or two fragments in common with all of their respective offspring. The probability of fragment sharing between full siblings, calculated as $(4 + 5q - 6q^2 + q^3)/4(2 - q)$ (Jeffreys *et al.*, 1985b), yielded a probability of 0.65, which was somewhat higher than most of the D-value averages actually obtained: 0.59,

0.56, 0.80, 0.58, 0.69 and 0.56, probably due to the majority of the offspring being only half-sibs. The offspring of B4 stand out by instead being considerably more homogeneous than predicted (D = 0.80). This homogeneity appears to be caused mainly by high levels of homozygosity in the maternal parent B4, as described above. However, assortative mating may also be involved.

An association was found between sampling distances and D-values in the insect-pollinated and dioecious box elder, with D = 0.64–0.68 for comparisons among closely growing trees and D = 0.48–0.58 for comparisons among trees occurring further apart (Nybom and Rogstad, 1990). It seems likely that the sampling distances here reflect the degree of genetic relatedness. The size range of the evaluated box elder fragments has some effect on the results obtained. Thus taking only fragments larger than 6 kb into consideration resulted in the same overall pattern but with considerably lower D-values (Nybom and Rogstad, 1990). Similar findings have been reported in man (Helminen *et al.*, 1988) and other mammals (Jeffreys and Morton, 1987). This effect is probably, at least to some part, due to the increased crowding of small-sized fragments, resulting in some being erroneously scored as identical. Great care should thus be taken in the hybridization procedure so that large-sized, and usually somewhat fainter, fragments are made clearly visible.

A similar association between minisatellite DNA variation and sampling distances has also been suggested in the insect-pollinated North American pawpaw. Plants collected from within small areas yielded D-values of 0.75–0.84 whereas a set of plants collected from several USA states resulted in D = 0.68–0.70 (Rogstad, Wolff and Schaal, unpubl. data). Flowers of this species are strongly protogynous but geitonogamy is possible. Thus genetic selfing might be an explanation for the overall comparatively large D-values.

Several future applications of genetic relatedness studies by use of minisatellite analysis may be visualized. Thus DNA fingerprints can provide markers for tagging the entire genome, as already demonstrated in ibex and goats (Hillel *et al.*, 1990). Such markers could become very useful in gene introgression breeding programs. Assessments of inbreeding may also be performed, as shown in a study using defined strains of chicken (Kuhnlein *et al.*, 1990).

References

Apuya, N. R., Frazier, B. L., Keim, P., Roth, E. J., and Lark, K. G. (1988) Restriction fragment length polymorphisms as genetic markers in soybean, *Glycine max* (L.) Merrill. *Theor. Appl. Genet.* 75: 889–901.

Bournival, B. L., and Korban, S. S. (1987) Electrophoretic analysis of genetic variability in the apple. *Sci. Hortic.* 31: 233–243.

Braithwaite, K. S., and Manners, J. M. (1989) Human hypervariable minisatellite probes detect DNA polymorphisms in the fungus *Colletotrichum gloeosporioides*. *Current Genet.* 16: 473–476.

Burke, T. (1989) DNA fingerprinting and other methods for the study of mating success. *Trends Ecol. Evol.* 4: 139–144.

Burke, T. and Bruford, M. W. (1987) DNA fingerprinting in birds. *Nature* 327: 149–152.

Cousineau, J. C., and Donnelly, D. J. (1989) Identification of raspberry cultivars *in vivo* and *in vitro* using isoenzyme analysis. *HortScience* 24: 490–492.

Dallas, J. F. (1988) Detection of DNA "fingerprints" of cultivated rice by hybridization with a human minisatellite probe. *Proc. Natl. Acad. Sci. USA* 85: 6831–6835.

Evola, S. V., Burr, F. A., and Burr, B. (1986) The suitability of restriction fragment length polymorphisms as genetic markers in maize. *Theor. Appl. Genet.* 71: 765–771.

Figdore, S. S., Kennard, W. C., Song, K. M., Slocum M. K., and Osborn, T. C. (1988) Assessment of the degree of restriction fragment length polymorphism in *Brassica*. *Theor. Appl. Genet.* 75: 833–840.

Gebhardt, C., Blomendahl, C., Schachtschabel, U., Debener, T., Salamini, F., and Ritter, E. (1989) Identification of 2n breeding lines and 4n varieties of potato (*Solanum tuberosum* ssp. *tuberosum*) with RFLP-fingerprints. *Theor. Appl. Genet.* 78: 16–22.

Georges, M., Lathrop, M., Hilbert, P., Marcotte, A., Schwers, A., Swillens, S., Vassart, G., and Hanset, R. (1990) On the use of DNA fingerprints for linkage studies in cattle. *Genomics* 6: 461–474.

Gilbert, D. A., Lehman, N., O'Brien, S. J., and Wayne, R. K. (1990) Genetic fingerprinting reflects population differentiation in the California Channel Island fox. *Nature* 344: 764–767.

Gill, P., Sullivan, K., and Werett, D. J. (1990) The analysis of hypervariable DNA profiles: problems associated with the objective determination of the probability of a match. *Hum. Genet.* 85: 75–79.

Hayward, N., Chen, P., Nancarrow, D., Kearsley, J., Smith, P., Kidson, C., and Ellem, K. (1990) Detection of somatic mutations in tumours of diverse types by DNA fingerprinting with M13 phage DNA. *Int. J. Cancer* 45: 687–690.

Helentjaris, T., King, G., Slocum, M. K., Siedenstrang, C., and Wegman, S. (1985) Restriction fragment polymorphisms as probes for plant diversity, and their development as tools for applied plant breeding. *Plant Mol. Biol.* 5: 109–118.

Helminen, P., Ehnholm, C., Lokki, M.-L., Jeffreys, A. J., and Peltonen, L. (1988) Application of DNA "fingerprints" to paternity determinations. *Lancet* i: 574–576.

Hillel, J., Plotzky, Y., Haberfeld, A., Lavi, U., Cahaner, A., and Jeffreys, A. J. (1989) DNA fingerprints of poultry. *Anim. Genet.* 20: 145–155.

Hillel, J., Schaap, T., Haberfeld, A., Jeffreys, A. J., Plotzky, Y., Cahaner, A., and Lavi, U. (1990) DNA fingerprints applied to gene introgression in breeding programs. *Genetics* 124: 783–789.

Huey, B., and Hall, J. (1989) Hypervariable DNA fingerprinting in *Escherichia coli*: minisatellite probe from bacteriophage M13. *J. Bacteriol.* 171: 2528–2532.

Jeffreys, A. J., and Morton, D. B. (1987) DNA fingerprints of dogs and cats. *Anim. Genet.* 18: 1–15.

Jeffreys, A. J., Wilson, V., and Thein, S. L. (1985a) Individual-specific 'fingerprints' of human DNA. *Nature* 316: 76–79.

Jeffreys, A. J., Brookfield, J. F. Y., and Semeonoff, R. (1985b) Positive identification of an immigrant test-case using human DNA fingerprints. *Nature* 317: 818–819.

Kuhnlein, U., Dawe, Y., Zadworny, D., and Gavora, J. S. (1989) DNA fingerprinting: a tool for determining genetic distances between strains of poultry. *Theor. Appl. Genet.* 77: 669–672.

Kuhnlein, U., Zadworny, D., Dawe, Y., Fairfull, R. W., and Gavora, J. S. (1990) Assessment of inbreeding by DNA fingerprinting: development of a calibration curve using defined strains of chickens. *Genetics* 125: 161–165.

Landry, B. S., Kesseli, R., Hei Leung, and Michelmore, R. W. (1987) Comparison of restriction endonucleases and sources of probes for their efficiency in detecting restriction fragment length polymorphisms in lettuce (*Lactuca sativa* L.). *Theor. Appl. Genet.* 74: 646–653.

Lynch, M. (1988) Estimation of relatedness by DNA fingerprinting. *Mol. Biol. Evol.* 5: 584–599.

310

Martienssen, R. A., and Baulcombe, D. C. (1989) An unusual wheat insertion sequence (WISI) lies upstream of an alpha-amylase gene in hexaploid wheat, and carries a "minisatellite" array. *Mol. Gen. Genet.* 217: 401–410.

McCouch, S. R., Kochert, G., Yu Z. H., Khush, G. S., Coffman, W. R., and Tanksley, S. D. (1988) Molecular mapping of rice chromosomes. *Theor. Appl. Genet.* 76: 814–829.

Nagamine, T., Todd, G. A., McCann, K. P., Newbury, H. J., and Ford-Lloyd, B. V. (1989) Use of restriction fragment length polymorphisms to fingerprint beets at the genotype and species levels. *Theor. Appl. Genet.* 78: 847–851.

Nybom, H. (1990a) DNA fingerprints in sports of 'Red Delicious' apples. *HortScience* 25: 1641–1642.

Nybom, H. (1990b) Genetic variation in ornamental apple trees and their seedlings (*Malus*, Rosaceae) revealed by DNA 'fingerprinting'. *Hereditas* 113: 17–28.

Nybom, H. (1990c) Evaluation of DNA fingerprints by computer vs occular inspection. *Fingerprint News* 2(4): 10–15.

Nybom, H., and Hall, H. K. (1991) Minisatellite DNA "fingerprints" can distinguish *Rubus* cultivars and estimate their degree of relatedness. *Euphytica* 53: 107–114.

Nybom, H., and Rogstad, S. H. (1990) DNA "fingerprints" detect genetic variation in *Acer negundo. Plant Syst. Evol.* 173: 49–56.

Nybom, H., Rogstad, S. H., and Schaal, B. A. (1990) Genetic variation detected by use of the M13 "DNA fingerprint" probe in *Malus, Prunus*, and *Rubus* (Rosaceae). *Theor. Appl. Genet.* 79: 153–156.

Nybom, H., and Schaal, B. A. (1990a) DNA "fingerprints" applied to paternity analysis in apples (*Malus* × *domestica*). *Theor. Appl. Genet.* 79: 763–768.

Nybom, H., and Schaal, B. A. (1990b) DNA "fingerprints" reveal genotypic distributions in natural populations of blackberries and raspberries (*Rubus*, Rosaceae). *Amer. J. Bot.* 77: 883–888.

Nybom, H., Schaal, B. A., and Rogstad, S. H. (1989) DNA "fingerprints" can distinguish cultivars of blackberries and raspberries. 5th International Symposium on *Rubus & Ribes*, *Acta Hortic.* 262: 305–310.

Reeve, H. K., Westneat, D. F., Noon, W. A., Sherman, P. W., and Aquadro, C. F. (1990) DNA "fingerprinting" reveals high levels of inbreeding in colonies of the eusocial naked mole rat. *Proc. Natl. Acad. Sci. USA* 87: 2496–2500.

Rogstad, S. H., Patton II, J. C., and Schaal, B. A. (1988a) M13 repeat probe detects DNA minisatellite-like sequences in gymnosperms and angiosperms. *Proc. Natl. Acad. Sci. USA* 85: 9176–9178.

Rogstad, S. H., Patton II, J. C., and Schaal, B. A. (1988b) A human minisatellite probe reveals RFLPs among individuals of two angiosperms. *Nucleic Acids Res.* 18: 1081.

Rogstad, S. H., Herwaldt, B. L., Schlesinger, P. H., and Krugstad, D. J. (1989) The M13 repeat probe detects RFLPs between two strains of the protozoan malaria parasite *Plasmodium falciparum. Nucleic Acids Res.* 17: 3610.

Ryskov, A. P., Jincharadze, A. G., Prosnyak, M. I., Ivanov, P. L., and Limborska, S. A. (1988) M13 phage DNA as a universal marker for DNA fingerprinting of animals, plants and microorganisms. *FEBS Letters* 233: 388–392.

Saghai-Maroof, M. A., Soliman, K. M., Jorgensen, R. A., and Allard, R. W. (1984) Ribosomal DNA spacer-length polymorphism in barley: Mendelian inheritance, chromosomal location, and population dynamics. *Proc. Natl. Acad. Sci. USA* 81: 8014–8018.

Striem, M. J., Spiegel-Roy, P., Ben-Hayyim, G., Beckmann, J., and Gidoni, D. (1990) Genomic DNA fingerprinting of *Vitis Vinifera* by the use of multi-locus probes. *Vitis* 29: 223–227.

Turner, B. J., Elder, J. F., Laughlin, T. F., and Davis, W. P. (1990) Genetic fingerprinting in clonal vertebrates detected by simple sequence DNA fingerprinting. *Proc. Natl. Acad. Sci. USA* 87: 5653–5657.

Vassart, G., Georges, M., Monsieur, R., Brocas, H., Lequarré, A.-S., and Christophe, D. (1987) A sequence in M13 phage detects hypervariable minisatellites in human and animal DNA. *Science* 235: 683–684.

Weising, K., Weigand, F., Driesel, A. J., Kahl, G., Zischler, H., and Epplen, J. T. (1989) Polymorphic simple GATA/GACA repeats in plant genomes. *Nucleic Acids Res.* 17: 10128.

Weising, K., Fiala, B., Ramloch, K., Kahl, G., and Epplen, J. T. (1990) Oligonucleotide fingerprinting in angiosperms. *Fingerprint News* 2(2): 5–10.

Weising, K., and Kahl, G. (1990) DNA fingerprinting in plants, the potential of a new method. *Biotech-Forum/Europe* 7: 230–235.

Westneat, D. F., Noon, W. A., Reeve, H. K., and Aquadro, C. F. (1988) Improved hybridization conditions for DNA 'fingerprints' probed with M13. *Nucleic Acids Res.* 16: 4161.

Wetton, J. H., Carter, R. E., Parkin, D. T., and Walters, D. (1987) Demographic study of a wild house sparrow population by DNA fingerprinting. *Nature* 327: 147–149.

Woodson, W. R. (1989) DNA fingerprinting for cultivar identification in carnation. 86th Annual Meeting Amer. Soc. Hortic. Sci., USA July 29–Aug 3, 1989 *Hort Science* 0 (Suppl.): 100.

Zimmerman, P. A., Lang-Unnasch, N., and Cullis, C. A. (1989) Polymorphic regions in plant genomes detected by an M13 probe. *Genome* 32: 824–828.

DNA Fingerprinting: Approaches and Applications
ed. by T. Burke, G. Dolf, A. J. Jeffreys & R. Wolff
© 1991 Birkhäuser Verlag Basel/Switzerland

Oligonucleotide Fingerprinting in Plants and Fungi

K. Weising[1], J. Ramser[1], D. Kaemmer[1], G. Kahl[1], and J. T. Epplen[2]

[1]*Pflanzliche Molekularbiologie, Fachbereich Biologie, Johann Wolfgang Goethe-Universität Frankfurt, Germany;* [2]*Max-Planck-Institut für Psychiatrie, Martinsried, Germany*

Summary
Synthetic oligonucleotides complementary to simple repetitive DNA sequence motifs are now routinely applied for multilocus DNA fingerprinting of humans and a large variety of animal species. Most recently, these probes have also been used successfully for the analysis of plant and fungal genomes. All simple motifs investigated to date (CA-, CT-, GATA-, GACA-, GAA-, GTG-, GGAT- and TCC-multimers) are present and repeated to various extents throughout the plant and fungal kingdoms. Usually, these probes reveal intra- and interspecific genetic variability resulting in polymorphic or even hypervariable banding patterns. Depending on the combination of species and oligonucleotide probe, species- variety-, accession-, strain- or individual-specific "fingerprints" were obtained in plants and fungi. Somatic stability was observed. For their successful application to DNA fingerprinting, the optimal probe/species-combinations that give distinct banding patterns have to be developed empirically. Various applications of plant DNA fingerprinting using oligonucleotide probes are suggested: (1) characterization of the extent of genetic variability within races, (2) assessment of the "purity" of inbred lines, (3) selection of the recurrent parental genome in backcross breeding programs, (4) identification of crop cultivars and fungal strains, (5) characterization of fusion hybrids, (6) evaluation of the extent of somaclonal variation at the molecular level.

Introduction

DNA fingerprinting makes use of the presence of hypervariable repetitive DNA sequences in the eukaryotic genome. Although highly polymorphic loci in the human genome had been detected as early as 1980 (Wyman and White, 1980), it was only in 1985 that genetic individualization of humans was achieved by hybridization to DNA sequences derived from multiple hypervariable DNA loci (Jeffreys *et al.*, 1985a, b). The human "minisatellites" 33.6 and 33.15, characterized in detail by Jeffreys and coworkers, contain a 10–15 bp GC-rich core sequence that was shown to cross-hybridize to a variety of mammalian (e.g. Jeffreys and Morton, 1987), avian (e.g. Burke and Bruford, 1987) and even plant genomes (Dallas, 1988; Rogstad *et al.*, 1988b), thereby revealing individual-specific patterns in several animal species. The polymorphic character of minisatellite-like sequences is mainly due to a variable number of tandem repeats, but also to some internal restriction site heterogeneity (Jeffreys *et al.*, 1990). Surprisingly, an internal repeat

Table 1. Probes used for DNA fingerprinting plant and fungal genomes

Probe & species	References
Human minisatellites (33.6; 33.15)	
Colletotrichum gloeosporioides	Braithwaite and Manners, 1989
Oryza sativa	Dallas, 1988
Oryza glaberrima	Dallas, 1988
Populus deltoides	Rogstad *et al.*, 1988b
Populus tremuloides	Rogstad *et al.*, 1988b
M13 repeat	
Fusarium sp.	Monastyrskii *et al.*, 1990
Pinus torreyana	Rogstad *et al.*, 1988a
Asimina triloba	Rogstad *et al.*, 1988a
Polyalthia glauca	Rogstad *et al.*, 1988a
Populus tremuloides	Rogstad *et al.*, 1988a
Lycopersicum esculentum	Rogstad *et al.*, 1988a
Arabidopsis thaliana	Zimmerman *et al.*, 1989
Linum bienne	Zimmerman *et al.*, 1989
Linum usitatissimum	Zimmerman *et al.*, 1989
Medicago sativa	Zimmerman *et al.*, 1989
Malus domestica	Nybom and Schaal, 1990a
	Nybom *et al.*, 1990
Prunus serotina	Nybom *et al.*, 1990
Rubus (several species)	Nybom and Schaal, 1990b
Random potato cDNA and genomic clones	
Solanum tuberosum	Gebhardt *et al.*, 1989
Minisatellite-like wheat sequence	
Triticum aestivum	Martienssen *et al.*, 1989
Random beet cDNA clone	
Beta vulgaris	Nagamine *et al.*, 1989
Simple repetitive sequences	
a) Fungi	
Saccharomyces cerevisiae	Walmsley *et al.*, 1989
Penicillium	Meyer *et al.*, 1991
Trichoderma	Meyer *et al.*, 1991
Aspergillus	Meyer *et al.*, 1991
Candida	Meyer *et al.*, 1991
Saccharomyces	Meyer *et al.*, 1991
Ascochyta rabiei	Weising *et al.*, submitted
b) Plants	
Cicer arietinum	Weising *et al.*, 1989, 1990, 1991
Lens culinaris	Weising *et al.*, 1989, 1990, 1991
Brassica napus	Weising *et al.*, 1989, 1990, 1991
Solanum tuberosum	Weising *et al.*, 1989, 1990, 1991
Nicotiana (several species)	Weising *et al.*, 1989, 1990, 1991
Beta vulgaris	Weising *et al.*, 1989, 1990, 1991
Hordeum vulgare	Weising *et al.*, 1989, 1990, 1991
Hordeum spontaneum	Weising *et al.*, 1989, 1990, 1991
Musa acuminata	Weising *et al.*, 1989, 1990, 1991
Beta vulgaris	Beyermann *et al.*, 1991
Hordeum vulgare	Beyermann *et al.*, 1991

sequence from bacteriophage M13 also detected polymorphic and hypervariable loci in a variety of organisms including vertebrates (Vassart *et al.*, 1987; Georges *et al.*, 1988), plants (Ryskov *et al.*, 1988; Rogstad *et al.*, 1988a; Zimmerman *et al.*, 1989; Nybom *et al.*, 1989, 1990; Nybom and Schaal, 1990a, b; see Tab. 1), fungi (Monastyrskii *et al.*, 1990) and even some bacteria (Huey and Hall, 1989).

Whereas minisatellite monomers are usually 10–35 bp long, tandem repeats of shorter motifs also exist in the DNA of most eukaryotes. These so-called "simple sequences" or "microsatellites" are ubiquitous components of all eukaryotic genomes analyzed so far (Tautz and Renz, 1984; Greaves and Patient, 1985; Epplen, 1988; Weising *et al.*, 1991). This particular class of repetitive DNA is made up from tandemly repeated short motifs of 2–10 bp, forming more or less monotonous stretches of variable length. Simple sequences are a major source of genomic variation (Tautz *et al.*, 1986), and it was therefore not surprising that probes complementary to simple motifs could be successfully applied to DNA fingerprinting of humans (Ali *et al.*, 1986; Schäfer *et al.*, 1988). The latter authors used in-gel hybridization with labeled oligonucleotide probes, a method exhibiting several advantages over the conventional blotting techniques: it is faster; prehybridization steps can be omitted; and a "100% transfer efficiency" is obtained. Oligonucleotide probes complementary to simple repetitive sequences, e.g. $(CAC)_5$ or $(GATA)_4$, are now increasingly used as multilocus probes revealing hypervariable target regions in many eukaryotic organisms. Meanwhile, more than 150 animal and plant species have been successfully fingerprinted with oligonucleotide probes (Epplen *et al.*, 1991). The present review summarizes the available results on oligonucleotide fingerprinting in plant and fungal genomes. Some applications of the technique in plant biology and breeding are discussed.

Simple Repetitive Sequences Are Ubiquitous Components of Animal, Plant, and Fungal Genomes

Since 1982 several systematical surveys on the occurrence of simple repetitive motifs in human, animal and yeast genomes have been undertaken (Hamada *et al.*, 1982; Tautz and Renz, 1984; Tautz *et al.*, 1986; Greaves and Patient 1985; Miklos *et al.*, 1989). The general outcome of these studies was that all simple sequences tested were represented to some degree in all eukaryotic species. Their abundance and organization, however, varied widely. Several bands or a smear over the whole molecular weight range were usually observed. Such hybridization patterns suggested that in contrast to classical satellite DNAs, simple sequences are dispersed rather than clustered in the genomes. DNA sequences exhibiting either pure or cryptic simplicity are thought to originate mainly from slipped-strand mispairing during replication (Levinson and Gutman, 1987). Some motifs, however, can also be generated by telomerase activities (Zakian, 1989). Once a stretch of simple sequences is present, heterogeneity may be created by slipping combined with point mutations, unequal crossover, and recombinational events (Tautz and Renz, 1984; Tautz *et al.*, 1986; Levinson and Gutman, 1987).

Since the discovery of simple sequences, their functional impact has been investigated. Whereas simple sequences comply with the concept of "selfish" DNA (Orgel and Crick, 1980) in being able to amplify and propagate in the absence of counterselective pressure, nevertheless possible functions have been implicated and discussed. Among these are the replication of telomeres and protection of chromosomal ends (Zakian, 1989), gene conversion and recombination (Hentschel, 1982; Rogers, 1983), transcriptional regulation of adjacent genes (e.g. Hamada *et al.*, 1984b), and sex determination especially in the case of GATA/GACA repeats (reviewed by Epplen, 1988). These functions may relate to the ability of several simple sequence stretches to form special conformations such as H-DNA, Z-DNA or hairpin structures. Under specific *in vitro* conditions, at least two types of simple sequences are able to adopt non-B-DNA conformation. (1) Pure homopurine-homopyrimidine stretches such as $(GA)_n$ or $(G)_n$ may form H-DNA, a hinged DNA structure being composed of a triple helix and single-stranded regions (Htun and Dahlberg, 1989). H-DNA is hypersensitive to the single-strand-specific S1 nuclease in supercoiled plasmids and sometimes also *in vivo* (see e.g. Weintraub, 1983; Margot and Hardison, 1985). These so-called S1-hypersensitive sites have been implicated in transcriptional regulation (Weintraub, 1983) and recombination (Hentschel, 1982). (2) Alternating purine/pyrimidine repeats such as $(CA)_n$ or $(CG)_n$, on the other hand, may form left-handed Z-DNA which is also thought to

---

Figure 1. Simple repetitive DNA motifs are present and organized distinctly in the genomes of fungi, lower and higher plants. HinfI-digested DNA was electrophoresed and consecutively hybridized in the gel to six different ^{32}P-labelled oligonucleotide probes as indicated. Positions of molecular weight markers are given in kb.

Fungi:	a) *Phycomyces blakesleeanus*, b) *Coprinus comatus*
Algae:	c) *Oedogonium spec.*, d) *Stenogramme interrupta*
Mosses:	e) *Polytrichum formosum*
Ferns:	f) *Equisetum arvense*, g) *Polypodium vulgare*, h) *Osmunda regalis*, i) *Dodia caudata*
Gymnosperms:	k) *Juniperus communis*
Monocots:	l) *Echinodorus osiris*, m) *Hordeum spontaneum* n) *Musa acuminata*, o) *Asparagus densiflorus*, p) *Chamaedorea cataracterum*
Dicots:	q) *Helleborus niger*, r) *Silene alba*, s) *Rumex acetosella*, t) *Urtica dioica*, u) *Humulus lupulus*, v) *Ficus benjamina*, w) *Lens culinaris*, x) *Cicer arietinum*, y) *Simmondsia sinensis*, z) *Aruncus dioicus*, a') *Brassica napus* var. oleifera, b') *Camellia sinensis*, c') *Solanum tuberosum*, d') *Nicotiana tabacum* var. atropurpurea, e') *Lactuca sativa*.

316

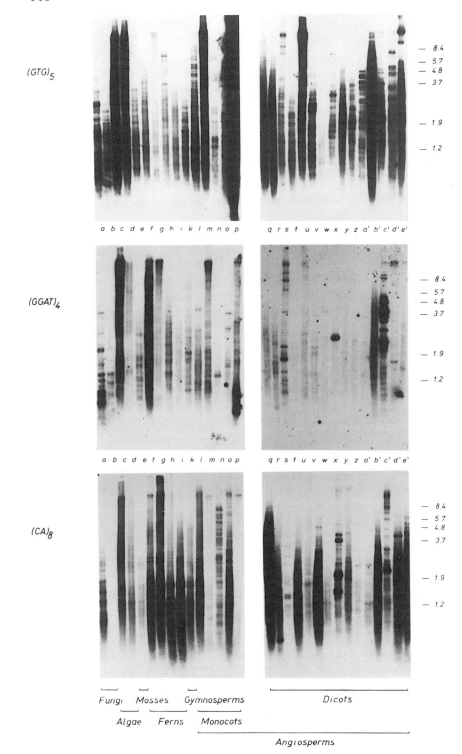

$(GTG)_5$

$(GGAT)_4$

$(CA)_8$

— 8.4
— 5.7
— 4.8
— 3.7

— 1.9

— 1.2

a b c d e f g h i k l m n o p   q r s t u v w x y z a' b' c' d' e'

Fungi   Mosses   Gymnosperms   Dicots

Algae   Ferns   Monocots

Angiosperms

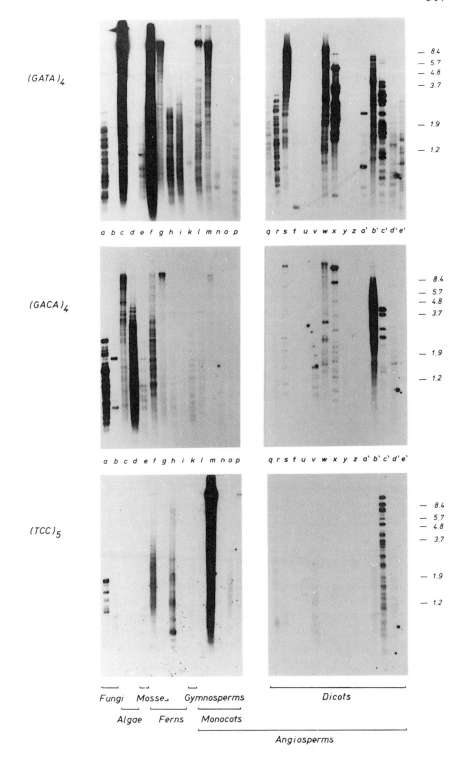

(GATA)₄

(GACA)₄

(TCC)₅

— 8.4
— 5.7
— 4.8
— 3.7

— 1.9

— 1.2

a b c d e f g h i k l m n o p   q r s t u v w x y z a' b' c' d' e'

Fungi   Mosses   Gymnosperms   Dicots

Algae   Ferns   Monocots

Angiosperms

influence the expression of cellular genes (Nordheim and Rich, 1983; Rich *et al.*, 1984). Indeed, poly(CA) was found to enhance transient gene expression when placed upstream of an SV-40 promoter and a reporter gene in a transfection experiment (Hamada *et al.*, 1984b). It was, however, also shown that the majority of $(CA)_n$ sequences found in animal genomes are not likely to be in the Z-conformation *in vivo* (Gross *et al.*, 1985). All in all, these results are consistent with the notion that the majority of simple sequences do indeed behave "selfishly". Some motifs, however, could perhaps acquire specific functions when located at certain positions within the genome, e.g. close to genes or within introns.

In contrast to the many data obtained within animal genomes, until recently essentially nothing was known about the occurrence and organization of simple sequences in plant genomes, the only exception being a study on telomeric repeats in *Arabidopsis* (Richards and Ausubel, 1988). This is somewhat surprising, since repetitive DNA sequences are even more abundant genomic elements in plants than in animals (Flavell, 1986). Most studies on plant repetitive DNA performed to date, however, were dedicated to the characterization of satellite DNA, transposons, and ribosomal DNA (reviewed in Flavell, 1986). This apparent lack of information prompted us to investigate the occurrence of simple sequences in plant and fungal genomes. Since preliminary experiments had provided evidence for the presence of repetitive GATA- and GACA-motifs in the genomes of banana, barley, rapeseed, potato, lentil and chickpea (Weising *et al.*, 1989; 1990), we undertook a more systematic study. HinfI-digested DNAs from a large number of species including fungi, algae, mosses, ferns, gymnosperms and angiosperms were screened for the presence and organization of various simple repetitive motifs by in-gel hybridization to the corresponding oligonucleotide probes (Weising *et al.*, 1991). Representative results obtained with the probes $(GATA)_4$, $(GACA)_4$, $(TCC)_5$, $(GTG)_5$, $(GGAT)_4$ and $(CA)_8$ are shown in Fig. 1. All sequence motifs were found to be present in the majority of tested species. Some motifs were so highly abundant in some species that a smear resulted on the autoradiograms, a situation reminiscent to the distribution of $(CA)_n$ repeats in human genomes (Hamada and Kakunaga, 1982; Hamada *et al.*, 1984a). Even very short exposure times did not reveal any distinct bands in these cases. This kind of pattern can be interpreted as resulting from a high copy number of short simple sequence stretches dispersed randomly all over the genome. Other motifs occured relatively rarely in some species, faint bands being only observed upon prolonged exposure of the autoradiograms (e.g. TCC-repeats in most species). In many cases, however, distinct banding patterns were revealed by the probes, sometimes superimposed on a smear. The occurrence of a limited number of strong bands suggests the presence of long, more or less homogeneous stretches of the respective motifs at a limited number of loci.

In summary, these results showed all simple sequences tested to be present and repetitive to various extents throughout the plant and fungal kingdoms, thereby confirming and extending the results obtained with animal and yeast genomes (e.g. Hamada *et al.*, 1982; Tautz and Renz, 1984; Greaves and Patient, 1985). Obviously, simple repetitive motifs are ubiquitous components of most, if not all, eukaryotic genomes. The fact that the relative abundance and the organization of specific motifs vary considerably between species and do not bear any obvious correlation to phylogenetic categories strengthens the hypothesis that simple sequences may have arisen independently throughout evolution (Levinson *et al.*, 1985).

Polymorphic Simple Repetitive Sequences in Plant and Fungal Genomes May Be Subjected to Oligonucleotide Fingerprinting

Within the last few years, several groups reported on the presence and variability of sequences complementary to the human minisatellites 33.6 and 33.15 and the M13 repeat sequence in plant genomes (see Tab. 1 and Nybom, 1991). The general outcome of these studies was that minisatellites provide useful tools for revealing genetic polymorphisms, i.e. for DNA fingerprinting, within a wide variety of plant species. In order to expand the collection of informative probes, we introduced selected oligonucleotide probes complementary to simple repetitive motifs to plant DNA fingerprint analyses. Since simple sequences are polymorphic or even hypervariable in animal genomes, we expected them to behave similarly in plants. This was indeed the case. Initial studies revealed various degrees of intraspecific variability within six species of angiosperms (Weising *et al.*, 1989; 1990). A more detailed study on the variability between different accessions of chickpea (*Cicer arietinum*), a Mediterranean crop plant, showed that the observed extent of polymorphism largely depended on the probe, and only slightly on the enzyme (Fig. 2; Weising, K., Weigand, F., Kahl, G., and Epplen, J. T., in preparation). In this study, hybridization of $(GATA)_4$ and $(CA)_8$ to TaqI-digested chickpea DNA reveals highly variable patterns, and $(GACA)_4$, $(GGAT)_4$ and $(TCC)_5$ show limited heterogeneity. In contrast, $(GTG)_5$ provides no intraspecific information at all. The "fingerprint" exhibited by $(GTG)_5$ is likely to be a species-specific one, since patterns obtained with other species from the genus *Cicer* look different (Fig. 3).

That distinct oligonucleotide probes exhibit different levels of informativeness within one and the same species was also observed in the case of barley and sugarbeet (Beyermann *et al.*, 1991), tobacco, lentil

320

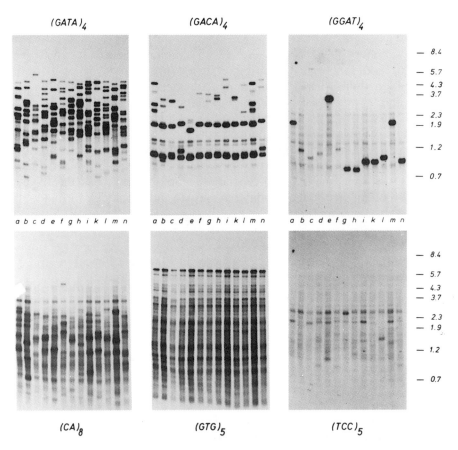

Figure 2. Different oligonucleotide probes reveal various levels of intraspecific polymorphism within chickpea (*Cicer arietinum*). DNA from thirteen chickpea accessions derived from the chickpea germplasm collection of ICARDA, Aleppo (Syria), were screened for genetic polymorphisms by in-gel hybridization with the indicated ^{32}P-labeled oligonucleotide probes. Digestion was performed with TaqI. Positions of molecular weight markers are given in kilobases.

and rapeseed (Weising *et al.*, 1991). It is possibly a general phenomenon that some sequences are turned over more rapidly than others. However, a high or low level of informativeness is not a property of the probe sequence *per se*. To give an example, GATA-repeats are polymorphic in chickpea (Fig. 2), but not in tobacco. The reverse is true for GTG-repeats (Weising *et al.*, 1991). Moreover, (GTG)$_5$ is the least informative probe in chickpea but the most informative probe in man (Schäfer *et al.*, 1988). The reasons for this completely unpredictable behaviour are unkown. In any case, the variability-creating mechanisms probably act also on adjacent sequences interspersed with the simple repeats to generate large-scale differences.

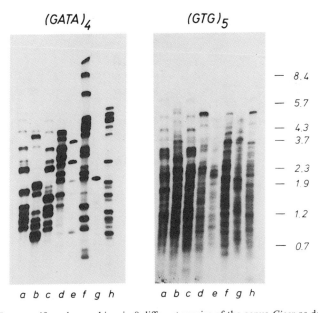

Figure 3. Interspecific polymorphism in 8 different species of the genus *Cicer* as demonstrated by oligonucleotide fingerprinting with the (GATA)₄ and (GTG)₅ probes, respectively. Note that (GTG)₅ differentiates between species, but not between accessions. Digestion was performed with TaqI. Positions of molecular weight markers are given in kilobases.
a) *C. yamashitae* b) *C. pinnatifidum* c) *C. judaicum* d) *C. reticulatum* e) *C. chorassanicum* f) *C. bijugum* g) *C. cuneatum* h) *C. echinospermum.*

In the case of some probe/species-combinations, not only accessions and varieties, but also individuals can be distinguished. For example, Fig. 4 shows that (GATA)₄ and (CA)₈ reveal differences between individuals belonging to the same chickpea accessions. Whereas patterns are more or less conserved within accessions, at least in the case of (GATA)₄, many fragments show slight, but individual-specific differences in mobility. This is possibly caused by slight variation in the number of tandemly repeated simple sequence units among alleles at some loci.

In general, individual-specific patterns are more frequently observed in animals than in plants. This may be explained by the occurrence of different reproductive strategies. Whereas sexual reproduction and outcrossing are the rule in most animal populations, selfing and vegetative propagation are widespread phenomena in plants. The highest levels of variability observed to date were obtained by hybridizing (GATA)₄ with DNA from different individuals of campion (*Silene alba*), a wildflower belonging to the pink family (Ramser, J., unpubl. observation). In this species, outcrossing is ensured by dioecy. On the other hand, within highly inbred tomato lines almost no variability was detected upon hybridization to (GATA)₄, which is an informative

322

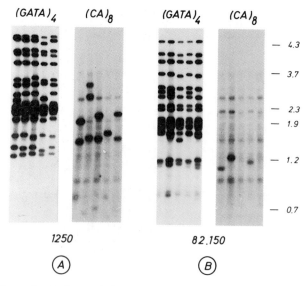

Figure 4. Oligonucleotide fingerprinting with (GATA)$_4$ and (CA)$_8$ detects variability between individual plants derived from the same chickpea accession. Digestion was performed with TaqI. Positions of molecular weight markers are given in kilobases.
A: 5 plants of *Cicer arietinum*, accession ILC 1250.
B: 5 plants of *Cicer arietinum*, accession FLIP 82.150.

probe for tomato (Beyermann, B., Weising, K., Kaemmer, D., Kahl, G., and Epplen, J. T.; in preparation). Whether individual-specificity in plant DNA fingerprinting is detected or not is obviously dependent upon at least three factors: (1) the reproductive strategies exhibited by the species under investigation (selfing, outcrossing, vegetative propagation), (2) the breeding history in the case of crop plants; e.g. the extent of inbreeding, (3) the probe itself (see above).

Oligonucleotide fingerprinting has so far not only been successfully applied in plants, but also in fungi. Walmsley *et al.* (1989) demonstrated the use of synthetic poly (dG-dT) for distinguishing between a collection of diverse strains of yeast (*Saccharomyces cerevisiae*). Meyer *et al.* (1991) analyzed the fingerprinting patterns of several genera of yeast and filamentous fungi. These authors found that genera, species, and strains of *Aspergillus*, *Penicillium*, *Candida*, *Trichoderma* and *Saccharomyces* could be discriminated by hybridization to (CT)$_8$, (GTG)$_5$, (GATA)$_4$, (GACA)$_4$, and the M13 repeat. Different clones obtained from one and the same strain yielded similar patterns in this study. We investigated six different pathotypes of *Ascochyta rabiei*, a fungal pathogen for chickpea. Figure 5 shows that four out of six single spore-derived mycelia yield clearly distinguishable fingerprint patterns (Weising, K., Kaemmer, D., Weigand, F., Epplen, J. T., and Kahl, G., submitted for publication). Clearly, differentiation of fungal strains by

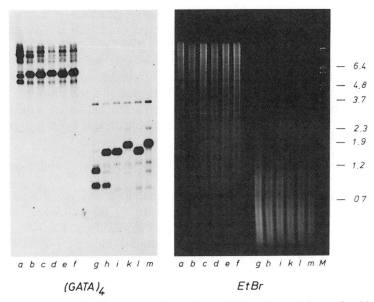

Figure 5. "Pathotyping" different races of *Ascochyta rabiei*, a pathogenic fungus for chickpea, by oligonucleotide fingerprinting. DNA was isolated from single spore-derived mycelia of 6 different fungal isolates. The isolates represent 6 races differing in their level of pathogenicity against chickpea. After digestion with EcoRI (lanes a–f) or HinfI (lanes g–m), fungal DNA was electrophoresed and hybridized in the gel to the (GATA)₄ oligonucleotide. Four different patterns were obtained. The patterns of races 3 and 5, and 4 and 6, respectively, are in each case indistinguishable. EtBr: ethidium bromide staining of the gel. Positions of molecular weight markers are given in kilobases.

oligonucleotide fingerprinting will be of particular use for brewers, bakers, and plant pathologists.

In summary, it can be concluded that the pattern complexity and variability obtained strongly depend on the repeated sequence motif used for hybridization in the species under investigation. The optimal combination of probe and species has to be determined empirically for each purpose. Using different probes, species-, variety- or individual-specific patterns can be obtained. Taking into account that simple sequences generally tend to be hypervariable (Tautz *et al.*, 1986; Tautz, 1989) and that even random 14 bp repeats revealed polymorphisms in the human genome (Vergnaud *et al.*, 1989), informative probes for oligonucleotide fingerprinting are probably not so rare.

Oligonucleotide Fingerprinting May Be Applied to Plant Biology and Breeding

In recent years, molecular marker techniques have gained widespread applications in many fields of plant genetics and breeding. Isozymes and restriction fragment length polymorphisms (RFLPs) have provided valuable tools for linkage analysis and the establishment of genetic maps in all major crop plants (reviewed by Tanksley et al., 1989). Multilocus DNA fingerprinting of hypervariable genomic regions will complement the classical RFLP-analyses of plants and fungi. A variety of probes are already available (Tab. 1): human minisatellites, the M13 repeat, and a probably unlimited collection of simple repetitive sequences. In plant breeding programs oligonucleotide fingerprinting may be used for:

(1) the molecular characterization of the extent of variability within races and cultivars;
(2) the assessment of the extent of inbreeding, as suggested by Kuhnlein et al. (1990); the genetic uniformity of highly inbred tomato lines has been observed (Beyermann, B., Weising, K., Kaemmer, D., Kahl, G., and Epplen, J. T., in preparation);
(3) DNA fingerprinting-assisted gene introgression by repeated backcross breeding (Hillel et al., 1990; the offspring individuals may be selected for maximal similarity to the recipient line and minimal similarity to the donor line, thereby reducing the required number of backcross generations);
(3) the identification of varieties and cultivars. DNA fingerprinting has already made possible the genetic characterization of plant species, varieties, cultivars, accessions, and of fungal strains. Thus it may serve as a tool for variety identification and protection of breeder's rights. However, several limitations may emerge in this field. First, depending on the extent of inbreeding, slight variations of patterns may occur between individuals (see Fig. 4), thereby making the classification of an unknown sample difficult. Second, a characteristic property of hypervariable sequences is, of course, their high mutation rate. Consequently, the "typical" pattern of a cultivar may change as a function of time and geographical distribution. It remains to be proven that a specific cultivar exhibits a constant pattern all over the world. Probes that reveal limited polymorphism rather than hypervariability are surely preferable for this kind of application.

Similar limitations have to be excluded experimentally for the estimation of genetic relatedness within populations by DNA fingerprinting. Paternity testing is, of course, possible and has been successfully done in apple trees (Nybom and Schaal, 1990a). This kind of study might be especially promising in the analysis of, for example, reproductive strategies of tropical rainforest trees; demographic studies on natural plant populations have already been initiated (raspberry and blackberry; Nybom and Schaal, 1990b). Third- or fourth-order relationships within sibships, however, are not easily detected by DNA fingerprinting with hypervariable probes (Lynch, 1988). This is even more true for taxonomic studies. Evolutionary relatedness within and between plant species and genera has been investigated by conventional RFLP analysis of chloroplast (Hosaka *et al.*, 1988; Dally and Second, 1990) and nuclear DNA (Debener *et al.*, 1990; Song *et al.*, 1990). It remains to be tested whether DNA fingerprinting using simple sequences of *low* variability (such as the GTG-repeat in chickpea) will provide an adequate tool in this area of research.

Another application of plant DNA fingerprinting implicates the molecular characterization of cell hybrids. Somatic hybridization of plant cells offers new opportunities to plant breeders by allowing the combination of properties from distantly related species via the transfer of advantageous traits (Harms, 1983). Characterization of hybrid plants produced by somatic hybridization is usually performed by morphological and isozyme markers. These techniques can be complemented directly at the DNA level by DNA fingerprinting. Direct analysis of the DNA will circumvent errors originating from modified transcription and translation levels of marker enzymes. More or less the whole genome is investigated using a collection of different oligonucleotide probes.

Finally, the use of DNA fingerprinting in plant tissue culture and genetic engineering can be encouraged. The technique of direct gene transfer usually involves the transformation of protoplasts and the regeneration of transgenic plants via a callus stage (reviewed by Weising *et al.*, 1988; Davey *et al.*, 1989). Recent evidence has shown that substantial molecular changes may occur in the course of regeneration from protoplasts (Brown *et al.*, 1990). This phenomenon, usually referred to as "somaclonal variation" (Lee and Phillips, 1988), is probably induced by stress during tissue culture. Primary regenerants often show little variation (mutations are heterozygous and recessive), whereas a proportion of their offspring exhibit phenotypic effects due to their switch to homozygosity. Undetected somaclonal variation may act as a severe limitation in the application of transformation procedures to breeding programs. Oligonucleotide fingerprinting may detect molecular changes and may thus serve as a fast screening method for the extent of somaclonal variation in different primary transformants and in the comparison of different transformation/regeneration protocols. Optimal protocols and the least affected transformants could be selected for subsequent projects.

Now that a variety of multilocus probes are available for plant DNA fingerprinting, the development of new probes should focus on multiallelic *single-locus* probes. These will be advantageous for several purposes such as linkage analysis and genome mapping. The most obvious strategy for obtaining single-locus probes is the cloning of hypervariable loci and the characterization of flanking regions that exhibit locus-specific hybridization. This has been done with human minisatellites (e.g.

326

Wong *et al.*, 1986), and also with the simple sequence $(CAC)_n$ (Epplen *et al.*, 1991). Flanking regions can also be used as locus-specific primers for DNA amplification in the polymerase chain reaction (Jeffreys *et al.*, 1988).

Most of the applications for plant DNA fingerprinting summarized above have to be tested experimentally. However, keeping the above-mentioned possible restrictions in mind, there is reason to believe that the technique will be of considerable use for plant breeders, plant pathologists, and population biologists. The availability of nonradioactive techniques will be of special importance in the applied fields. Digoxigenin-labeled oligonucleotide probes (Zischler *et al.* 1989) are already available and chemiluminescence methods might be developed in the near future. These methods can then be routinely used in institutions lacking radioisotope facilities, for example in plant breeding institutes.

Acknowledgements
We appreciate the help of S. Kost for preparation of the figures. K. Weising acknowledges a fellowship from DECHEMA (Frankfurt, FRG). The simple repeated oligonucleotides are subject to patent applications. Commercial inquiries should be directed to Fresenius AG, Oberursel, FRG.

References

Ali, S., Müller, C. R., and Epplen, J. T. (1986) DNA fingerprinting by oligonucleotide probes specific for simple repeats. *Hum. Genet.* 74: 239–243.

Beyermann, B., Nürnberg, P., Weihe, A., Meixner, M., Epplen, J. T., and Börner, T. (1991) Fingerprinting plant genomes with oligonucleotide probes specific for simple repetitive DNA sequences. *Theor. Appl. Genet.*, (in press).

Braithwaite, K. S., and Manners, J. M. (1989) Human hypervariable minisatellite probes detect DNA polymorphisms in the fungus *Colletotrichum gloeosporioides*. *Curr. Genet.* 16: 473–475.

Brown, P. T. H., Kyozuka, J., Sukekiyo, Y., Shimamoto, K., and Lörz, H. (1990) Molecular changes in protoplast-derived rice plants. *Mol. Gen. Genet.* 223: 324–328.

Burke, T., and Bruford, M. W. (1987) DNA fingerprinting in birds. *Nature* 327: 149–152.

Dallas, J. F. (1988) Detection of DNA "fingerprints" of cultivated rice by hybridization with a human minisatellite DNA probe. *Proc. Natl. Acad. Sci. USA* 85: 6831–6835.

Dally, A. M., and Second, G. (1990) Chloroplast DNA diversity in wild and cultivated species of rice (genus *Oryza*, section *Oryza*). Cladistic-mutation and genetic-distance analysis. *Theor. Appl. Genet.* 80: 209–222.

Davey, M. R., Rech, E. L., and Mulligan, B. J. (1989) Direct DNA transfer to plant cells. *Plant Mol. Biol.* 13: 273–285.

Debener, T., Salamini, F., and Gebhardt, C. (1990) Phylogeny of wild and cultivated *Solanum* species based on nuclear restriction fragment length polymorphisms (RFLPs). *Theor. Appl. Genet.* 79: 360–368.

Epplen, J. T. (1988) On simple repeated GATA/GACA sequences in animal genomes: a critical reappraisal. *J. Hered.* 79: 409–417.

Epplen, J. T., Ammer, H., Epplen, C., *et al.* (1991) Oligomideotide fingerprinting using simple repeat motifs: a convenient, ubiquitously applicable method to detect hypervariability for multiple purposes. In: Burke, T. Dolf, G., Jeffreys, A. J. and Wolff R. (eds) DNA Fingerprinting: Approaches and Applications. Birkhäuser, Basel pp 50–69 (This volume).

Flavell, R. (1986) Repetitive DNA and chromosome evolution in plants. *Phil. Trans. Roy. Soc. London, series B* 312: 227–242.

Gebhardt, C., Blomendahl, C., Schachtschabel, U., Debener, T., Salamini, F., and Ritter, E. (1989) Identification of 2n breeding lines and 4n varieties of potato (*Solanum tuberosum* ssp. tuberosum) with RFLP fingerprints. *Theor. Appl. Genet.* 78: 16–22.

Georges, M., Lequarré, A.-S., Castelli, M., Hanset, R., and Vassart, G. (1988) DNA fingerprinting in domestic animals using four different minisatellite probes. *Cytogenet. Cell Genet.* 47: 127–131.

Greaves, D. R., and Patient, R. K. (1985) (AT)$_n$ is an interspersed repeat in the *Xenopus* genome. *EMBO J.* 4: 2617–2626.

Gross, D. S., Huang, S.-Y., and Garrard, W. (1985) Chromatin structure of the potential Z-forming sequence (dT-dG)$_n$ (dC-dA)$_n$. *J. Mol. Biol.* 183: 251–265.

Hamada, H., and Kakunaga, T. (1982) Potential Z-DNA forming sequences are highly dispersed in the human genome. *Nature* 298: 396–398.

Hamada, H., Petrino, M. G., and Kakunaga, T. (1982) A novel repeated element with Z-DNA-forming potential is widely found in evolutionary diverse eukaryotic genomes. *Proc. Natl. Acad. Sci. USA* 79: 6465–6469.

Hamada, H., Petrino, M. G., Kakunaga, T., Seidman, M., and Stollar, B. D. (1984a) Characterization of genomic poly (dT-dG) poly (dC-dA) sequences: structure, organization, and conformation. *Mol. Cell. Biol.* 4: 2610–2621.

Hamada, H., Seidman, M., Howard, B. H., and Gorman, C. M. (1984b) Enhanced gene expression by the poly(dT-dG) poly(dC-dA) sequence. *Mol. Cell. Biol.* 4: 2622–2630.

Harms, C. T. (1983) Somatic hybridization by plant protoplast fusion. In: Potrykus, I., Harms, C. T., King, P. J., and Shillito, R. D. (eds.) Protoplasts 1983: Lecture proceedings. Birkhäuser, Basel, pp. 69–84.

Hentschel, C. C. (1982) Homocopolymer sequences in the spacer of a sea urchin histone gene repeat are sensitive to S1 nuclease. *Nature* 295: 714–716.

Hillel, J., Schaap, T., Haberfeld, A., Jeffreys, A. J., Plotzky, Y., Cahaner, A., and Lavi, U. (1990) DNA fingerprints applied to gene introgression in breeding programs. *Genetics* 124: 783–789.

Hosaka, K., Ogihara, Y., Matsubayashi, M., Tsunewaki, K. (1988) Phylogenetic relationship between the tuberous *Solanum* species as revealed by restriction endonuclease analysis of chloroplast DNA. *Jpn J. Genet.* 59: 349–369.

Htun, H., and Dahlberg, J. E. (1989) Topology and formation of triple-stranded H-DNA. *Science* 243: 1571–1579.

Huey, B., and Hall, J. (1989) Hypervariable DNA fingerprinting in *Escherichia coli*: Minisatellite probe from bacteriophage M13. *J. Bacteriol.* 171: 2428–2532.

Jeffreys, A. J., Wilson, V., and Thein, S. L. (1985a). Individual-specific "fingerprints" of human DNA. *Nature* 316: 76–79.

Jeffreys, A. J., Wilson, V., and Thein, S. L. (1985b). Hypervariable "minisatellite" regions in human DNA. *Nature* 314: 67–73.

Jeffreys, A. J., and Morton, D. B. (1987) DNA fingerprints of dogs and cats. *Anim. Genet.* 18: 1–5.

Jeffreys, A. J., Wilson, V., Neumann, R., and Keyte, J. (1988) Amplification of human minisatellites by the polymerase chain reaction: towards DNA fingerprinting of single cells. *Nucleic Acids Res.* 16: 10953–10971.

Jeffreys, A. J., Neumann, R., and Wilson, V. (1990) Repeat unit sequence variation in minisatellites: a novel source of DNA polymorphism for studying variation and mutation by single molecule analysis. *Cell* 60: 473–485.

Kuhnlein, U., Zadworny, D., Dawe, Y., Fairfull, R. W., and Gavora, J. S. (1990) Assessment of inbreeding by DNA fingerprinting: development of a calibration curve using defined strains of chickens. *Genetics* 125: 161–165.

Lee, M., and Phillips, R. L. (1988) The chromosomal basis of somaclonal variation. *Ann. Rev. Plant Physiol.* 39: 413–437.

Levinson, G., Marsh, L., Epplen, J. T., and Gutman, G. A. (1985) Cross-hybridizing snake satellite, *Drosophila*, and mouse DNA sequences may have arisen independently. *Mol. Biol. Evol.* 2: 494–504.

Levinson, G., and Gutman, G. A. (1987) Slipped-strand mispairing: A major mechanism for DNA sequence evolution. *Mol. Biol. Evol.* 4: 203–221.

Lynch, M. (1988) Estimation of relatedness by DNA fingerprinting. *Mol. Biol. Evol.* 5: 584–599.

328

Margot, J. B., and Hardison, R. C. (1985) DNase I and nuclease S1 sensitivity of the rabbit alpha globin gene in nuclei and in supercoiled plasmids. *J. Mol. Biol.* 184: 195–210.

Martienssen, R. A., and Baulcombe, D. C. (1989) An unusual wheat insertion sequence (WIS1) lies upstream of an alpha-amylase gene in hexaploid wheat, and carries a "minisatellite" array. *Mol. Gen. Genet.* 217: 401–410.

Meyer, W., Koch, A., Niemann, C., Beyermann, B., Epplen, J. T., and Börner, T. (1991) Differentiation of species and strains of filamentous fungi by DNA fingerprinting. *Curr. Genet.*, (in press).

Miklos, G. L. G., Matthaei, K. I., and Reed, K. C. (1989) Occurrence of the $(GATA)_n$ sequences in vertebrate and invertebrate genomes. *Chromosoma* 98: 194–200.

Monastyrskii, O. A., Ruban, D. N., Tokarskaya, O. N., and Ryskov, A. P. (1990) DNA fingerprints of some *Fusarium* isolates differentiated toxicogenically. *Genetika* 26: 374–377.

Nagamine, T., Todd, G. A., McCann, K. P., Newbury, H. J., and Ford-Lloyd, B. V. (1989) Use of restriction fragment length polymorphism to fingerprint beets at the genotype and species level. *Theor. Appl. Genet.* 78: 847–851.

Nordheim, A., and Rich, A. (1983) Negatively supercoiled simian virus 40 DNA contains Z-DNA segments within transcriptional enhancer sequences. *Nature* 303: 674–670.

Nybom, H., Schaal, B. H., and Rogstad, S. H. (1989) DNA "fingerprints" can distinguish between cultivars of blackberries and raspberries. *Acta Hort.* 262: 305–310.

Nybom, H., and Schaal, B.H. (1990a) DNA "fingerprints" applied to paternity analysis in apples (*Malus* × *domestica*). *Theor. Appl. Genet.* 79: 763–768.

Nybom, H., and Schaal, B. H. (1990b) DNA "fingerprints" reveal genotypic distributions in natural populations of blackberries and raspberries (*Rubus*, Rosaceae). *Amer. J. Bot.* 77: 883–888.

Nybom, H., Rogstad, S. H., and Schaal, B. A. (1990) Genetic variation detected by use of the M13 "DNA fingerprint" probe in *Malus*, *Prunus*, and *Rubus* (Rosaceae). *Theor. Appl. Genet.* 79: 153–156.

Nybom, H. (1991) Applications of DNA fingerprinting in plant breeding. In: Burke, T., Dolf, G., Jeffreys, A. J., and Wolff, R. (eds) DNA Fingerprinting: Approaches and Applications. Birkhäuser, Basel pp 294–311. (This volume).

Orgel, L. E., and Crick, F. H. C. (1980) Selfish DNA: The ultimate parasite. *Nature* 284: 604–607.

Rich, A., Nordheim, A., and Wang, A. H. (1984) The chemistry and biology of left-handed Z-DNA. *Ann. Rev. Biochem.* 53: 791–846.

Richards, E. J., and Ausubel, F. M. (1988) Isolation of a higher eukaryotic telomere from *Arabidopsis thaliana. Cell* 53: 127–136.

Rogers, J. (1983) CACA sequences – the ends and the means? *Nature* 305: 101–102.

Rogstad, S. H., Patton, J. C., and Schaal, B. A. (1988a) M13 repeat probe detects DNA minisatellite-like sequences in gymnosperms and angiosperms. *Proc. Natl. Acad. Sci. USA* 85: 9176–9178.

Rogstad, S. H., Patton, J. C., and Schaal, B. A. (1988b) A human minisatellite probe reveals RFLPs among individuals of two angiosperms. *Nucleic Acids Res.* 16: 11378.

Ryskov, A. P., Jincharadze, A. G., Prosnyak, M. I., Ivanov, P. L., and Limborska, S. A. (1988) M13 phage DNA as a universal marker for DNA fingerprinting of animals, plants, and microorganisms. *FEBS Letters* 233: 388–392.

Schäfer, R., Zischler, H., Birsner, U., Becker, A., and Epplen, J. T. (1988) Optimized oligonucleotide probes for DNA fingerprinting. *Electrophoresis* 9: 369–374.

Song, K., Osborn, T. C., and Williams, P. H. (1990) *Brassica* taxonomy based on nuclear restriction fragment length polymorphisms (RFLPs). 3. Genome relationships in *Brassica* and related genera and the origin of *B. oleracea* and *B. rapa* (syn. *campestris*). *Theor. Appl. Genet.* 79: 497–506.

Tanksley, S. D., Young, N. D., Paterson, A. H., and Bonierbale, M. W. (1989) RFLP mapping in plant breeding: new tools for an old science. *Bio/Technology* 7: 257–264.

Tautz, D. (1989) Hypervariability of simple sequences as a general source for polymorphic DNA markers. *Nucleic Acids Res.* 17: 6463–6471.

Tautz, D., and Renz, M. (1984) Simple sequences are ubiquitous repetitive components of eukaryotic genomes. *Nucleic Acids Res.* 12: 4127–4138.

Tautz, D., Trick, M., and Dover, G. A. (1986) Cryptic simplicity in DNA is a major source of genetic variation. *Nature* 322: 652–656.

Vassart, G., Georges, M., Monsieur, R., Brocas, H., Lequarre, A. S., and Christophe, D. (1987) A sequence in M13 phage detects hypervariable minisatellites in human and animal DNA. *Science* 235: 683–684.

Vergnaud, G. (1989) Polymers of random short oligonucleotides detect polymorphic loci in the human genome. *Nucleic Acids Res.* 17: 7623–7630.

Walmsley, R. M., Wikinson, B. M., and Kong, T. H. (1989) Genetic fingerprinting for yeasts. *Bio/Technology* 7: 1168–1170.

Weintraub, H. (1983) A dominant role for DNA secondary structure in forming hypersensitive structures in chromatin. *Cell* 32: 1191–1203.

Weising, K., Schell, J., and Kahl, G. (1988) Foreign genes in plants: transfer, structure, expression, and applications. *Ann. Rev. Genet.* 22: 421–477.

Weising, K., Weigand, F., Driesel, A., Kahl, G., Zischler, H., and Epplen, J. T. (1989) Polymorphic GATA/GACA repeats in plant genomes. *Nucleic Acids Res.* 17: 10128.

Weising, K., Fiala, B., Ramloch, K., Kahl, G., and Epplen, J. T. (1990) Oligonucleotide fingerprinting in angiosperms. *Fingerprint News* 2(2): 5–8.

Weising, K., Beyermann, B., Ramser, and Kahl, G. (1991) Plant DNA fingerprinting with radioactive and digoxigenated oligonucleotide probes complementary to simple repetitive DNA sequences. *Electrophoresis*, (in press).

Wong, Z., Wilson, V., Jeffreys, A .J., and Thein, S .L. (1986) Cloning a selected fragment from a human DNA "fingerprint": isolation of an extremely polymorphic minisatellite. *Nucleic Acids Res.* 14: 4605–4616.

Wyman, A. R., and White, R. (1980) A highly polymorphic locus in human DNA. *Proc. Natl. Acad. Sci. USA* 77: 6754–6758.

Zakian, V. A. (1989) Structure and function of telomeres. *Ann. Rev. Genet.* 23: 579–604.

Zimmerman, P. A., Lang-Unnasch, N., and Cullis, C. A. (1989) Polymorphic regions in plant genomes detected by an M13 probe. *Genome* 32: 824–828.

Zischler, H., Nanda, I., Schäfer, R., Schmid, M., and Epplen, J. T. (1989) Digoxigenated oligonucleotide probes specific for simple repeats in DNA fingerprinting and hybridization *in situ*. *Hum. Genet.* 82: 227–233.

DNA Fingerprinting: Approaches and Applications
ed. by T. Burke, G. Dolf, A. J. Jeffreys & R. Wolff
© 1991 Birkhäuser Verlag Basel/Switzerland

The Isolation and Characterisation of Plant Sequences Homologous to Human Hypervariable Minisatellites

A. Daly, P. Kellam, S. T. Berry, A. J. S. Chojecki, and S. R. Barnes

I.C.I Seeds, Jealott's Hill Research Station, Bracknell, Berkshire RG12 6EY, Great Britain

Summary

We have isolated DNA probes, homologous to the human hypervariable minisatellite sequence 33.15, from the genome of rice (*Oryza sativa*). These probes are capable of producing a multilocus rice DNA fingerprint. The rice sequence has a tandem repeating structure based on a 12 bp GC-rich repeat which shows homology to its human counterpart. This probe detects up to 30 loci which are at a number of unlinked chromosomal sites. The GC-rich sequence is invariably associated with an open reading frame (ORF) of unknown function. The ORF is probably a member of a small multigene family.

Introduction

The use of molecular probes to detect polymorphisms at the DNA level has permitted the generation of detailed genetic maps in a number of major crop species. Such maps have hitherto been based upon variation at single loci, detected as restriction fragment length polymorphisms (RFLP's), using single or low copy sequences such as cDNA's (Helentjaris *et al.*, 1985) or hypomethylated genic regions (McCouch *et al.*, 1988). For most applications at least fifty such probes, evenly spaced at 20 cM intervals, would be needed to provide adequate marking for a complex plant genome, such as maize. In practice the use of this number of probes is time-consuming and relatively inefficient because many fail to reveal useful polymorphism for a given cross.

The discovery of minisatellite sequences in the human genome (Jeffreys *et al*, 1985), also described as variable numbers of tandem repeats (VNTRs) (Nakamura *et al.*, 1987), has overcome these limitations by providing a series of highly polymorphic markers scattered throughout the genome. Minisatellites appear to be present in all vertebrate genomes (Jeffreys *et al.*, 1987) and consist of small repeating units arranged in tandem arrays which can vary in length from a few tens of base pairs up to twenty kilobases. Polymorphisms are detected as changes in the length of the arrays by probing restriction digests with the core repeat. These variations in the number of repeat units are possibly due to unequal sister chromatid exchange and/or replication slippage (Jeffreys *et al.*, 1990). In humans, the polycore probes 33.6 and 33.15 are capable of detecting many polymorphic loci simultaneously, producing a DNA fingerprint (Jeffreys *et al.*, 1985). The application of this technology to plant breeding would have many benefits and in this paper we describe the isolation of plant sequences homologous to the human 33.15 sequence and their behaviour in a segregating population.

Materials and Methods

Unless otherwise stated, methods were essentially as described by Maniatis *et al.*, 1982.

Plant Material

Seed of *Oryza sativa* was obtained from the International Rice Research Institute, Los Baños, Philippines, or from the University of Nottingham. Plants were grown under greenhouse conditions in soil and harvested for DNA when about 6 weeks old.

DNA Isolation

DNA was isolated from leaf material using the modified CTAB method of Saghai-Maroof *et al.* (1984). DNA for the preparation of genomic libraries was further purified on CsCl gradients.

Restriction Digests, Electrophoresis and Southern Blotting

Restriction digests were performed under the manufacturer's recommended conditions (Northumbria Biologicals Ltd.). Electrophoresis was in horizontal agarose gels in TBE buffer at constant voltage. Agarose concentrations were either 0.8% or 1.4% depending on the size range of DNA fragments to be separated. Gels were treated as described by Southern (1975) and the DNA blotted onto Hybond-N membranes (Amersham).

DNA Labelling and Hybridisation

DNA probes were radiolabelled with α^{32}P-dCTP (3000 Ci/mM, Amersham) by random priming (Feinberg and Vogelstein, 1984) to high specific activities ($10^8 - 10^9$ cpm/μg). Filters were prehybridised in

5 × SSC, 50% formamide, 1% skimmed milk powder and 25 mM sodium phosphate (pH 6.8) for 6 hours at 42°C. Filters were hybridised with the labelled probe (5 ng/ml) overnight in fresh hybridisation solution incorporating 10% dextran sulphate. Unbound probe was removed by washing at 65°C in 2 × SSC (2 washes of 30 minutes), followed by a single 30 minute wash in 0.1 × SSC at the same temperature. Autoradiography was at −80°C, using Kodak XAR film with a single intensifying screen.

Preparation and Screening of Rice Genomic Libraries

A Sau3A partial digest of 250 μg of rice DNA was size-fractionated by velocity sedimentation in a linear sucrose gradient (10–40%). Fractions containing DNA in the size-range 15–20 kb were identified by gel electrophoresis and pooled. The DNA was ligated to BamHI/EcoRI arms of λEMBL3 (Stratagene) at 4°C for 48 hours and packaged *in vitro* using a Giga-Pack Gold kit (Stratagene). Phage libraries were plated on *E. coli* strain NM621, plaques transferred to nylon membranes and screened with the human minisatellite probe 33.15 (Jeffreys *et al.*, 1985).

Phage DNA Purification

Phage were isolated from 500 ml cultures by PEG precipitation, the particles banded on a CsCl gradient and the DNA recovered by phenol extraction and ethanol precipitation.

Subcloning and Sequencing

Restriction fragments were purified from agarose gels by electrophoresis onto DE81 paper (Dretzen *et al.*, 1981). Fragments were then ligated into the appropriate cloning site in M13mp18/19. Nested sets of deletions were generated using exonuclease III (Henikoff, 1984). Single-stranded templates were sequenced using Sequenase (United States Biochemical Corp.) and the ^{35}S-labelled sequencing products separated on 40 cm buffer gradient acrylamide gels.

Results

Isolation and Subcloning of Rice Sequences Homologous to 33.15

Screening of 10,000 primary plaques from the rice genomic library with 33.15 yielded 7 positives, of which 5 were successfully purified. DNA

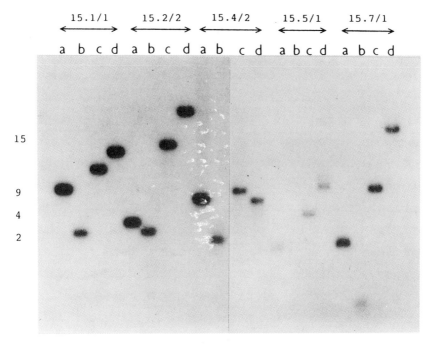

Figure 1. A Southern blot of the five 33.15 positive rice genomic clones cut with four different restriction enzymes (a-HinfI; b-HaeIII; c-DraI; d-HindIII) and probed with 33.15.

from these 5 recombinants was digested with a range of enzymes and probed with 33.15 in order to identify the smallest homologous fragment for subcloning (Fig. 1). The BamHI digests of the clones, which gave the smallest clonable fragments are not shown in Fig. 1.

All attempts to subclone these BamHI fragments using plasmid vectors failed; however, it was possible to recover a full-length clone of the BamHI fragment from lambda 15.2/2 in one orientation only in M13mp18. Sub-fragments from the other positive clones also showed the same preference as to orientation, but were far less stable.

Nucleotide Sequencing of the 33.15 Homologous Sequences

The 2 kb BamHI fragment from lambda clone 15.2/2 was sequenced using a combination of ExoIII deletions and one subclone. Certain deletions were never found and subclones to permit the sequencing of one region were never made successfully. We can only assume that the sequence in question, in specific arrangements with vector DNA, is inviable or lost through recombination. Similar problems have been encountered with human minisatellites (Jeffreys, pers. comm.). Attempts to sequence the missing 400 bp using synthetic oligonucleotide

```
         10         20         30         40         50         60
GGATCCTTAA GGGTTATGTA CGGAACCATT CACACCCCGA GGGGAGTATC ATCGAGAGTT
         70         80         90        100        110        120
ACACCACCGA AGAGGTCATC GATTTTTGTG TGGACTAGAT GTCAGAAACA TCTTCAATTG
        130        140        150        160        170        180
GATTACCACG ATCTCATCAT GAAGGGAGGC TTGATGGTGT TGGTACTGTT GGAAGAAAGA
        190        200        210        220        230        240
CTGTTAGGTT GGATCGGAAA GGTGTACGAT AAAGCTCATT TCACGGTACC GCAGCATATG
        250        260        270        280        290        300
ACTGAGGTGG TGCCGTATGT TGACGAACAC CTTGCAGTTC TCCGACAGGA GAACCCAGAC
        310        320        330        340        350        360
CGATTAGAGA GTTGGGTCAG AAATAAGCAC ATGTCTTCTT TCAACGAGTG GCTGAAGAAC
        370        380        390        400        410        420
CGAATTGCTA GGTTGCAGAA CTTGTCTAGT AAAACACTTC AGTGGTTGTC ACAGGTCCTG
        430        440        450        460        470        480
AATGGAGTGT CACCACCTGG CAAGGATATG ACATAAATGG ATACATCTTT CACACGGTAA
        490        500        510        520        530        540
AGCAAGACAG CAAATGCACA GTGCACAACA GTGGGTTACG CATCGAGGTT GCTAGTGACG
        550        560        570        580        590        600
GTGGTCGTCG TGATAAATAC TATGGTAGAG TTGAGCAAAT ATTGGAGCTA GATTACTTGA
        610        620        630        640        650        660
AGTTCAAAGT CTCGTTGTTT CGTTGTCGAT AGGTCGATCT TCGCAATGTA AAAGTTGACA
        670        680        690        700        710        720
ATGAAGGTTT CACCACTGTC AACTTGGCTA ACAATGCGTA CAAGGATGAA CCGTTCGTTC
        730        740        750        760        770        780
TCGCCAAACA AGTTGTTCAA GTGTTCTACA TAGTTGACCC GTGTAACAAG AAACTACATG
        790        800        810        820        830        840
TTGTTCGTGA AGGGAAAAGG AGAATTGTTG GATTGGACAA TATTGCAGAC GAGGATGATT
        850        860        870        880        890        900
ACAACCAGCA CGTCCACGGC ATAGGTCAAG AAATACCTCT AGAAGAGGAG GAAGAAGAAG
        910        920        930        940        950        960
ATGACGTTCA ATATGCACGC ATCGACCATG AGGAAGGATT ATTTTTGTAA TTTATGTAAG
        970        980        990       1000       1010       1020
TAATTTATGT CTACATGTTT GTTATCTGTA TGACTCTAAT ATCTTCACAA TGATCAATAG
       1030       1040       1050       1060       1070       1080
TATATTGCTT ATGAAATAGT CAAATAAAAT AATTAAGTGA AGCACCAACT AGAAGTGAAA
       1090       1100       1110       1120       1130       1140
TAAAATAATT AAGCAAGTAT AACCATTGGA AATAAAATAA TAAAATTAGT AAAATAATGA
       1150       1160       1170       1180       1190       1200
AGTGCAATTA CTACTGTTAG TGAAATAAAA TAATTAAGTA TAACAAAAAT ATGTAGTGCA
       1210       1220
TGTGGAAATA TGTTAAGGAA AATAAAA--UNSEQUENCED GAP OF APPROX. 400bp--
         10         20         30         40         50         60
TAGGTTGGAC CTCTTCTAGA TCGATCCAGA TTCGGCACCC ACTCTCACTC ACTCACTCTC
         70         80         90        100        110        120
TCGACCGCCG CCGTCCTCTC CGCCGCCATC TCCACCCTCG CCGCCTCGAC CTCTCCGCCG
        130        140        150        160        170        180
CCCTCTCCGC TGCCATCTCC GCCGCCCTCT CTGCCGCCAT CTCCGCCCTA TCCGCCGCCC
        190        200        210        220        230        240
TCTCCACCGC CGCCTCGACC TCTCCACCGC CGCCCTCGCC GCCTCCCTCT CCACCGCCGC
        250        260        270        280        290        300
CGCCTCACTC ACTCTCTCGA CCGCCGCCGT CCTCTCCGCC GTCCGTCCCC GACCTCCCGA
        310
CACCGGGATC C
```

Figure 2. The partial nucleotide sequence of the 2 kb BamHI fragment from the genomic rice clone 15.2/2.

primers also failed due to difficulties in finding a unique sequence adjacent to the missing region. The nucleotide sequence is shown in Fig. 2 and the schematic representation of the clone in Fig. 3 divides the subclone up into 3 distinct regions:

1) GC-rich Repeating Motif

Between positions 1645 and 1895 there is an imperfect array of a 12 bp repeat. Figure 4 shows the rice consensus sequence and compares it to the human 33.15 core repeat. The human and plant repeats show clear homology, yet differ significantly in the spacing of the (G) blocks. Careful examination of the plant sequence also reveals a simpler repeat structure based upon $(GGN)_n$.

2) AT-rich Repeating Motif

Between positions 670 and 1260 there are 8 imperfect repeats of an 18 bp AT-rich sequence. No clear secondary structural features are visible, but it is possible to speculate that this region might be involved in the regulation of the adjacent open reading frame, only 210 bp upstream.

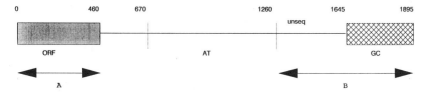

Figure 3. A schematic representation of the 2 kb BamHI fragment from the genomic rice clone 15.2/2. Probe regions defining the ORF (A) and GC-rich regions (B) are designated by the underlying double-headed arrows.

Rice 15.2/2 consensus (12bp)

A G A G G G C G G c **G G**

Human 33.15 consensus (16bp)

A G A g g t **G G G C** a **G G** t **G G**

Figure 4. The consensus sequences of the core repeats from the rice and human minisatellites 15.2/2 and 33.15 (homologies between the two repeats are shown in large bold type).

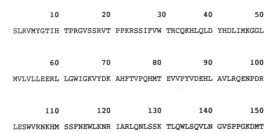

```
         10        20        30        40        50
SLRVMYGTIH TPRGVSSRVT PPKRSSIFVW TRCQKHLQLD YHDLIMKGGL

         60        70        80        90        100
MVLVLLEERL LGWIGKVYDK AHFTVPQHMT EVVPYVDEHL AVLRQENPDR

         110       120       130       140       150
LESWVRNKHM SSFNEWLKNR IARLQNLSSK TLQWLSQVLN GVSPPGKDMT
```

Figure 5. Amino acid sequence of the C-terminus of the protein coded for by the ORF in the rice clone 15.2/2.

336

3) Open Reading Frame

From positions 1 to 460 is an open reading frame (ORF) capable of encoding 150 amino acids at the C-terminus of a protein (Fig. 5). Sequence analysis of the other four 33.15 positive rice clones showed that they also contained the ORF. The N-terminal part of this putative gene has not been studied, nor has a cDNA clone been isolated. Searches of the EMBL and Genbank databases have shown that there is no known DNA sequence/protein homology to this ORF.

Organisation and Variation of 15.2/2 Homologous Genomic Sequences

Figure 6 shows the result of hybridising radiolabelled fragments containing the ORF and GC regions to BamHI and EcoRI digests of 5

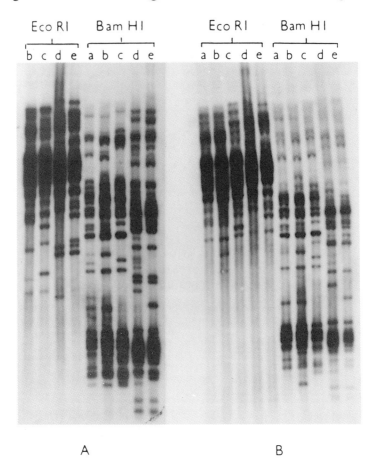

A   B

Figure 6. Two Southern blots of genomic rice DNA from five different inbred lines (a–e), digested with the restriction enzymes BamHI and EcoRI and probed with the ORF region (A) and the GC-rich region (B) of the rice clone 15.2/2. Lanes: a-Silewah; b-0.56; c-T909; d-IR45; e-IR36.

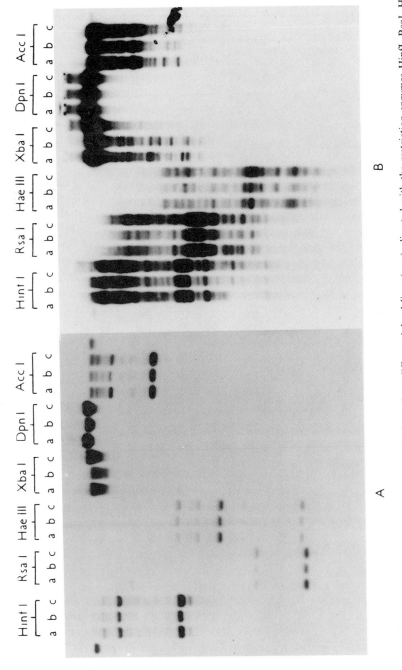

Figure 7. Two Southern blots of genomic rice DNA from three different inbred lines (a–o), digested with the restriction enzymes HinfI, RsaI, HaeIII, XbaI, DpnI and AccI and probed with the ORF region (A) and the GC-rich region (B) of the rice clone 15.2/2.

Table 1. Recombination fractions between the pairs of bands which segregated in a 3:1 ratio in the F2 population between IR39357-91-3-2-3 (parent 5) and Kwang Lu Ai 4 (parent 8). Bands from parent 5 are labelled 5.1, 5.5 etc. Standard errors are shown in parentheses

	5.1	5.5	5.6	5.7	8.1	8.2	8.3	8.4	8.5
5.1		50(8)	44(9)	46(8)	40(9)	28(10)	40(9)	34(10)	21(10)
5.5			51(10)	52(8)	43(9)	38(10)	44(9)	44(9)	56(8)
5.6				36(8)	57(9)	52(10)	43(11)	42(11)	47(11)
5.7					49(9)	52(8)	51(9)	49(9)	60(8)
8.1						55(9)	50(8)	49(8)	51(8)
8.2							54(9)	41(7)	47(8)
8.3								43(8)	56(9)
8.4									21(5)

inbred rice lines (the regions of the subclone covered by the probes are shown in Fig. 3). The autoradiograph shows that both fragments can detect a clear multi-banded pattern which varies between the rice lines tested. When the genomic DNA is digested with enzymes that cut between the ORF and GC-rich regions (Fig. 7), only the GC repeats give a polymorphic fingerprinting pattern. The exposure times of the autoradiographs probed with the GC-rich probe (Figs. 6 and 7) were only a few hours at room temperature.

Segregation Analysis of a Rice F2 Population

DNA from one hundred individuals from an F2 population originating from a cross between IR39357-91-3-2-3 (parent 5) and Kwang Lu Ai 4 (parent 8) were digested with BamHI and fingerprinted using the rice GC probe. Pairwise combinations of 9 bands (four bands from parent 5 and five bands from parent 8) which showed normal Mendelian segregation within the F2 were tested for linkage using the method of Allard (1956). The recombination fractions for all possible combinations of the 9 bands are shown in Tab. 1.

From this analysis it can be shown that none of the bands are allelic, 3 pairs of loci are linked (5.1/8.2, 5.1/8.5, and 8.4/8.5) 3 pairs show loose linkage (5.1/8.4, 5.5/8.2, 5.6/5.7), while the remainder are unlinked.

Discussion

The results presented here clearly demonstrate that sequences homologous to the human minisatellite sequence 33.15 are present in the rice genome. This extends the observations made by Dallas (1988) who showed that the polycore probe 33.6 gave a clear fingerprinting pattern on rice; whereas 33.15 gave bands, but with a background smear of hybridisation. This latter result is typical of many, using either the human probes (Rogstad *et al.*, 1988) or the protein III gene of M13 (Nybom *et al.*, 1990) on plant DNA; however some high molecular weight polymorphisms can be resolved. Rice, it appears, possesses fewer 33.6/33.15 cross-hybridising sequences which make the fingerprinting patterns in other plant species illegible.

It may, however, be possible by the present approach – of cloning homologous plant sequences – to overcome such problems and even to extend the range of species amenable to DNA fingerprinting, as is the case with bird minisatellites (Gyllensten *et al.*, 1989). Another advantage of cloning species-specific minisatellites is that the signals obtained when probing Southern blots are considerably stronger and banding patterns sharper. When 33.6 is used to probe rice genomic digests, exposure times of 3–6 days are typical (Dallas, 1988); whereas the rice minisatellite clone 15.2/2 only requires exposure times of a few hours.

By studying the molecular organisation of the rice minisatellite clone 15.2/2 (Fig. 3) and other 33.15 positive clones (data not shown) three distinct regions become apparent: a GC-rich region, an AT-rich region and an open reading frame (ORF). Southern blot analysis clearly demonstrated that the polymorphic fingerprinting pattern was due to the GC-rich region (Fig. 7), which consists of a tandemly repeated 12 bp repeat related to 33.15 core sequence (Fig. 4). Presumably as in humans, variation is occurring in the numbers of these repeats within each array, but the number of loci detected in rice (approx. 30) is considerably less than in an equivalent human DNA fingerprint (Jeffreys *et al.*, 1985).

Although the isolated rice GC-rich sequences show homology to 33.15 they differ considerably in their genomic flanking sequences. The Southern blots shown in Fig. 6 show that the rice minisatellite is invariably associated with an ORF of unknown function. This evidence is in agreement with the partial sequence data obtained from the other 33.15 positive lambda clones (data not shown). In humans however, minisatellites tend to be clustered with other tandem and dispersed repeats (Armour *et al.*, 1989), although they were originally identified in genic regions (Weller *et al.*, 1984). We have no evidence so far for this type of organisation in the rice genome.

The simplest explanation for the linkage between GC-rich and ORF sequences is that the ORF is a member of a small multigene family during whose evolution, minisatellite sequences were also multiplied. Whether the GC-rich sequences have any functional significance with regard to the ORF is open to speculation and no evidence is provided by the current study. However, it is worth pointing out that if such sequences are genuinely recombinogenic (Wahls *et al.*, 1990), they could provide sites for the initiation of recombination or gene conversion with non-homologous loci and hence offer an opportunity for the homogenisation of the linked members of a multigene family. This could therefore provide a mechanism for the phenomenon of "concerted evolution" observed in a number of similar situations (Dover, 1982).

The linkage of the rice minisatellite to a multigene family poses the question of whether these sequences are clustered within the rice genome. In order to answer this, 100 individuals from an F2 population were fingerprinted to allow a crude genetic analysis of segregation patterns of pairs of bands. The recombination fractions given in Tab. 1 can be used to derive a linkage map. This map shows that while some loci are indeed clustered within the genome, the remainder are at dispersed locations.

Conclusions

This paper demonstrates that hypervariable minisatellite-type repeats are present in the genomes of plants. The exploitation of such sequences in plant breeding programmes could greatly accelerate and improve the selection of material made by the plant breeder, for example in recovering the recurrent parent genotype in backcrossing experiments. Multilocus probes could also be used for testing the purity of inbred lines and hybrid seed, as well as for varietal protection in plant breeders rights.

Acknowledgements
We are grateful to Dr. A. Jeffreys for useful discussions and to Dr. G. Khush (IRRI) for the supply of the F2 rice population.

References

Allard, R. W. (1956) Formulas and tables to facilitate the calculation of recombination values in heredity. *Hilgardia* 24: 235–279.

Armour, J. A. L., Wong, Z., Wilson, V., Royle, N. J., and Jeffreys, A. J. (1989) Sequences flanking the repeat arrays of human minisatellites: association with tandem and dispersed repeat elements. *Nucleic Acids Res.* 17: 4925–4935.

Dallas, J. F. (1988) Detection of DNA "fingerprints" of cultivated rice by hybridisation with a human minisatellite DNA probe. *Proc. Natl. Acad. Sci.* 85: 6831–6835.

Dover, G. (1982) Molecular drive: a cohesive mode of species evolution. *Nature* 299: 111–117.

Dretzen, G., Bellard, M., Sassons-Corsi, P., and Chambon P. (1981) A reliable method for the recovery of DNA fragments from agarose and acrylamide gels. *Anal. Biochem.* 112: 295–298.

Feinberg, A. P., and Vogelstein B. (1983) A technique for radiolabelling DNA restriction endonuclease fragments to high specific activity. *Anal. Biochem.* 132: 6–13.

Gyllensten, U. B., Jakobsson, S., Temrin, H., and Wilson, A. C. (1989) Nucleotide sequence and genomic organization of bird minisatellites. *Nucleic Acids Res.* 17: 2203–2214.

Helentjaris, T., King, G., Slocum, M., Siedenstrang, C., and Wegman, S. (1985) Restriction fragment polymorphisms as probes for plant diversity and their development as tools for applied plant breeding. *Plant Molecular Biology* 5: 109–118.

Henikoff, S. (1984) Unidirectional digestion with exonuclease III creates targeted breakpoints for DNA sequencing. *Gene* 28: 351–359.

Jeffreys, A. J., Wilson, V., and Thein, S. L. (1985) Hypervariable "minisatellite" regions in human DNA. *Nature* 314: 67–73.

Jeffreys, A. J., Wilson, V., Wong, Z., Royle, N., Patel, I., Kelly, R., and Clarkson, R. (1987) Highly variable minisatellites and DNA fingerprints. *Biochem. Soc. Symp.* 53: 165–180.

Jeffreys, A. J., Neumann, R., and Wilson, V. (1990) Repeat unit variation in minisatellites: A novel source of DNA polymorphism for studying variation and mutation by single molecule analysis. *Cell* 60: 473–485.

McCouch, S. R., Kochert, G., Yu, Z. H., Wang, Z. Y., Khush, G. S., Coffman, W. R., and Tanksley, S. D. (1988) Molecular mapping of rice chromosomes. *Theor. Appl. Genet.* 76: 815–829.

Maniatis, T., Fritsch, E. F., and Sambrook, J. (1982) Molecular cloning: A Laboratory Manual (Cold Spring Harbor Lab., Cold Spring Harbor, N. Y.).

Nakamura, Y., Leppert, M., O'Connell, P., Wolff, R., Holm, T., Culver, M., Martin, C., Fujimoto, E., Hoff, M., Kumlin, E., and White, R. (1987) Variable number of tandem repeat (VNTR) markers for human gene mapping. *Science* 235: 1616–1622.

Nybom, H., Rogstad, S. H., and Schaal, B. A. (1990) Genetic variation detected by use of the M13 "DNA fingerprint" probe in *Malus*, *Prunus* and *Rubus* (Rosaceae). *Theor. Appl. Genet.* 79: 153–156.

Rogstad, S. H., Patton, J. C., and Schaal, B. A. (1988) A human minisatellite probe reveals RFLP's among individuals of two angiosperms. *Nucleic Acid Res.* 23: 11378.

Saghai-Maroof, M. A., Soliman, K. M., Jorgensen, R. A., and Allard, R. W. (1984) Ribosomal DNA spacer-length polymorphism in barley: Mendelian inheritance, chromosomal location and population dynamics. *Proc. Natl. Acad. Sci. USA* 81: 8014–8018.

Southern, E. M. (1975) Detection of specific sequences among DNA fragments by gel electrophoresis. *J. Mol. Biol.* 98: 503–517.

Wahls, W. P., Wallace, L. J., and Moore, P. D. (1990) Hypervariable minisatellite DNA is a hotspot for homologous recombination in human cells. *Cell* 60: 95–103.

Weller, P., Jeffreys, A. J., Wilson, V., and Blanchetot, A. (1984) Organisation of the human myoglobin gene. *EMBO J.* 3: 439–446.

DNA Fingerprinting: Approaches and Applications
ed. by T. Burke, G. Dolf, A. J. Jeffreys & R. Wolff
© 1991 Birkhäuser Verlag Basel/Switzerland

DNA Fingerprinting; a Biotechnology in Business

P. G. Debenham

Cellmark Diagnostics, Blacklands Way, Abingdon Business Part, Abingdon,
Oxfordshire OX14 1DY, Great Britain

Summary
Since its discovery by Professor Alec Jeffreys, first published in 1985, DNA fingerprinting has
never been far from the headlines. Cellmark, as the leading commercial DNA fingerprinting
enterprise, has had the challenge of establishing its business in this high profile environment.
The challenge for such a business has had three elements. Firstly and primarily to establish a
laboratory system that consistently provides results to the highest standards possible. Secondly
to communicate the science to its broad customer base which ranges from the general public
to the experts of legal, scientific and governmental systems. Then finally there is the challenge
of setting up the business in the context of the volatile, venture-capital based biotechnology
market where the requirements for assured quality spar with those for cut prices. Given this
backdrop it is not surprising that DNA fingerprinting has had its detractors, yet the consensus
of opinion is that this technology has become a near routine business with the challenges
effectively met and the success that was hoped for.

Setting Up a DNA Testing Service

Setting up a DNA testing laboratory with molecular biologists could
have its problems. Molecular biologists have an inexorable desire to
change protocols for a particular technique, as if there is a need to
stamp their personal mark upon the method. Probably not all changes
are intentional as, for instance, when an extended coffee break causes an
extended enzyme incubation which in turn provides an improved result.
Such lack of attention to a protocol is not necessarily to the discredit of
the scientist, but often reflects his knowledge that many processes have
broad optima so that any timing of a step within limits is probably
without detriment. Neither a lack of attention to detail for whatever
reason, nor the drive to innovate, are inevitable in molecular biology
but such factors are confounding elements in the provision of a DNA
fingerprinting laboratory service. In such service one of the key impera-
tives is reproducibility. Faith in the result obtained hinges on the proven
nature of a given protocol that can be seen to have been followed
exactly. Many DNA tests, because of limiting amounts of samples,
cannot be repeated. Thus, painstaking documentation of every step is
essential if the results are to be trusted.

Given that there are many slightly different ways that a molecular
biologist can get a DNA fingerprint from, say, a blood sample, the
focus for a DNA diagnostic business must be to adapt a tried and tested

technical protocol and stick to it. Such a protocol should be optimised for each laboratory customised to its particular equipment, supply of chemicals, etc. Thus a universal methodology may not be exactly achievable though it is often put forward as a holy grail. The essential requirement is that the methodology is tried, well tested and most importantly produces results within internationally monitored limits (see later). At Cellmark we started with the procedures utilised by Jeffreys and optimised them for use in a streamlined biotechnological service. Whilst the heart of such a protocol can be published for all to compare (Smith *et al.*, 1990) the fine detail inevitably requires documentation that runs into tens of pages and cannot be published.

Defining what exactly constitutes an appropriate level of attention to detail in the technical preparation of a DNA fingerprint is a matter of opinion. From day one at Abingdon we imposed a witnessing procedure so that every key step involving a sample or information transfer is checked and documented to minimise the opportunity for human error. This imposed attention to detail is arguably essential and the consequent financial burden it causes is the cost of gaining confidence in the results.

Rigorous attention to detail within the diagnostic laboratory could be wasted effort if the protocol is sub-optimal or the quality of the reagents purchased varied. The use of the many chemicals and enzymes in DNA fingerprinting tends to limit the scientist's ability to apportion blame correctly should a case study fail to yield analysable results. Thus, for example, the failure of a probe radio-labelling step, the weak transfer of DNA from gel to blot or the curving of bands on the final autoradiograph may have a host of explanations. Incorrect identification of the cause of error could lead to failure of subsequent DNA fingerprints which would be disastrous for forensic investigations and financially burdensome to the commercial laboratory. Unlike the research laboratory, the diagnostic laboratory must impose detailed checking of all new ingredients. This ranges from batch-to-batch optimisation of nylon membranes for the UV exposure required to bind DNA, to activity assays on new restriction enzyme batches which can exclude those which can be the cause of curved bands etc. The use of detailed, computerised stock and solution coding systems further lets the laboratory pinpoint the one faulty solution preparation amongst many required in the complex process of producing a DNA fingerprint.

Variability is natural in the actual samples to be processed. Therefore it is important that steps are in place to monitor both the quantity and quality of the DNA obtained so that samples are matched for quantity or adjusted for quality. Normally, and particularly with blood samples, only a percentage of any sample is initially extracted for

DNA. A further aliquot can be extracted from samples with insufficient DNA and combined with the original preparation. Laboratory documentation must be able to trace such samples so as to eliminate any mixing of samples. Equally, documentation must be appropriate to trace the repeated restriction digestion of a sample yielding abnormally high yields of DNA. With high yields of DNA further digestion should be performed, to avoid partial digestion results.

Partial digestion of a DNA sample could yield misleading results in a DNA fingerprint analysis. Unassigned DNA bands in a paternity analy- ? sis, or unmatched bands in an identity analysis could be mistakenly taken to be evidence of a lack of relationship or identity match when in fact the result is an artefact of partial digestion of a critical sample in the case. For HinfI DNA digests a hypervariable single locus probe, MS51, has been developed which identifies a locus that is one of the last to be completely digested in multi-locus probe analysis of restriction digest time course studies (using 33.15 as a probe, manuscript in preparation). Thus with experience of the digest patterns obtained with MS51, MS51 can be used to analyse the presence of a partial digest. Partial restriction digestion can occur with every restriction enzyme particularly when cutting DNA in forensic cases, where digestion conditions can be unknowingly "poisoned" by components of the material from which the sample was extracted. MS51 can act as a digestion monitor only for HinfI digests and does not produce informative results with other restriction enzyme digests, such as HaeIII, the enzyme utilised by the FBI.

A related technical problem for a DNA fingerprinting service is that of "band shifting". In such situations two identical samples will have DNA fingerprint bands slightly shifted from each other. There is no mystique in such a result, it is purely an indication that DNA from one of the two samples has not been completely purified away from "components" in the material from which the sample was extracted. Such components have not been fully documented but it is commonly recognised that shifts can be imposed on a sample by deliberately adjusting laboratory components in the sample such as its salt content or ethidium bromide concentration prior to electrophoresis. Shifts are rarely observed in an experienced laboratory where optimised DNA extraction procedures minimise the presence of "components".

However, the availability of a monomorphic probe to technically substantiate the presence of a shift is an important tool for a diagnostic laboratory to have in its repertoire. One such probe (pCTS100, manuscript in preparation) has now been characterised specifically for HinfI digests. This probe intriguingly has tandemly repeated DNA elements yet it identifies a constant band of 4.2Kb in unrelated individuals. pCTS100 contains numerous HaeIII cut sites and is not suitable for use with this enzyme.

Public Confidence in DNA Testing

The maintenance of a DNA fingerprinting laboratory to the highest standards is essential for public confidence in DNA testing. The public perception of this technology, as mirrored in the press and media, has moved from an initial phase of unqualified praise for this ultimate identity test to one of confused concern in some countries because of a few headlines "errors" in court. The association of doubt with DNA testing has been a notable success of the defence lawyers in the USA who to their credit have ensured that high standards must be maintained. To their discredit it seems that if great care is not taken then disproportionate emphasis on trivial technical detail can inhibit the true course of justice and rapists, etc., may go free. Similarly the merit of DNA results as evidence can be misleadingly diminished by a mis-reading of the ongoing discussion between scientists with respect to the statistics of DNA analysis as a debate on the validity of DNA testing itself. The complex scientific discussions on this issue are a debate on evidential weight. It should be remembered that no-one as yet has published evidence, in a peer reviewed scientific journal, showing incorrect matching of samples due to an artefact of DNA analysis. It might be counter argued that insufficient control studies, say on DNA degradation, are published. Yet it must be realised that control studies, which certainly must be performed, often show the obvious and thus do not meet the criteria of scientific journals for novel knowledge. The credibility of DNA fingerprinting is however extended by the hundreds of publications showing the ever increasing diversity of DNA fingerprinting applications. It is therefore appropriate to inform the public that the balance of scientific argument is clearly in DNA fingerprinting's favour and that the very few, well-aired, errors are the exception rather than the rule.

In the UK, DNA fingerprinting has stood the test of time with more than 30,000 individual tests reported by Cellmark for the broad range of categories from paternity testing to forensic casework. Large numbers of samples have also been reported by Cellmark, the FBI and Lifecodes in the USA and by the array of forensic laboratories in Europe. In the UK, DNA fingerprinting has been reviewed and approved by the government for its use in appeal cases as evidence in immigration family analyses (Home Office, 1988) and the government has changed English law to provide a legal status to DNA evidence in paternity testing (Blood Test Regulations (Amendment) 1989 of the Family Law Reform Act of 1969). The government is expected soon to be announcing contracts for ongoing immigration casework in the UK to be resolved by DNA fingerprinting. Probably the UK is more advanced in its commitment to DNA testing than other countries at this time because of the residence of Professor Jeffreys here and to some extent the presence of Cellmark. It would, however, not be surprising to see aspects of the UK's commitment to be taken up by many countries in the foreseeable future.

Quality Assurance and its Cost

The high profile nature of DNA fingerprint testing has kept the public eye on the testing laboratories. The testing laboratories must sustain confidence through reasoned arguments for each question on the technology or on assured quality. The definition of what is the correct operating technology should be set by experts internal to an operation, but should be evaluated by independent scientific experts; and the management systems should be separately evaluated by independent quality assessors. "Experts" need extensive experience of this "hands-on" technology to assess the technology and its results and not just a scientific background. The option taken by Cellmark has been to set their own procedural and management systems on the basis of tried and tested technology. The scientific results are constantly evaluated both internally, through technical audits, but also externally by the critical eye of experts for the defence when DNA evidence is utilised in court. The management systems and documentation are externally evaluated to the International Quality standard ISO 9002 (BS5750 in the UK). Cellmark is the first DNA testing laboratory in the world to achieve certification to this high standard.

Given the complexity of the technology, the numerous technical variations between operating laboratories and a number of differing guidelines published by important scientific bodies, the question must be addressed as to whether the same DNA result would be obtained by differing laboratories if given the same samples. This question has been briefly addressed by EDNAP (European DNA Profiling group) and CACLD (Californian Association of Crime Labs Directors) in limited inter-laboratory trials. Both trials indicated principally a consensus of results with minor variations between the small numbers of participating laboratories. Over the next few years however, the numbers of laboratories operating DNA testing services (either government forensic laboratories or private laboratories) will number in their tens if not hundreds and it is important that public confidence in DNA testing is not damaged by the malpractice or "cut-corners" technology employed by any one DNA laboratory. On the other hand how would any one laboratory necessarily know that its results were of poor standard or even inaccurate? The answer can be provided by an international quality assessment scheme which will not set out to arbitrarily set standards but which will compare those achieved by any voluntarily participating laboratory. In this way practitioners of DNA fingerprinting can be seen to be addressing, in practical terms, the issues at the heart of the technology requiring the confidence of the public.

Commercially, the imposition of a continual auditing or assessment scheme has a cost penalty that must be met. The problem of course is whether the public are prepared to pay the premium for the assured quality or will turn to cut price services. Many companies in the molecular genetics field are principally product manufacturers and distributors, the business commitment only extending to the shelf-life of the products. DNA fingerprinting, as primarily a service business, requires a completely different philosophy as DNA results are not akin to a product from a catalogue. A service business therefore must plan a solid and long-term business strategy which is both compatible with customer sensitivities and assured quality.

Future Technologies in Practice

For many laboratories PCR technology is not seen as a replacement for "blot and probe" DNA fingerprinting but as an adjunct, particularly for forensic cases when minimal DNA is available for analysis. DNA amplification may soon provide the unique identification, or near-unique identification, that can be obtained by multilocus or single locus probe technologies. However, while the development of "ampflps" (amplified fragment length polymorphisms) using PCR to pick out DNA loci of limited repeat length polymorphisms, should avoid the debate of allele definition associated with single locus probes, it may be subject to the issue of population sub-grouping instead. Therefore it may not be surprising to see a combination of DNA amplification and probe technologies working side by side in the forensic and commercial laboratories.

For some laboratories probe technologies have not been readily available because until recently it has required radio-active labelling facilities to provide sufficiently sensitive DNA probes. Non-isotopic labelling methods whilst not new, have previously not provided the same detection sensitivities obtained by phosphorous-32 labelling. Now, however, it is possible to combine non-isotopic technologies with oligo-probe systems to provide rapid hybridisations and film exposures (Giles *et al.*, 1990). With chemiluminescent technology the probes are detected by X-ray film providing the same "hard-copy" of evidence obtained previously by isotopic methods. The non-isotopic technologies can be expected to improve greatly the prospects for the numerous small DNA laboratories who have been unable to enter the market previously because of the lack of isotope handling facilities.

The scene is therefore set for a great growth in DNA testing on a truly international scale. DNA fingerprinting has brought DNA testing to the fore but around the corner there are the real possibilities of diagnostic laboratories offering a wide range of DNA tests, from inherited disorders to even complex multi-genic traits. If DNA finger-printing can be seen to have been introduced in a careful and concerned manner, the public will have the reassurance required to accept a broader range of DNA tests in the future.

References

Giles, A. F., Booth, K. J., Parker, J. R., Garman, A. J., Carrick, D. T., Akhavan, H., and Schaap, A. P. (1990) Rapid, simple, non-isotopic probing of Southern blots for DNA fingerprinting. In: Polesky, H. F., and Mayr, W. R. (eds), Advances in Forensic Haemogenetics 3. Springer-Verlag, Berlin, pp. 40–42.

Home Office (1988). DNA profiling in DNA immigration casework. Home Office Publication. ISBN 086252–349–4–1988.

Smith, J. C., Newton, C. R., Alves, A., Anwar, R., Jenner, D., and Markham, A. F. (1990) Highly polymorphic minisatellite DNA probes. Further evaluation for individual identification and paternity testing. *J. For. Sci. Soc.* 30: 3–18.

DNA Fingerprinting: Approaches and Applications
ed. by T. Burke, G. Dolf, A. J. Jeffreys & R. Wolff
© 1991 Birkhäuser Verlag Basel/Switzerland

DNA Fingerprinting: Its Application in Forensic Case Work

W. Bär[1] and K. Hummel[2]

[1]*Institute of Forensic Medicine, University of Zürich, Zürichbergstrasse 8, CH-8028 Zürich, Switzerland;* [2]*Institut für Blutgruppenserologie, Postfach 880, D-7800 Freiburg i. Br., Germany*

Summary
Forensic serology deals with cases of disputed paternity and criminal stains. The spectacular improvement using DNA-profiling is best demonstrated with cases of criminal stains where with the same ease as a suspect can be identified, an innocent person can be excluded. In cases with band-shifting, a statistical definition of a match applying Bayes theorem as a decision making tool seems mandatory since matching or non-matching cannot be treated as a binary event. Good postmortem DNA stability is found in brain cortex, lymph nodes and psoas muscle. DNA fingerprinting is also a perfect tool to investigate disputed identity of blood alcohol samples. In paternity cases, we recommend both multi locus and single locus probes in kinship cases, e.g. in mother-child-putative father cases as well as father-daughter and brother-sister incest cases, grandparent cases and two-men cases. For the biostatistical evaluation of SLP patterns the formal genetics for a system of multiple allelism is used and for multi locus probes the model of multiple diallelism is applicable.

Introduction

The major task of forensic medicine is to investigate medico-legal matters in order to establish objective evidence and then make it available to the judges in their search for the truth.

Forensic serology – a subfield of forensic medicine – mainly deals with cases of disputed paternity and criminal stains, usually body secretions such as blood, semen or salivia. In both situations, the aim of the investigation is to identify or exclude a person as father or donor of a stain. The success rate of these investigations is strongly dependant on the power of the available tools of discrimination.

Genetic polymorphisms are very suitable for investigating these kinds of problems. The rate of discrimination is a function of the number of markers used and of the frequencies of the alleles of each system. However, in forensic stain work, limiting factors to the full exploration of some of these systems are their weak stability and the often sparse amount of stain material available. In practice therefore, only a few markers are reliably detectable and the value of the examination is limited.

When dealing with paternity cases, the amount of material is usually not a problem. But cases with deficient parties, i.e. where a mother or a putative father is deceased or missing, are becoming more frequent. With the increasing numbers of systems used and with the constant development of powerful investigative tools, problem cases can now be resolved.

The spectacular improvement when dealing with stains using DNA-profiling (Jeffreys et al., 1985a) is best demonstrated with cases of criminal stains where the power of discrimination has indeed increased dramatically (Gill et al., 1986).

Visual interpretation of a multi-locus-profile is a simple procedure: when the multi-locus-profile of the stain and the donor's blood show identical patterns, a suspect can be identified. With the same ease, an innocent person can be excluded. Thus, one result of the substantial progress of DNA-profiling is a significant increase in the unequivocal administration of the law.

The number of DNA probes commercially available has constantly increased and many forensic laboratories have started to use DNA techniques in daily routine work. The technique is still quite time-consuming and not always easy. Both isotopic and non-isotopic labelling reactions are used, but greater sensitivity of isotopic markers will make it necessary to continue to employ ^{32}P in forensic laboratories for some time to come. Most laboratories use single locus probes because of their better sensitivity and easier working conditions of high stringency. However, the application of 4 to 5 single-locus probes consecutively is very time-consuming and labour-intensive; hence, the attraction of the one-step investigation when using a multi-locus probe, despite its more difficult techniques.

Of late, mistakes or hasty interpretations of DNA profiling results have been given a lot of publicity (Lander, 1989). Nevertheless, in the USA as well as recently in Germany, the courts have accepted DNA profiles as legally admissable evidence.

Analysis of Semen Stains

About 200,000 to 1 million sperm heads contained in about 25 μl semen are usually needed to produce a multi-locus-profile of a semen stain. However, for a single locus probe only about one twentieth of this amount is necessary, so the increase in sensitivity is quite considerable. In 99 rape cases analyzed at the Institute of Forensic Medicine of Zürich, the amount of high molecular DNA extracted from semen stains varied considerably, ranging from less than 10 ng up to 10 μg. In rape cases, vaginal swabs are inevitably contaminated by female DNA and this tends to obscure the suspect's DNA pattern. Preferential lysis of vaginal cells by preliminary incubation in a SDS/Proteinase K mixture, a treatment which does not affect the sperm nuclei due to their high content of disulphide bonds, allows the separation of the male DNA by simple centrifugation of sperm heads (Gill et al., 1986).

Conventional serological typing of semen stains is usually only capable of detecting two polymorphic markers – the ABH-group substances and the phosphoglucomutase (PGM_1) isozymes. Contamination by female vaginal secretions is not manageable, unlike cases dealing with DNA, and often makes an interpretation of the serological results impossible. The discrimination value of the examination can be calculated by multiplying the phenotype frequencies of the markers revealed, if independence is observed. Conventional techniques lead to values usually in the order of 1 out of 100 nonrelated individuals. Identical calculations can be performed based on results of DNA profiles provided that the frequencies of the bands are known, a prerequisite not yet always fufilled for the DNA probes available for practical use. Another requirement – underestimated at the beginning of DNA profiling and now probably overestimated – is the definition of a band match, particularly when band shifts – quite common in forensic material – do occur.

These problems are far from being solved and one can have doubts whether a solution will ever be possible. Gill *et al.* (1990) proposed a statistical definition of a match applying Bayes theorem as a decision-making tool, realizing that in the case of band shifts, matching or non-matching cannot be treated as a binary event. The application of their statistical method requires, however, a detailed analysis of the characteristics and errors of the electrophoretic system used, parameters that each laboratory must determine for its own locally employed method and equipment.

Strategy in Stain Analysis Using DNA Profiling

In complex cases, (e.g. when several men are involved in a rape case or when less than 200 to 500 ng of high molecular weight DNA is available), single locus probes should be used in the first instance. If a stain mixture can be excluded, a multi-locus probe can be used in a second and usually final identifying or excluding test. In all other cases, sequential hybridizations with a number of single-locus DNA probes is necessary. In the event of band shifts, the number of SLPs needs to be increased. The value of discrimination is calculated by multiplying the phenotype frequencies. The estimate of the frequency of a single band is based on its length, measured either manually or by the aid of a computerized analysis system. When suspects of other racial origin are involved and data of the frequencies of the probe used is lacking, we use a constant frequency of 10% per band.

A match of two bands is interpreted visually. In cases with discernable band shifting, the number of probes used is increased to further evaluate the small probability of a true exclusion, possibly being masked by the band shift. By increasing the number of probes in these cases we take into account the reduced statistical power of discrimination, which can be overcome by examining the case with additional probes.

Postmortem Stability of DNA

Data about stability of DNA and RNA in our bodies is sparse (Pääbo, 1985; Wood et al., 1986; Madison et al., 1987). High molecular weight DNA is a prerequisite to obtaining reliable RFLP-patterns (Jeffreys, 1984; Goelz et al., 1985; Hughes et al., 1986). Therefore, it is important to know the postmortem stability of DNA under various conditions. Postmortem decay of human bodies is a complex and not yet fully understood process beginning with autolysis and putrefaction, followed by aerobic and bacterial decomposition of organic material. These processes show a maximum of activity at temperatures between 34–40°C. Humidity of the air influences the rate of autolysis (Mackie, 1929). Loss of enzyme regulation and lactic acidosis in autolysis enhances the activity of some enzymes, for example, the hydrolases (Bradley, 1938; Gössner, 1955). Most of these enzymes are remarkably resistant to autolysis itself but are rapidly destroyed by bacteria. DNA in dead cells is degraded both by endonucleases and exonucleases. Yields of undegraded DNA by extraction of specimens of various tissues from human bodies at varying postmortem periods were reported by Bär et al. (1988). High molecular weight DNA is recoverable postmortem in sufficient quantities from various human organs as well as from blood, although not all organs are equally suitable. Good DNA stability is found in brain cortex, lymph nodes and psoas muscle over a period of three weeks post-mortem. Spleen and kidney show good DNA stability up to 5 days postmortem but after longer periods, rapid degradation is observed. Yields of DNA from blood tend to be inconsistent because of the non-homogeneity of samples. Blood clots are rich with DNA. Generally, the amount of degraded DNA correlates fairly well with the length of the postmortem period. However in some cases, DNA degradation is already prominent after a short postmortem period. In DNA fingerprints using the min-isatellite probe 33.15 gradual disappearance to complete loss of the long fragments (15–23 kb) can be observed. No extra bands were noted, thus excluding erroneous conclusions. However, the evidentiary value of older samples is lower.

DNA fingerprints is also a perfect tool to investigate disputed identity of blood alcohol samples (Bär et al., 1989). However, blood samples stored at ambient temperature for longer periods can show considerable degradation of high molecular weight DNA, diminishing the value of the fingerprint investigation because of the loss of the less frequent bands formed by the longer DNA fragments. Addition of the complexing agent EDTA can retard such a degradation and does not affect the chemical analysis of blood alcohol by enzymatic methods. The determination of sex with DNA probes in the blood alcohol sample can increase the confidence in the investigation (Gill, 1987).

DNA Fingerprinting and Paternity

By the early 1980s, experts on parentage and kinship generally felt the existing comprehensive range of polymorphisms and the efficiency of available physico-chemical separation techniques were optimal. The same applied to computer programmes developed for the biometric evaluation of findings. And indeed, the state of blood group expertise as practised in Central Europe had hardly changed in the past decade. Notwithstanding, 1985 brought a surprising advance: DNA analysis. The bands of a DNA profile segregate in a Mendelian fashion and the method of DNA profiling is therefore very suitable for the investigation of disputed paternity (Jeffreys *et al.*, 1985, 1986). Both multi-locus and single-locus probes lead to conclusive results in simple trio cases. The power of exclusion of nonfathers is high for both the multi-locus as well as a battery of 4 to 5 single-locus probes and in simple trio cases one can assume paternity for a non-excluded man. But without any doubt and since we are dealing with a scientific method, there remains a small chance of error. It is common practice to evaluate multi-locus profiles by the method of band-sharing. Its result is mainly based on paternal male and child (Lansmann *et al.*, 1981; Burke and Bruford, 1987); maternal information is not considered.

Essen-Möller (1938) introduced a biostatistical method to evaluate blood group findings in paternity cases by calculating the so-called W-Value. This permits a scientific evaluation of the two hypothesis: paternity versus non-paternity and provides the rate of error for the decision taken (1-W). It therefore supplies more complete information than the paternity index of Gürtler (1956) or the band sharing method. Ihm and Hummel (1975) finally described an algorithm for the biostatistical evaluation of cases with any degree of kinship. For the evaluation of SLP patterns the formal genetics for a system of multiple allelism is used and for multi-locus probes the model of multiple diallelism is applicable (Hummel *et al.*, 1990).

In deficiency cases – cases in which the putative father or another cohabitant or the child's mother is deceased, and relatives of the deceased may be accessible – blood-group opinions often produce only insufficient probability values. DNA analysis is particularly useful for unequivocally solving such cases.

The only deficiency cases that normal blood group opinions settle satisfactorily, either through a high W-valve or through an exclusion, are the so-called grandparent cases. However, should the W-value be only moderately high, one must consider the possibility that the respective grandfather may not be related. In such cases the W-value from the DNA analysis can substantiate or disprove kinship between the child and the grandmother or grandfather.

Apart from the above-mentioned grandparent cases, other types of deficiency occur in practice, especially when a deceased or missing man can be "replaced" by kin, e.g. siblings, own children plus their mother etc. Because there is usually little chance of exclusion, a very high or a very low probability is required for a decision pro or contra. Hence at least one multi-locus probe will be necessary, supported by several single-locus probes. In cases with deficient parties it is often necessary to use as many polymorphic genetic systems as possible, although HLA opinion usually is not of much help in such cases. The combination of the findings of the blood groups and DNA analysis and their combined W-value do not always show values above 99.73%, but in these cases values of 99% and above are very useful for the courts in conjunction with other evidence.

In cases without findings from the mother or father single locus probes are best used because of their ability to produce exclusions. Disputed sibship without accessible parents is treated differently since the null hypothesis cannot be refuted by "exclusion". Addition of one or more multi-locus probes to the set of investigative tools is useful in these cases and W-values of 95% and above (or 5% and below) are acceptable.

In conclusion, we recommend both multi-locus and single-locus probes in kinship cases, e.g. in mother-child-putative father cases and two-males cases. A set of 3 to 5 single-locus probes and/or one multi-locus probe are usually sufficient when combined with the blood group findings.

References

Bär, W., Kratzer, A., Mächler, M., and Schmid, W. (1988) Postmortem Stability of DNA. *Forensic. Sci. Intern.* 39: 59–70.

Bär, W., and Kratzer, A. (1989) Abklärung strittiger Identität von Blutalkoholproben mit DNA-Fingerprinting. *Z. Rechtsmed.* 102: 263–270.

Bradley, H. C. (1938) Autolysis and atrophy. *Physiol. Rev.* 18: 173.

Burke, T., and Bruford, M. W. (1987) DNA fingerprinting in birds. *Nature* 327: 149–152.

Essen-Möller, E. (1938) Die Beweiskraft der Aehnlichkeit im Vaterschaftsnachweis; theoretische Grundlagen. *Mitt. Anthrop. Ges.* (Wien) 68: 9–53.

Gill, P. (1987) A new method for sex determination of the donor of forensic samples using a recombinant DNA probe. *Electrophoresis* 8: 35–38.

Gill, P., Jeffreys, A. J., and Werret, D. J. (1985) Forensic application of DNA "fingerprints". *Nature* 318: 577–579.

Gill, P., Werrett, D. J., and Evett, I. W. (1990) Problems associated with the determination of band match probabilities. Advances Forens Haemogenetics, vol. 3., Polesky, H. F. and Mayr, W. R. (eds.), Springer Verlag Berlin Heidelberg, p. 63–67.

Goelz, S. E., Hamilton, S. R., and Vogelstein, B. (1985) Purification of DNA from formaldehyde fixed and paraffin embedded human tissue. *Biochem. Biophys. Res. Comm.* 130: 118–126.

Gössner, W. (1955) Untersuchungen über das Verhalten der Phosphatasen und Esterasen während der Autolyse. *Virchows Arch.* 327: 304.

Hughes, Margaret A., and Jones, D. S. (1986) Body in the bog but no DNA. *Nature* 323: 208.

Hummel, K., Fukshansky, N., Bär, W., and Zang, K. (1990) Biostatistical approaches using minisatellite DNA patterns in paternity cases (mother-child-putative father trios). Advances Forensic Haemogenetics, vol. 3., Polesky, H. F., and Mayr, W. R. (eds.), Springer Verlag Berlin Heidelberg, p. 17–22.

Ihm, P., and Hummel, K. (1975) Ein Verfahren zur Ermittlung der Vaterschaftswahrscheinlichkeit aus Blutgruppenbefunden unter beliebiger Einbeziehung von Verwandten. *Z. Immun. Forsch.* 149: 405–416.

Jeffreys, A. J. (1984) Raising the dead and buried. *Nature* 312: 198.

Jeffreys, A. J., Wilson, V., and Thein, S. L. (1985) Hypervariable "minisatellite" regions in human DNA. *Nature* 314: 67–73.

Jeffreys, A. J., Wilson, V., Thein, S. L., Weatherall, D. J., and Ponder, B. A. J. (1986) DNA "Fingerprints" and segregation analysis of multiple markers in human pedigrees. *Am. J. Hum. Genet.* 39: 11–24.

Lander, E. S., (1989) DNA Fingerprinting on trial. *Nature* 339: 501–505.

Lansman, R. A., Shade, R. O., Shapira, J. F., and Avise, J. C. (1981) The use of restriction endonucleases to measure mitochondrial DNA sequence relatedness in natural populations. *J. Mol. Evol.* 17: 214–226.

Mackie, F. P. (1929) The microscopical changes occuring in organs after death. *Ind. J. Med. Res.* 16: 827.

Madisen, L., Hoar, D. I., Holroyd, C. D., Crisp, M., and Hodes, M. E. (1987) DNA banking: The effects of storage of blood and isolated DNA on the integrity of DNA. *Am. J. Med. Genet.* 27: 379–390.

Pääbo, S. (1985) Molecular cloning of ancient Egyptian mummy DNA. *Nature* 314: 644–645.

Wood, T. L., Frantz, G. D., Menkes, J. H., and Tobin, A. J. (1986) Regional distribution of messenger RNAs in postmortem human brain. *J. Neurosci. Res.* 16: 311–324.

DNA Fingerprinting: Approaches and Applications
ed. by T. Burke, G. Dolf, A. J. Jeffreys & R. Wolff
© 1991 Birkhäuser Verlag Basel/Switzerland

Tracking the Violent Criminal Offender through DNA Typing Profiles – a National Database System Concept

F. S. Baechtel, K. L. Monson, G. E. Forsen, B. Budowle, and J. J. Kearney

Forensic Science Research and Training Section, FBI Laboratory, FBI Academy, Quantico, Virginia, U.S.A

Summary
Implementation of standard methods for the conduct of restriction fragment length polymorphism analysis into the protocols of United States crime laboratories offers an unprecedented opportunity for the establishment of a national computer database system to enable interchange of DNA typing information. The FBI Laboratory, in concert with crime laboratory representatives, has taken the initiative in planning and implementing such a database system. The Combined DNA Index System (CODIS) will be composed of three sub-indices: a statistical database, which will contain frequencies of DNA fragment alleles in various population groups; an investigative database which will enable linkage of violent crimes through a common subject; and a convicted felon database that will serve to maintain DNA typing profiles for comparison to profiles developed from violent crimes where the suspect may be unknown.

Introduction

Within recent years, a number of procedures have become available that enable the genetic typing of individuals at the level of their DNA. Of these methods, DNA typing by restriction fragment length polymorphism (RFLP) analysis has been adopted by many crime laboratories. The widespread acceptance and implementation of the RFLP procedures in the crime laboratories of the U.S. offer a unique opportunity for developing national computer databases that can assist law enforcement officials in the investigation of violent crimes and the apprehension of criminal offenders.

The need for assistance in these areas is compelling when one considers that many of the individuals who commit violent crimes are repeat offenders. A U.S. Department of Justice report (1989) indicates that of the 108,580 individuals who were released from prison in 11 of the United States in 1983, 67,898 were rearrested for a felony or serious misdemeanor within three years of release. The bulk of these arrests was for homicide, sexual assault, or robbery. Significantly, more than one in eight of the rearrests was in a state other than the state of earlier

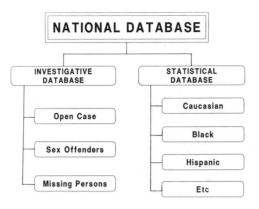

Figure 1. Structure of national database system.

conviction and imprisonment. In addition, only 70% of the homicide cases and 50% of the rape cases committed in the U.S. are closed by arrest. Given the magnitude of violent crime and the mobility of felons, a national database structure that could link crime scene evidence with convicted felons through DNA typing procedures would be a powerful asset to law enforcement officials. This document describes the initiative currently underway in the FBI Laboratory to bring such a national database concept to fruition. The database has been named CODIS, which stands for COmbined DNA Index System. In its initial configuration, CODIS has been structured to contain a statistical database and an investigative support database (Fig. 1). These databases are not designed to hold detailed information concerning a DNA profile submission; rather they will serve as indices to reference the laboratories that contributed the DNA profiles.

Statistical Database

Estimates of DNA fragment allele frequency at various loci for human DNA specimens can only be derived through reference to appropriate population data. Since all laboratories conducting RFLP analyses on DNA specimens presumably will have need for allele frequencies, the establishment of a statistical database is a function appropriate to a national database. The statistical database initially will contain fragment profiles at several loci obtained through analysis of DNA from randomly chosen individuals from major population groups within North America. Profiles from Caucasians, U.S. Blacks, Hispanics, Orientals, and American Indians as well as profiles from individuals representing sub-populations within any of these major groups will be included in this database. DNA profiles placed into this database will be developed on a collaborative basis by cooperating crime laboratories and maintained by the FBI. All crime laboratories in North America would have access to the information in this database.

There are several advantages to having a statistical database at the national level: (1) Because it is composed of data submitted by numerous agencies, it will be larger in scope than would be possible for each participating agency to assemble by itself. Large population databases can more accurately reflect the true population distribution of fragment alleles, especially those that occur at low frequencies; (2) laboratories will gain access to allele frequencies for population groups not represented in their local areas due to unequal population distributions; and (3) a central repository for population data compels the standardization of methodologies if data are to be compatible.

The statistical database will be seeded initially with DNA data obtained from random donors, and continually expanded with additional data from this donor source. No information that might be used to identify a specimen source will be attached to the DNA profiles in the statistical database.

Investigative Support Databases

Open Case DNA Profiles

An open case DNA profile database can serve to link violent crime cases through a common subject, usually an unidentified suspect(s). DNA profile information entered into this database would be derived from case evidence specimens (e.g. semen, blood, or saliva stains). Profiles developed from live or dead victims would also be entered into this database. Open case data submitted to the database would be compared with open case data already residing within the database as well as data in the convicted violent offenders database (see below). If no matches were found with existing data, the new submission would be added to the open case data files for future reference. In the event of a profile match between entered data and resident data, the exchange of information pertinent to the match would be coordinated with the agency holding the referenced DNA case evidence profile. The actual storage of detailed case records will be the responsibility of the submitting agency.

Information that linked separate cases would be of value in tracking the activities of serial offenders throughout different jurisdictions in North America. Identification of the offender in one jurisdiction could lead immediately to the solution of linked cases elsewhere.

Convicted Violent Offender DNA Profiles

A number of jurisdictions in the U.S. have passed statutes that compel the DNA profiling of individuals convicted of sex offences. The convicted violent offenders database would receive DNA profiles developed by state and local laboratories under these mandates. This database would be used to determine if the convicted individual had been involved in similar crimes elsewhere, and also would be accessible for reference by open case submissions. As with the open case file database, detailed records of offences would be kept by the contributing agencies, with the CODIS acting only as an index to the source of those records.

Missing Persons/Unidentified Bodies DNA Profiles

The missing persons/unidentified body database would serve to identify missing persons through genetic linkage with their close biological relatives, or to link found human remains with their source. Appropriately close biological relatives of missing individuals would have their DNA profiles prepared and entered into this database for comparison with DNA profiles obtained from found bodies, body parts, or body fluids.

Criteria for Data Acceptance

Genetic information in the form of RFLP profiles can be successfully interchanged among laboratories through databases only when certain essential procedural criteria are observed. Failure to follow these procedural guidelines will jeopardize the value of the RFLP results as they relate to other DNA profiles resident in the databases.

In order to interchange data in the CODIS system, participating laboratories must agree to the following standard practices:

(1) DNA must be fragmented using the restriction endonuclease HaeIII;
(2) DNA fragments must be separated electrophoretically in gels of a composition that yield data compatible with that resident in the database;
(3) Employ DNA probes to a standard array of loci;
(4) Employ standard reference fragments that enable accurate sizing estimates and employ DNA specimens to satisfy quality control standards; and
(5) Submit to periodic quality assurance procedures.

360

Growth Assumptions

A number of variables are likely to influence the growth of the CODIS system over the next five years. These would include:

(1) The number of sex offenders required to give blood samples for DNA profiling;
(2) The number of violent crimes producing body fluid specimens amenable to DNA profiling;
(3) Acceptability of DNA testing procedures and examination results in the courts;
(4) State, local and federal funding for implementation of DNA profiling procedures; and
(5) Availability of training in RFLP testing procedures.

Perhaps the most influential factors in limiting the growth of the CODIS system will be the rate at which crime laboratory scientists can be trained in the methodology of RFLP analysis and the extent to which agencies can afford the costs of this new technology. Currently, the FBI Laboratory, through its facilities at the FBI Academy in Quantico, Virginia, provides most of the training of domestic crime laboratory personnel in DNA analysis methods. During the past 21 months, 200 individuals from 67 crime laboratory agencies (including 9 persons from 8 foreign countries) have been trained in DNA profiling technology by the FBI Laboratory. Despite the magnitude of this training effort, fewer than 10 public crime laboratories in the U.S. are conducting DNA analyses on actual case evidence as of this symposium. Clearly, the initial rate of submission of DNA profiles to CODIS will be low. However, many of the agencies that are staffed by individuals already trained in the technology should be examining cases and submitting data to CODIS within two to three years.

Reference

United States Department of Justice, (1989) Recidivism of prisoners released in 1983. Bureau of Justice Statistics report. NCJ 116261.

DNA Fingerprinting: Approaches and Applications
ed. by T. Burke, G. Dolf, A. J. Jeffreys & R. Wolff
© 1991 Birkhäuser Verlag Basel/Switzerland

The Quality Control of Cell Banks Using DNA Fingerprinting

G. N. Stacey, B. J. Bolton, and A. Doyle

European Collection of Animal Cell Cultures, PHLS Centre for Applied Microbiology and Research, Porton Down, Salisbury, Wiltshire SP4 0JG, Great Britain

Summary
Reproducibility in animal cell culture technology requires careful preparation and characterisation of banks of cell cultures. The two standard techniques used in the quality control of such banks are isoenzyme analysis and cytogenetics which require complex and time-consuming procedures to enable cell line identification. However, DNA fingerprinting is potentially a more powerful method of analysis which can detect mutation and intra-species cross-contamination. At the European Collection of Animal Cell Cultures (ECACC) multilocus fingerprint analysis using probes 33.6 and 33.15 has been assessed in the quality control of cell banks. This method has confirmed consistency between master and working banks, has proven useful over a wide species range and can differentiate closely related cell lines. The key advantage of this method is its ability to detect cross-contamination by cell lines from a wide range of species using a straightforward and economical test. In addition the reproducibility of DNA fingerprints indicates their possible role in cell line authentication procedures which are important for patent and product licence applications.

In the field of animal cell technology it is important to have available a reliable source of authenticated cultures to enable the production of consistent experimental data in research and the reproducibility required in industrial scale processes.

Currently, two of the most commonly used methods for the authentication of cell lines are cytogenetic and isoenzyme analysis. Guidelines are set for the characterisation of cell lines used in the production of vaccines and other therapeutic products by WHO (WHOSG, 1987), the US Food and Drug Administration (OBRR, 1987) and the Commission of the European Communities (CPMP, 1987; 1988). To a large extent these have recognised conventional methods which have several disadvantages, not least of which is the interpretation of the data. Multilocus DNA fingerprinting, using probes 33.6 and 33.15 discovered by Alec Jeffreys, is a new technique which has been extensively tested in the analysis of animal populations and has been recently accredited with a British Standard (BS 5750) for use in human paternity testing and forensic analysis. A recent development is the use of the technique in the quality control of cell lines, where distinct and very important advantages over the existing quality control procedures have been shown (Gilbert *et al.*, 1990; Stacey *et al.*, 1990).

Two events which can significantly affect the defined features of a cell line are accidental cross-contamination with other lines and genetic mutation. Cell line cross-contamination may go undetected unless unexpected experimental data or obvious morphological differences give reasons to suspect the line. At the beginning of the 1980's, extensive Hela cell contamination of cell lines was discovered in many North American laboratories following cytogenetic and isoenzyme analysis (Nelson-Rees, 1981). However, cases of cross-contamination, which would probably not have been detected by such techniques, have been readily resolved by DNA fingerprinting (Van Helden et al., 1988).

Mutation of cell lines may also go undetected without appropriate quality control procedures since the effects of mutation on the characteristics of the cell line may be insignificant in the early stages. DNA fingerprinting has been used to identify mutations in lines of the same origin, cultured at different laboratories (Thacker et al., 1988). In this case no changes in the characteristics of the cells were recognised until DNA fingerprinting was employed. Mutation rates in somatic cells have been estimated at 10^{-6} per gene per generation (Alberts et al., 1989). Hence, if cell lines have a similar level of mutability then mutations would be expected to occur frequently in exponentially growing cell cultures which can exceed 10^6 cells per ml. Obviously, the chances of developing a stable mutation with a significant effect on a cell line will be dramatically increased in continuous culture and large scale cultures.

In order to avoid the problems discussed above, cell "banking" systems have been devised whereby large quantities of homogeneous cultures are stored in a viable but non-dividing state in liquid nitrogen at $-196°C$. Such banks provide cultures which are genetically uniform and as close as possible to the original cell culture. A hierarchy of cell banks is generally used (as indicated in Fig. 1) to ensure that the number of population doublings (pd's) between the initial stock and the cells in use is kept to a minimum. This is achieved by replacing a depleted bank with a sample expanded from the next banking level towards the initial stock. Maintaining a minimum number of pd's between initial stock and cell banks reduces the chance of mutation, but is also of critical importance when dealing with cell lines of limited lifespan (e.g. Human Diploid Cell Strains MRC-5 and WI 38). Where cell lines have an infinite life expectancy limits are usually set for the number of pd's permitted for cell lines used for industrial or clinical purposes, particularly those with a highly differentiated function (Hayflick and Hennesen, 1988). For quality assurance, it is important to establish the consistency of each banked cell line both vertically through the tiers of banks and horizontally within each bank. However, horizontal analysis is most important in banks most distant from the initial stock in terms of cell doublings. For example, it is usual to test cells at

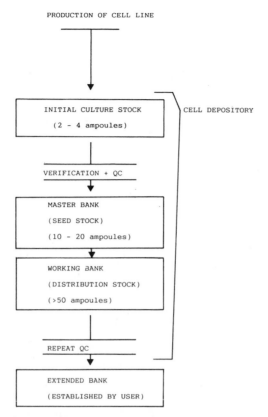

Figure 1. Cell banking procedure. The initial deposited sample is expanded and preserved to produce the master bank which is carefully characterised. The working bank is generated and replenished from the master bank. QC, quality control procedures.

10 pd's beyond production level in cell lines where therapeutic protein products are derived (Hayflick and Hennesen, 1988).

Quality control of a cell line by analysis of a specific product or function is unlikely to identify cross-contamination and genetic aberrations outside the gene sequences related to that particular feature of the cell line. Characterisation of a cell line by techniques such as cytogenetic or isoenzyme analysis enables it to be monitored for chromosomal defects (cytogenetics) and contamination with other cell types (cytogenetics and isoenzymology), which may not have immediately obvious effects on the performance of the cell line. However, cytogenetics is a difficult and time-consuming technique which may not identify intra-species cross-contamination and is not suitable for cell lines showing cytogenetic variation over short term passage (Miele *et al.*, 1989). Furthermore, whilst new kits for isoenzyme analysis have simplified this procedure, a minimum of thirteen separate analyses may be required to give a degree of confidence in specific identification of cell lines (Hay, 1988). Conversely, techniques with extreme sensitivity which could detect minor fluctuations in the state of a cell culture would be too sensitive to be of any practical value.

Thus, quality control of cell lines needs a method with an appropriate balance between the ability to specifically identify a cell line and the ability to recognise contamination by other cell lines. Multilocus DNA fingerprinting with probes 33.6 and 33.15 appears to represent such a method and is proving to have great potential in the quality control of cell banks.

At the ECACC this technique is in the process of assessment in the quality control of cell banks. A two tier banking system is operated at the ECACC with a "master" bank produced from the initial stock and a "working" or distribution bank generated from the master (see Fig. 1). The DNA fingerprinting method used is performed under licence from ICI Cellmark Diagnostics and the general methods and conditions have been developed from those reported previously (Jeffreys et al., 1985). In an initial investigation, fingerprints and isoenzyme profiles were produced from samples of the master and working banks for eight cell lines of human (five lines), murine (two lines) and primate (one line) origin. This data showed identical fingerprint patterns for samples of cells from master and working banks and examples are given in Fig. 2. In these experiments, samples were compared on the same Southern membrane to eliminate variables in the fingerprinting process. However, our data shows a high degree of reproducibility between membranes, which indicates that it may be possible to produce a catalogue of cell line DNA fingerprints for identification purposes. Useful cell lines have been generated from a diverse range of animals, including humans, rodents, bats, ungulates, birds, reptiles, amphibians and insects. Thus, in cell banking it is important that the fingerprinting method used should be applicable to the wide range of species encountered in animal cell technology.

The Jeffreys probes 33.6 and 33.15 have proved to be as useful in differentiating individual cats, birds and dogs as in human populations and have been applied to many other species (Hill, 1987). Species fingerprinted at the ECACC are given in Tab. 1 and examples of fingerprints from mouse, hamster and human lines are shown in Fig. 3. Obviously, the capacity to differentiate individuals within a population depends on the genetic diversity of that population. Therfore, it is important to assess carefully the use of probes 33.6 and 33.15 in cell lines where the origin of cells may be restricted to a few strains of each species (e.g. murine hybridomas) or where a number of derivative lines may share a common ancestor. In our experience diverse fingerprint profiles are generally obtained for cell lines from individual species. Furthermore, our investigations of closely related hybridomas from the

M W   M W   M W

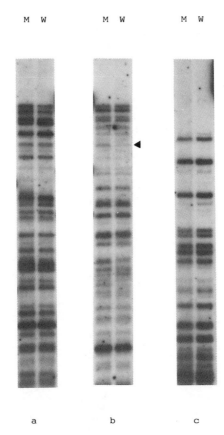

a   b   c

Figure 2. Probe 33.6 DNA fingerprint profiles of master (M) and working (W) cell bands. a, K562 a human myelogenous leukaemic cell line. b, MOLT4 a human T cell line. c, P3/X63Ag8.653 a murine myeloma cell line. The one case of a relative band intensity difference is indicated by an arrow (◄).

BALB/c mouse strain show differences in their fingerprints. This is illustrated in Fig. 4 which shows differences in fingerprint profiles found in different BALB/c × BALB/c hybridomas sharing one identical fusion partner. However, identical fingerprints have been found in closely related but phenotypically distinct cell lines, as found with the lymphotropic virus susceptible lines HUT 78 and H9 (Mann *et al.*, 1989). These lines have identical multilocus fingerprints, as shown in Fig. 5, but can be distinguished using single locus probes (Mann *et al.*, 1989).

The wide range of species to which this technique can be applied means that it is very useful in the detection of cell line cross-contamination. However, it is important to establish the sensitivity of this aspect of multilocus fingerprinting. Experiments at the ECACC have identified

Table 1. Cell lines from the European Collection of Animal Cell Cultures fingerprinted as part of quality control procedures. High levels of band sharing were detected only in some mouse cell lines e.g. the BALB/c lines. Nevertheless, they were still differentiated

Animal	Origin of cell line Species/strain	Tissue	Cell line name
Human		Transformed lung fibroblast.	MRC5
		Myeloid leukaemia	K562
		Cervical epithelial carcinoma	HelaB
		Promyelocytic leukaemia	HL60
		Thyroid lymphoma	Daudi
		T cell leukaemia	MOLT 4
		HUT 78 derivative	H9
		T cell leukaemia	HUT 78
		Histiocytic lymphoma	U937
		Rectal epithelium	HT29/219
		B lymphocytes	HLA defined – various
Primate	African green	Kidney	Vero
	monkey	Kidney	CV1
			BSC1
Dog		Kidney epithelium	MDCK
Cat		Kidney	CRFK
Rabbit		Kidney epithelium	Various
Hamster	Syrian	Kidney fibroblast	BHK
	Chinese	Ovary epithelium	CHO
Mouse	BALB/c	Myelome	P3/X63Ag8.653
	BALB/c	Myeloma	NS1
		Myeloma	NS0
	C3H/An	Areolar and adipose Tissue fibroblast	L929
	Swiss	Embryo fibroblast	3T3
Bat		Lung	Tb1Lu
Bird	Goose	Sternum	CGBQ
Lizard	Iguana	Heart	Ig-H2

DNA from one B-lymphoblastoid cell line diluted at between 5 and 10% ($\mu g/\mu g$) in DNA from another B-lymphoblastoid cell line. Further experiments in collaboration with the Institut Pasteur have investigated DNA from previously undisclosed mixtures of cells and for combinations of the cell lines BHK and CHO which indicate a level of detection of 1–5% (CHO cell/BHK cell). Thus, multilocus fingerprinting can readily detect low level contamination which might go unnoticed by the existing quality control techniques.

Apart from cross contamination and mutation there are many factors encountered in tissue culture procedures that might conceivably affect the fingerprint of a cell line. For example, mycoplasma contamination is a significant problem in animal cell culture and has been associated with the appearance of chromosome aberrations and even cell death (Aula

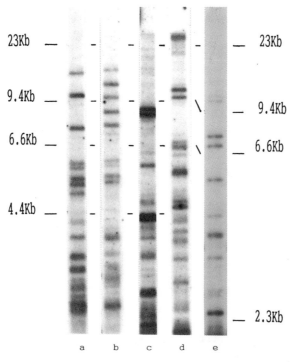

Figure 3. Probe 33.6 DNA fingerprints for cell lines from mouse, human, monkey and hamster cells lines. a) P3X63Ag, BALB/c mouse myeloma line. b) L929, C3H mouse lung fibroblast line. c) Vero monkey kidney cell line. d) MRC5 human lung fibroblast line. e) BHK hamster kidney fibroblast line. All fingerprints except e) were produced on the same Southern membrane.

and Nichols, 1967; Levine *et al.*, 1968). We have previously reported DNA fingerprint analysis of a Vero cell line which showed two marked band intensity changes following mycoplasma infection (Stacey *et al.*, 1990). In this case it appeared that mycoplasma contamination over a period of only 20 days had induced changes in the cell line's fingerprint when compared with an uninfected control. Consequently, it will be important to establish the effects of long-term mycoplasma contamination and subsequent eradication. Growth conditions, including the effects of growth phase, heat shock and chemical or hormonal stimulation, must also be investigated for their influence on cell line fingerprints. Such experiments are important to establish the level of sample standardisation required when DNA fingerprinting is used in quality control procedures. Alterations in culture supplements and adaptation to serum-free conditions (a desirable feature in production processes) do not appear to have any effect on the fingerprint profile. This is illustrated for a murine hybridoma cultured with different supplements and adapted to serum free culture as shown in Fig. 6.

368

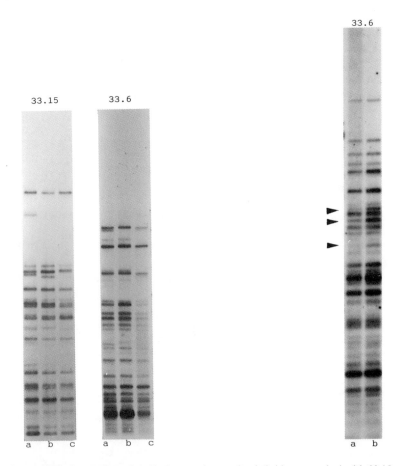

Figure 4. DNA fingerprints of antibody secreting murine hybridomas probed with 33.15 and 33.6. a) hybridoma CP3A5 produced by fusion of a BALB/c mouse spleen preparation with the myeloma cell line NS1. b) hybridoma CPeA101, a fusion of the same spleen preparation used with the myeloma line NSO. c) hybridoma StA19, a different BALB/c spleen preparation to that in a) and b) fused with the same immortal partner used in a) (Hybridomas were kindly provided by Dr. S. A. Clark, PHLS, CAMR, UK).

Figure 5. DNA fingerprints of the human T cell line HUT78 (b) and its derivative line H9 (a). The band patterns are identical although there are three bands in H9 with notably decreased intensities (▶).

A large number of fingerprinting methods have been developed using diverse types of DNA probe. The multilocus method described above and operated with Cellmark protocols provides a high quality of fingerprint which is a very important feature when testing for minor changes in fingerprint profile due to mutation or very faint superimposed profiles in cases of cross-contamination. Oligonucleotides homologous to hypervariable DNA core sequences are increasingly used and in the future may offer an alternative to multilocus fingerprinting of cell lines.

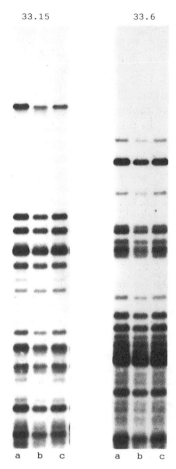

Figure 6. DNA fingerprints of a murine hybridoma cultured with different media supplements. a) media supplemented with foetal calf serum (10%). b) media supplemented with horse serum (10%). c) hybridoma adapted to serum free culture.

Other features such as patterns of DNA methylation may also prove useful in the analysis of cell line stability. However, at present the most important supplement to multilocus fingerprint analysis of cell lines would be a set of species specific, yet polymorphic, single locus probes. Multilocus probing could then be used as a screen for cross-contamination and identity; with confirmation possible using species specific single locus probes.

The potential effects of mycoplasma contamination on experimental data has led a number of scientific journals to require authors to provide evidence that cell lines used in publications are mycoplasma-free (Mowles and Doyle, 1990). In view of the difficulties in recognising cell line cross-contamination and mutation, it would seem appropriate to advocate similar measures using DNA fingerprinting to establish cell line identity and purity in order to promote scientific confidence when comparing data. Commercial procedures such as filing patents and product licensing may also benefit from the application of DNA fingerprint technology. While the influences of some tissue culture procedures on hypervariable DNA remains to be investigated, multilocus fingerprinting of samples taken under standardised conditions, offers an economical and effective method for the quality control of cell banks.

References

Alberts, B., Bray, D., Lewis, J., Raff, M., Roberts, K. and Watson J. D. (1989) Molecular Biology of the Cell. (2nd Edition). Garland Publishing.

Aula, P. and Nichols, W. W. (1967) The cytogenetic effects of mycoplasma in human leucocyte cultures. *J. Cell Physiol.* 70: 281–90.

Committee for Proprietary Medicinial Products AdHoc Working Party on Biotechnology/Pharmacy (CPMP). (1987) Guidelines on the production and quality control of medicinal products derived by recombinant DNA technology. *T.I.B. Tech.* 5: G1-3.

Committee for Proprietary Medicinal Products AdHoc Working Party on Biotechnology/Pharmacy (CPMP). (1988) Guidelines on the production and quality control of monoclonal antibodies of murine origin intended for use in man. *T.I.B. Tech.* 6: G5–8.

Gilbert, D. A., Reid, Y. A., Gail, M. H., Pee, D., White, C., Hay, R. J. and O'Brian, S. J. (1990) Application of DNA fingerprints for cell line individualisation. *Am. J. Hum. Gen.* 47: 499–514.

Hay, R. J. (1988) The seed stock concept and quality control for cell lines. *Analyt. Biochem.* 171: 225–237.

Hayflick, L. and Hennessen, W. (eds). (1988) *Develop. Biol. Stand.* 70: 153–157.

Hill, W. G. (1987) DNA fingerprints applied to animal and bird populations. *Nature* 327: 98.

Jeffreys, A. J., Wilson, V. and Thein, S. L. (1985) Individual specific "fingerprints" of human DNA. *Nature* 316: 76–79.

Levine, E. M., Thomas, L., McGregor, D., Hayflick, L. V. and Eagle, H. (1968) Altered nucleic acid metabolism in human cell cultures infected with mycoplasma. *Proc. Natl. Acad. Sci. USA* 60: 583–589.

Mann, D. L., O'Brien, S. J., Gilbert, D. A., Reid, Y. A., Popovic, M., Reid-Connole, E., Galle, R. and Gazdar, A. F. (1989) Origin of the HIV-susceptible human CD4 cell line H9. *AIDS Res. Retroviruses* 5: 253–255.

Miele, M., Bonatti, S., Menickini, P., Ottaggio, L. and Abbondandolo, A. (1989) The presence of amplified regions alters the stability of chromosomes in drug resistant Chinese hamster cells. *Mut. Res.* 219: 171–178.

Mowles, J. M. and Doyle, A. (1990) Cell culture standards – time for a rethink? *Cytotechnology* 3: 107–108.

Nelson-Rees, W. A., Daniels, D. W. and Flandermeyer, R. R. (1989) Cross-contamination of cells in culture. *Science* 212: 446–452.

Office of Biologics Research and Review (OBRR). (1987) Points to consider in the characterisation of cell lines used to produce biologicals. Food and Drugs Administration. Bethesda MD 20892.

Stacey, G. N., Lanham, S., Booth, S. J., and Bolton, B. J. (1991) Characterisation of cell banks by DNA fingerprinting. Proceedings of the 10th meeting of the European Society for Animal Cell Technology. Avignon, France. Butterworths-Heineman.

Thacker, J., Webb, M. B. T. and Debenham, P. G. (1988) Fingerprinting cell lines; use of human hypervariable DNA probes to characterise mammalian cell cultures. *Somat. Cell and Mol. Gen.* 14: 519–525.

Van Helden, P. D., Wiid, I. J. F., Albrecht, C. F., Theron, E., Thornley A. L. and Hoal-van Helden, E. G. (1988) Cross-contamination of human cell esophageal squamous carcinoma cell lines detected by DNA fingerprint analysis. *Can. Res.* 48: 5660–5662.

World Health Organisation Study Group (WHOSG) (1987) Cells, products, safety. *Dev. Biol. Stand.* 68: 1–81.

DNA Fingerprinting: Approaches and Applications
ed. by T. Burke, G. Dolf, A. J. Jeffreys & R. Wolff

Detection of Amplified VNTR Alleles by Direct Chemiluminescence: Application to the Genetic Identification of Biological Samples in Forensic Cases

R. Decorte and J.-J. Cassiman

Center for Human Genetics, University of Leuven, Campus Gasthuisberg O&N6, B-3000 Leuven, Belgium

Summary
Minisatellite or variable number of tandem repeat (VNTR) regions contain such a high degree of polymorphism that they allow one to construct an individual-specific DNA "fingerprint". Analysis of these sequences by Southern blot however, consumes much DNA and is not applicable to degraded DNA samples often recovered from body-fluid stains found at crime scenes. The polymerase chain reaction (PCR) technique may overcome these problems. With oligonucleotide primers flanking the repeat region, amplification of the VNTR alleles followed by direct visualization on ethidium bromide-stained agarose gels is possible. In those cases were the PCR yield is too low for direct visualization, the product can be blotted to a nylon membrane and hybridized with a labelled internal probe. Alternatively, the PCR product can be biotinylated during amplification and visualized by direct chemiluminescence after Southern transfer. The remarkable sensitivity of the PCR technique has allowed the detection of genetic polymorphisms in single cells, hair roots and single sperm. A drawback of this very high sensitivity however is that special precautions have to be taken to prevent accidental contamination resulting in erroneous interpretation of the results.

Introduction

At the end of the 19th century, the "fingerprint" patterns formed by the sweat glands at the tips of the fingers were recognized for the first time as unique identifiers of human individuals (Ceccaldi, 1989). Since then forensic scientists have looked for similar unique markers which would allow them to identify blood and body-fluid samples with the same degree of certainty. Up to 1985, biologic samples for forensic analysis were mainly typed for protein polymorphisms such as blood groups, serum proteins and red-cell enzyme systems (Sensabaugh, 1981). However, blood samples usually enter the laboratory as dried stains on different kinds of material and are sometimes contaminated, which reduces the number of markers which can be typed. For the typing of semen, the number of protein markers which can be used is even smaller and, as a result, the likelihood of finding evidence which could exclude a suspect is low. Moreover, semen samples from a sexual assault are often contaminated with vaginal secretions which carry enzyme and blood group activity.

Restriction fragment length polymorphisms (RFLPs) have become extremely useful for the study of genetic variation at the DNA level in man (Cooper *et al.*, 1985). While most polymorphic loci have two alleles with a maximum heterozygosity of 50%, a number of them are multi-allelic and have a high polymorphic information content. The use of these hypervariable (minisatellite or VNTR) regions to produce an individual-specific DNA "fingerprint" (Jeffreys *et al.*, 1985a, b) therefore meant a big step forward in the ultimate goal of the forensic scientist. Moreover, it has been shown that stable and reproducible 'fingerprint' or VNTR patterns can be produced from dried bloodstains (up to 4 years old), semen stains, hair roots and bloodstains exposed to different environmental influences (Gill *et al.*, 1985; Kanter *et al.*, 1986; Giusti *et al.*, 1986; Gill *et al.*, 1987; McNally *et al.*, 1989a). An additional breakthrough in the identification of rapists has been the development of a differential cell lysis method to separate sperm nuclei from vaginal cellular debris, obtained from semen-contaminated vaginal swabs (Gill *et al.*, 1985; Giusti *et al.*, 1986).

Analysis of VNTR Regions by Southern Blot Analysis

Until recently, typing of VNTR regions was done entirely by Southern blot analysis (Southern, 1975). Depending on the hybridization and washing conditions which were used in this assay, different results could be obtained. Under conditions of low-stringency hybridization, the minisatellite probes developed by Jeffreys *et al.* (1985a) detect a number of loci containing tandem repeats of similar sequence. Using two different multi-locus minisatellite probes which detect different subsets of minisatellite regions, it is possible to create a DNA "fingerprint" with a chance of identity between two non-related individuals of 5.4×10^{-21} except for identical twins (Jeffreys *et al.*, 1985b). However, this technique remains difficult to perform even for a competent molecular biologist and the interpretation of the generated pattern of bands requires great experience. Similar sized fragments in two individuals may represent the same alleles or may be derived from different loci (Lewin, 1989). The result of this is that two non-related individuals will share approximately 25% of their bands by chance (Jeffreys *et al.*, 1985b; Wong *et al.*, 1987). Hypervariable minisatellites tend to cluster in the proterminal regions of human autosomes and therefore may be linked to each other and not segregate as independently as previously assumed (Royle *et al.*, 1988). Finally, DNA "fingerprint" patterns have been revealed using human probes in several animals and plants including poultry (Hillel *et al.*, 1989), fish and cattle (Vassart *et al.*, 1987;

Georges *et al.*, 1988), dogs (Jeffreys and Morton, 1987) and rice (Dallas, 1988). This is particularly troublesome in the analysis of bloodstains in forensic cases where some stains may belong to a domestic or a farm animal.

In contrast to the multi-locus approach, a minisatellite or a VNTR probe will only detect its complementary locus on a Southern blot under conditions of high-stringency hybridization (Nakamura *et al.*, 1987). Whereas the multi-locus probes detect variability in different organisms, the single-locus human probe, as far as is known, only show cross-reaction with DNA from primates (Bowcock *et al.*, 1989; Wong *et al.*, 1990). Single-locus profiles have the advantage over multi-locus profiles to be less complex and to allow for the determination of locus-specific allele frequencies (Baird *et al.*, 1986; Balazs *et al.*, 1989; Chimera *et al.*, 1989; Odelberg *et al.*, 1989; Gasparini *et al.*, 1990). Moreover, the locus-specific probes can be pooled together and used to generate a DNA fingerprint similar to the multi-locus pattern. If these loci are not in linkage disequilibrium with each other, then the combined chance that two non-related individuals will have the same genotype approaches the chance obtained for the multi-locus probes (Wong *et al.*, 1987).

The single-locus probes also have the advantage of improving the sensitivity of DNA typing. Whereas DNA fingerprints can be obtained from at least 500 ng of DNA (Jeffreys *et al.*, 1985b), single-locus profiles require only 60 ng of DNA (Wong *et al.*, 1987), assuming that the DNA is relatively undegraded. However, forensic samples in crime cases do not always contain high-molecular-weight DNA for the analysis of VNTR regions. In fact, a recent report on the stability of DNA in bloodstains from actual casework revealed that only 41% of the samples contained high-molecular-weight DNA, while in 48% of the samples either degraded DNA or no-detectable DNA was obtained (McNally *et al.*, 1989b). Moreover, typing of VNTR regions by Southern blot analysis consumes a large amount of DNA and re-analysis of a forensic sample is not always possible.

Other potential problems seen in the analysis of VNTR regions by Southern blot analysis are the low resolution for small size differences between relatively large DNA fragments (Baird *et al.*, 1986; Boerwinkle *et al.*, 1989; Gill *et al.*, 1990) and the difference in migration of DNA across an agarose gel (e.g. band shifting; Lander, 1989). For hypervariable loci, some alleles may remain undetected because their size difference is between 11 and 70 base pairs. This could cause a heterozygous individual with alleles that differ by only one repeat unit to be typed as a homozygote. As a consequence, it will lead to incorrect population frequency estimates for the different alleles and to deviations from the Hardy-Weinberg equilibrium (Lander, 1989; Odelberg *et al.*, 1989).

Band shifting is a technical problem which involves a number of variables such as the concentration of sample DNA, contamination of the sample, the composition of the gel and the running conditions. Because of the small differences possible between two alleles, it is sometimes difficult to know whether the relative positions of the two bands arose from the size of the allele fragments or whether the small difference resulted from band shifting. This may lead to erroneous conclusions about the presence of matching band pattern. A number of procedures for the correction of this effect have been proposed including the use of monomorphic probes to calculate bandshift correction factors (Norman, 1989), declaring a match only when bands fall within a certain number of standard deviations (Baird et al., 1986; Gill et al., 1990), estimating the degree of match by using probability functions (Gjertson et al., 1988) and running a mixed sample of evidence DNA and suspect DNA (Lander, 1989). From a scientific point of view, the mixing experiment represents the most straightforward approach to eliminate the problem of band shifting between two samples of DNA. Unfortunately, most forensic samples do not contain enough DNA to allow such an experiment and therefore as an alternative the use of monomorphic probes is highly recommended (Lander, 1989).

Application of the Polymerase Chain Reaction Technique to the Analysis of VNTR Regions

The limitations observed with Southern blot analysis of VNTR regions (e.g. sensitivity of the probes, low resolution for small size differences) and the problems associated with DNA from forensic samples (e.g. low amounts of DNA, degraded DNA) may be overcome by applying the polymerase chain reaction (PCR) technique. This method which has been developed by Mullis and co-workers (1986), is based on the enzymatic amplification of a DNA fragment that is flanked by two oligonucleotide primers that hybridize to opposite strands of the target sequence. Repeated cycles of the reaction involves denaturation of the target DNA, annealing of the two primers to their complementary sequences and extending the annealed primers on the template with a DNA polymerase (Fig. 1). Since the extension product of each primer can serve as a template for the other primer, each cycle will double the amount of target sequence produced in the previous cycle. This results in an exponential increase of the target sequence and in 20–30 cycles the original sequence has been amplified by more than a million-fold. In early experiments, the Klenow fragment of *Escherichia coli* DNA polymerase I was used with an extension step at 37°C, but after each denaturation step fresh enzyme had to be added because it was not resistant to the high temperatures used for the denaturation of the

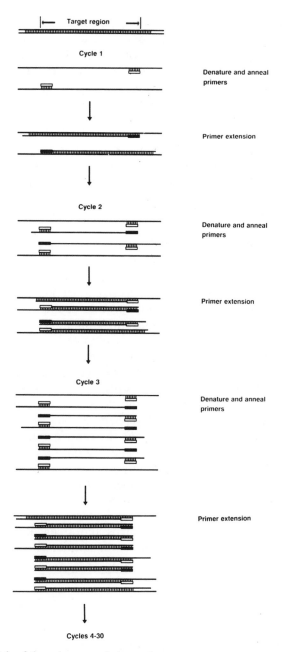

Figure 1. Principle of the polymerase chain reaction.

376

target DNA (Mullis *et al.*, 1986). The isolation and cloning of a thermostable DNA polymerase from the bacterium *Thermus aquaticus* made it possible to add the enzyme only in the first cycle of the reaction and even to automate the process in a temperature-cycling device simplifying the whole procedure (Saiki *et al.*, 1988; Lawyer *et al.*, 1989). In addition, the specificity of the reaction improved by the higher annealing and extension temperatures and the overall yield of amplification product increased.

A prerequisite for the PCR technique is that at least part of the target DNA sequence encompassing the region of interest must be known for the design of the amplification primers. This makes it possible to choose the primer sequences so that they flank the polymorphic site directly. For insertion/deletion polymorphisms, it implies that even small differences in length of one to two base pairs can be detected when the amplified fragment is small enough to allow discrimination by high resolution gel electrophoresis (Weber and May, 1989; Litt and Luty, 1989; Economou *et al.*, 1990). This "high resolution design" is also applicable to VNTR regions where the size difference can be between 11 and 70 base pairs. The two oligonucleotide primers flank the tandem repeat region and are complementary to a unique DNA sequence (Fig. 2). In this way a smaller fragment is obtained than after restriction enzyme digestion and conventional Southern blotting. The length variation between the alleles at a given VNTR locus can be readily detected by size fractionation of the PCR products on an analytical gel (Fig. 3).

Perhaps the main advantage of the PCR method is its capability to analyze a target sequence in a DNA sample whatever the quantity or quality of the DNA is, provided that the sample contains at least one intact DNA strand encompassing the region to be amplified. This sensitivity has allowed the typing of different human DNA polymorphism in single hair roots (Higuchi *et al.*, 1988), individual somatic cells (Jeffreys *et al.*, 1988) and single sperm (Li *et al.*, 1988). Even degraded DNA from formalin-fixed, paraffin-embedded archival tissues or extracted from ancient museum specimens and archaeological findings is suitable for the amplification reaction (Impraim *et al.*, 1987; Pääbo,

Figure 2. Localization of the two oligonucleotide primers for the amplification of VNTR regions. A and B are the two primers. R.E. site is the recognition site of a restriction endonuclease used for the analysis of VNTR regions by Southern blot and hybridisation with a labelled DNA probe.

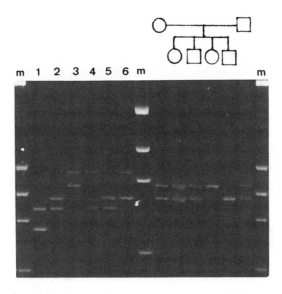

Figure 3. Amplification of the minisatellite 33.4 with direct detection of the PCR products on ethidium bromide-stained agarose gel. Lane 1 to 6 on the left side represent the amplified alleles from 6 unrelated Caucasians. On the right side, the Mendelian inheritance in a nuclear family is shown. Lane m contains molecular weight markers.

1989). Therefore, the PCR technique becomes a valuable tool for the analysis of polymorphic regions in human DNA samples, particularly in forensic cases where high amounts of undegraded DNA are frequently not available.

Development of PCR Procedures for the Amplification of VNTR Regions

The PCR technique has found wide application in genetic linkage studies or in the diagnosis of genetic diseases. In these cases base-pair substitutions are visualized with restriction enzymes or with allele specific oligonucleotides following amplification of substrate DNA. The amplified fragments contain almost no repeated sequences and do not contain extreme base-pair ratios. Contrary to these "single copy" sequences, VNTRs are composed of tandem repeat units and have either a high GC-content (e.g. p33.6; Jeffreys *et al.*, 1988) or a high AT-content (e.g. apolipoprotein B; Decorte *et al.*, 1990b). This could lead to a number of potential problems in the amplification reaction including secondary structure of the GC-rich repeat units, slippage of DNA polymerase in AT-rich regions, generation of recombinant alleles during the extension phase and preferential amplification of the smallest allele in a heterozygous individual. Some of these problems have been observed in the first report describing the amplification of minisatellite regions (Jeffreys et al., 1988). At high cycle numbers, a background of additional fragments appeared which could be removed by digestion with S1 nuclease. These background fragments are mainly incomplete single-stranded templates from previous cycles which can act as primers in the annealing phase at higher cycle numbers. Because of the complementarity between the tandem repeats, they can hybridize at several sites of the VNTR region and anneal out-of-register. This results in the generation of PCR products of abnormal length which can be visualized on ethidium bromide-stained agarose gels after a sufficient number of cycles (R. Decorte, unpubl. data). Additional parameters which can generate spurious fragments are the annealing and extension temperatures of the amplification cycles (Decorte et al., 1990b). For high AT-rich regions, such as the VNTR 3' of the apolipoprotein B gene, it is possible that the background fragments arise through DNA slippage, but this cannot explain the extreme differences between some fragments. Furthermore, studies on the fidelity of DNA synthesis by Taq DNA polymerase have reported mainly base substitution errors and frameshift errors of only one base (Keohavong and Thilly, 1989; Eckert and Kunkel, 1990). Therefore, out-of-register annealing is probably the only cause for these spurious fragments, while DNA slippage gives rise only to small deletions or insertions, which can be seen in the amplification of fragments with di-nucleotide polymorphism (Weber and May, 1989; Litt and Luty, 1989).

A second factor which affects the efficiency of VNTR amplifications is the preferential amplification of shorter alleles, resulting in a molar imbalance between short and long allele products. Since the length of VNTR alleles can be quite disparate, it could be that the longest alleles tend not to be detected. Although the yield of longer alleles can be improved by increasing the extension time and the amount of Taq DNA polymerase, alleles larger than 10.2 kb have not been detected (Jeffreys et al., 1988).

Finally, the secondary structure of GC-rich repeat units which may be difficult to overcome by the DNA polymerase, is partly destroyed by the high annealing and extension temperatures employed in the PCR reaction. Although the inclusion of 10% dimethylsulfoxide (DMSO) in the amplification reaction will inhibit DNA synthesis by 50% (Gelfand, 1989), some PCR assays for VNTR regions are facilitated by DMSO. The presence of DMSO may remove secondary structures formed by the repeat units and therefore increase the efficiency of the amplification (R. Decorte, unpubl. data).

Detection Procedures for Amplified VNTR Alleles

Initially, the detection of amplified minisatellite alleles could only be accomplished by Southern blot analysis with an internal probe (Jeffreys *et al.*, 1988). Subsequent reports of amplification procedures for VNTR regions, demonstrated the possibility of visualizing allelic fragments on an ethidium bromide-stained agarose gel (Boerwinkle *et al.*, 1989; Horn *et al.*, 1989; Bowcock *et al.*, 1989; Wu *et al.*, 1990 and Decorte *et al.*, 1990b). The alleles of these loci (except for c-Ha-ras) fall in a size range between 100 and 1000 base pairs (Tab. 1) and are amplified with equal efficiencies. Some of the minisatellite loci have comparable allele distributions and it has been possible to directly detect discrete amplified alleles for 33.6 (R. Decorte, unpub. data) and 33.4 (Fig. 3; Decorte *et al.*, 1990a). Additional improvements have been reported by Jeffreys and co-workers (1990) which has permitted the visualisation of amplified alleles up to 6 kb long on ethidium bromide-stained agarose gels. However, when using a standard number of 25 cycles, the minimum amount of total DNA required for amplification with direct detection on agarose gels is 60 ng (R. Decorte, P. Marynen & J.-J. Cassiman, in preparation). For lower amounts of DNA, Southern blot analysis with a radioactive internal probe must be performed in order to detect the PCR products. The final result can be obtained after several hours of autoradiography or even after several days for low yields of PCR product.

Table 1. VNTR regions which can be amplified by PCR

VNTR region	HGM locus[a]	Chromosome localization	Hetero-zygosity (%)	Allelic length range (kb)	Reference[b]
pλg3	D7S22	7q36->qter	97	0.6-20	[1]
λMS32	D1S8	1q42-q43	97	1.1-20	[1]
pMS51	D11S97	11q13	77	1.3-4.3	[1]
p33.1	—	—	66	1.1-2.5	[1]
p33.4	—	—	70	0.8-1.3	[1]
p33.6	D1S111	1q23[c]	67	0.5-1.0	[1]
apolipoprotein B	APOB	2p23-p24	78	0.54-0.87	[2] + [3]
				0.56-0.97	[4]
pYNZ22	D17S30	17p13	86	0.17-0.94	[5]
interleukin-6	IL6	7p21-p14	d	0.61-0.76	[6]
type II collagen	COL2A1	12q14.3	81	0.6-0.7	[7]
c-Ha-ras	HRAS	11p15.5	58	0.86-2.65	[3]
p3.4BHI	IGHJ	14q32.33	73	0.42-0.92	[3]
p602	DXYS17	Xp22.3/Yp11.3	87	0.52-1.24	[8]

[a]Human Gene Mapping 10 locus assignment (— = unassigned)
[b]References: [1] Jeffreys *et al.* (1988), [2] Boerwinkle *et al.* (1989), [3] Decorte *et al.* (1990b), [4] Ludwig *et al.* (1989), [5] Horn *et al.* (1989), [6] Bowcock *et al.* (1989), [7] Wu *et al.* (1990) and [8] R. Decorte, P. Marynen & J.-J. Cassiman (in preparation).
[c]Recently mapped by Chimini *et al.* (1989).
[d]Polymorphism information content of 0.37.

380

At present non-radioactive hybridization and detection procedures have become available which are a rapid and ultrasensitive alternative to the conventional ^{32}P-based systems (Bronstein *et al.*, 1990). The time required for an analysis is markedly reduced but a probe must still be labelled and it requires an overnight hybridization procedure. Therefore, we developed an alternative method for the detection of the amplified VNTR alleles after Southern blot transfer (Decorte *et al.*, 1990a). By including biotin-dUTP in the PCR reaction, the amplified alleles will be labelled by incorporation of this nucleotide. The PCR products can be detected immediately after Southern transfer by incubation with an avidin-alkaline-phosphatase complex and a chemiluminescent substrate. Upon dephosphorylation by alkaline-phosphatase, the substrate decomposes, resulting in an emission of light which can be imaged on x-ray film. From amplification to visualization of the results only 2 days are required, a reduction in time of at least 1 to several days in comparison to conventional Southern blot analysis. Furthermore, this method proved to be very sensitive allowing for the detection of amplified products starting from 15 pg of human DNA within 20 min. of exposure (R. Decorte, P. Marynen & J.-J. Cassiman, in preparation). Amplified products labelled by incorporation of nucleotides with biotin

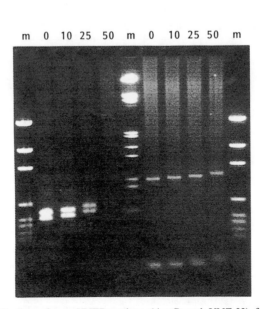

Figure 4. Amplification of two VNTR regions (ApoB and YNZ-22) for a heterozygous individual in the presence of different amounts (%) of biotin-11-dUTP relative to the total amount of dTTP and biotin-11-dUTP. Lane m contains molecular weight markers.

HVR-Ig

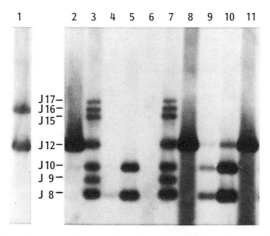

Figure 5. Exclusion of a suspect in a combined sexual assault–murder case for the hypervariable region (HVR-Ig) 5′ to the joining segments of the immunoglobulin heavy chain gene (IGHJ). Lane 1, DNA isolated from peripheral blood of the suspect. Lane 2, DNA isolated from a hair root of the victim. Lane 3 and 7, molecular weight marker composed of different alleles observed for HVR-Ig. Lane 4 and 5, DNA isolated from a semen stain (on clothing from the victim) with the differential lysis method without a reducing agent (lane 6) and in the presence of dithiotreitol (lane 4 and 5). Lane 8, DNA isolated from vaginal swab without differential lysis and with differential lysis (lane 9 to 11); lane 9 and 10, in the presence of DTT and lane 11 without.

or digoxigenin migrate slower through agarose gels than non-labelled fragments (Fig. 4; Lo *et al.*, 1988; Lanzillo, 1990). When molecular weight markers such as λ or pBR322 are used as size markers for the calculation of the different alleles, one could easily misclassify a particular fragment when the repeat unit size is very small. This problem can be resolved by using a mixture of labelled amplified alleles for each VNTR as a molecular weight marker (Fig. 5; Boerwinkle *et al.*, 1989). This standard makes the identification of known or new alleles in DNA samples straightforward. Secondly, a nomenclature for assigning the different alleles can be based on the number of repeats present in a DNA fragment (Odelberg *et al.*, 1989; Boerwinkle *et al.*, 1989). This makes it possible to standardize and to compare results from different laboratories.

Application of PCR Procedures to the Analysis of DNA from Biological Samples

It has been demonstrated that typing of DNA polymorphisms by PCR can be done on semen (Li *et al.*, 1988; Jeffreys *et al.*, 1990), bloodstains

(Williams *et al.*, 1988), single hairs (Higuchi *et al.*, 1988), buccal epithelial cells (Lench *et al.*, 1989) and even epithelial cells collected from urine (Gasparini *et al.*, 1989). Most of these biological samples are typically found at crime scenes. However, the samples usually enter the laboratory under far from ideal circumstances and are sometimes contaminated with bacteria or fungi. This can result in low amounts of genomic DNA and co-purification of non-human DNA or inhibitory substances. In addition to low recoveries, DNA from evidence samples is sometimes degraded (McNally *et al.*, 1989b). The sensitivity of the PCR technique may overcome most of these problems but low PCR product yields are not uncommon (R. Decorte, unpubl. data). A prerequisite for the typing of degraded DNA samples is that the VNTR alleles have a size distribution similar to the DNA fragments from the sample to be analyzed. Provided the DNA sample contains at least one intact DNA fragment comprising the two primer sites, it is possible to amplify the target region. Contaminating bacterial or fungal DNA will not amplify because the two oligonucleotide primers recognize only human sequences. Inhibitory substances are mainly co-purified from bloodstains and may affect the ability of *Taq* DNA polymerase to amplify certain regions. In some forensic samples, admixture of DNA from different individuals is present. This can be the case for bloodstains, semen or hair. In sexual assault cases, the semen samples are composed of a mixture of vaginal cells from the victim and sperm cells from the rapist. The protein coat of the sperm head contains disulphide bridges which can be broken by dithiotreitol (DTT). DNA from the epithelial cells can be isolated by a differential lysis procedure in the absence of this reducing agent leaving the sperm heads intact. Semen cells are recovered by centrifugation and subsequently lysed in the presence of DTT. Mixtures of blood and saliva are not separable by differential lysis but in mixed hair samples, individual hair roots can be typed (Higuchi *et al.*, 1988).

The remarkable sensitivity of the PCR technique has generated special concerns about accidental contamination by foreign DNA. Indeed, false positive results could lead to incorrect conclusions and even exclude suspects from being involved in a crime. These false positives can result from sample-to-sample contamination or from carry-over of amplified DNA from previous amplifications. Therefore, strict preventive measures must be taken to avoid such artifacts as described in the list of recommendations of Kwok and Higuchi (1989). Isolation of DNA from evidence samples and the setting up of the PCR reactions has to be performed in a separate room, from which tubes with PCR products are excluded. Carry-over of DNA can be prevented by using positive displacement pipetting devices and each series of amplification reactions has to include a proper positive control DNA (a DNA sample known to amplify under the conditions applied) and a negative control

(DNA replaced by water which was used for dilution of the samples). These precautions are relevant to all genetic analysis where PCR is used. However, they cannot exclude contamination of the evidence samples before they enter the forensic laboratory. Therefore, law enforcement agencies have to take the necessary steps to prevent contamination of the evidence by saliva and hair.

Finally, PCR can be done on minute DNA samples so that the DNA extracted from evidence material will not be exhausted after one analysis. This allows for a repeat amplification in case of failure. Only 10 to 20% of an amplification reaction is used for gel analysis, while the rest remains available for mixing experiments to exclude for bandshifting during agarose gel electrophoresis. Moreover, in view of the recent controversy about DNA fingerprinting and its admissibility in U.S. courts (Lander, 1989) defence attorneys will tend to request contra-expertises. Since only small samples of the evidence material are required for DNA extraction and PCR analyses, enough material should remain for another independent investigation.

PCR-based Typing of VNTR Regions in DNA from Forensic Evidence: Casework Examples

Before a VNTR marker can be used for genetic identification purposes, it must satisfy a number of criteria: it has to be very polymorphic with a high level of heterozygosity; PCR-based typing has to be uncomplicated with detection of all possible alleles; population data for the alleles must be available in order to allow a statistical analysis of the obtained results and finally, the VNTR marker should be inherited independently of other VNTR regions. A large number of VNTR regions for which PCR-based typing has been described (Tab. 1) fulfil these conditions. Alleles greater than the maximum limit of detection with PCR are present in the population only for pλg3 and λMS32, and for interleukin-6 and c-Ha-ras low heterozygosity exists. A practical procedure for genetic identification (Tab. 2) can be set up according to

Table 2. Procedure for genetic identification

Phase I	DXYS17
	YNZ22 (D17S30)
	minisatellite 33.6 (D1S111)
Phase II	apolipoprotein B (APOB)
Phase III	minisatellite 33.4
	HVR-Ig (IGHJ)

The different VNTR regions were ranked according to their information content and in each phase the same cycling parameters are used for amplification.

the degree of heterozygosity of the VNTR marker and the cycle profile used for PCR amplification of the VNTR alleles. This makes it possible to analyze simultaneously different VNTR regions. Although one could co-amplify several loci together with this procedure (Jeffreys et al., 1988; R. Decorte, unpub. data), the analysis of the "fingerprint" pattern would be more complicated certainly when the allele fragment distribution overlaps between the different VNTRs. Subsequent hybridization with the VNTR probes should resolve the different alleles but increases the time needed for the analysis of the DNA from biological samples.

At present the procedure, presented in Tab. 2, has been used in combination with typing of sequence polymorphisms at the HLA-DRB locus by the reversed-dot-blot system (Saiki et al., 1989; I. Buysse et al., in preparation) for the identification of biological samples in 6 sexual assaults, 3 homicide cases and bloodstains which were found after a street fight. The following examples will illustrate the application of the PCR technique for the analysis of VNTR regions in DNA from biological samples obtained from a rape and a homicide.

In the first case (Fig. 5), a woman was raped and murdered. A semen stain on the clothing of the victim (lane 4 to 6), a vaginal swab (lane 8 to 11) and a hair root from the victim (lane 2) were submitted for comparison with a bloodsample from a suspect (lane 1). DNA from the semen stain and the vaginal swab was obtained by the differential cell lysis method. Briefly, vaginal epithelial cell DNA was isolated in the absence of DTT (lane 6 and 11) after which sperm heads were collected by centrifugation and lysed with DTT (lane 4, 5, 9 and 10). One sample from the vaginal swab was used directly for DNA extraction without pre-lysis of the vaginal cells (lane 8). DNA from epithelial cell fractions, hair root, semen fractions and blood sample together with positive control DNA and a negative control (water) were amplified for the hypervariable region (HVR-lg) 5' to the joining segments of the immunoglobulin heavy chain gene on chromosome 14. A sample from the amplified DNAs was electrophoresed overnight through 3% composite agarose gels (Decorte et al., 1990b) and subsequently transferred to a nylon membrane. Detection was done with avidin-alkaline-phosphatase and the substrate AMPPD (Decorte et al., 1990a). The results were visualized after exposure to x-ray film for 15 min. The semen stain on the clothing of the victim was free of vaginal epithelial cells (lane 6) while in the vaginal swab both sperm and vaginal cells were present. Differential lysis of the vaginal swab could separate the bulk of epithelial cells from the sperm cells but still some traces of female DNA remained. The sample from the vaginal swab which was directly lysed (lane 8), revealed only alleles from the victim suggesting that the amount of sperm DNA present in the vaginal swab falls below the detection level. In both the semen stain (lane 5) on the clothing and in the vaginal swab (lane 9 and 10) identical alleles were found for the male DNA (J8–J10). The woman was homozygous for another allele (J12–J12) in the hair root (lane 2) and in the isolated vaginal epithelial cells (lanes 8 and 11). When the genotype for the suspect (J12–J16) was compared with the alleles found for the sperm DNA, no match was found. Based on these findings, the suspect could be excluded as being the donor of the sperm cells found on the clothing of the victim and in the vaginal swab.

The second case (Fig. 6), is a homicide. The suspect had bloodstains on his pants (2) and two towels containing bloodstains (1 and 3) were found in his apartment. Conventional blood group typing revealed that the bloodstains on the two towels contained a mixture of blood. From each exhibit, two samples (A and B) were taken for DNA extraction; in addition a peripheral blood sample from the suspect and the victim were extracted. The different alleles present in the isolated DNA were detected with direct chemiluminescence after amplification of the VNTR

ApoB

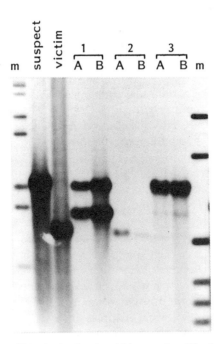

Figure 6. Identification of bloodstains in a homicide case. Amplification of the VNTR region (ApoB) 3′ of the apolipoprotein B gene. 1 and 3, DNA isolated from two bloodstains (A and B) from two different towels which were found in the apartment of the suspect. 2 is DNA isolated from two bloodstains (A and B) from the pants of the suspect. m indicates a molecular weight marker labelled with biotin-11-dUTP.

region 3′ of the apolipoprotein B gene (ApoB) and overnight size fractionation through 3% composite agarose gel (Decorte *et al.*, 1990b). Both the suspect and the victim appeared to be homozygous but for a different allele of ApoB. When the genotype of the victim was compared with the results for the bloodstains on the evidence material, a match was found with the bloodstains on the pants (2A and B). This genotype occurs in 4.5% of the Belgian population. The genotype of the suspect showed no match with the bloodstains on his pants but was compatible with one of the alleles found in the bloodstains on the two towels (1 and 3). The third allele on the two towels was different from the alleles both of the victim and of the suspect and therefore belonged to a third individual. It could also be that the two alleles in the bloodstains on the towels represent the genotype of this third individual and that by chance the suspect's genotype is compatible with one of these alleles. Similar results were obtained for all VNTRs (Tab. 2) used in the procedure and made it possible to construct a possible haplotype for the third individual.

Potential Developments in PCR-based Typing of VNTR Regions

The number of PCR-based typing procedures for VNTR regions will presumably increase over the next few years. In combination with a similar approach to the study of polymorphic HLA regions, it will allow the construction of highly individual-specific DNA "fingerprints". Higher resolution of small size differences will possibly increase the number of alleles presently known for each VNTR region. As a consequence more detailed allele distribution frequencies will become available, including those for different ethnic groups.

The observation that the repeat units of VNTR regions have internal polymorphisms (Minisatellite Variant Repeats) detectable with restriction endonucleases has added a new dimension to the allelic variability of minisatellites or VNTRs (Jeffreys *et al.*, 1990). These polymorphisms should allow the discrimination of similar sized alleles between two different individuals. For forensic testing, it would mean that when a match in allele size has been found between the biological sample and the suspect, more reliable conclusions could be drawn about the origin of the allele. However, the whole procedure for identification of the MVRs is at the present time very laborious and still would not be applicable to mixed biological samples in which different individuals have an allele in common.

At present, the detection of VNTR alleles is restricted to 6 kb directly on ethidium bromide-stained agarose gels and 10 kb after Southern blot analysis with an internal probe (Jeffreys *et al.*, 1988; 1990). These limitations might be overcome by other polymerases which are more heat-resistant and have higher processive synthesis than *Taq* DNA polymerase. As a consequence, it would become possible to detect the alleles of most VNTR loci by the PCR technique. Additional improvements on the detection of VNTR alleles could be obtained by the incorporation of fluorescent labelled primers into the PCR product during amplification with subsequent detection on agarose gels by an automatic laser system similar to automatic sequencers (Zimran *et al.*, 1989). Co-amplification of different VNTR regions could then be done with primers containing different fluorochromes, thereby allowing the discrimination of the different alleles from each VNTR. This will make the whole procedure from PCR reaction until detection completely automatic and will substantially reduce the amount of time needed for the construction of a reliable DNA "fingerprint".

Conclusion

While the use of VNTR amplification has only recently been introduced in the genetic identification of individuals or biological samples, the potentials of this approach are already quite obvious and the advantages over the classical DNA fingerprinting make it a valid alternative to the latter method, which still suffers important shortcomings. Exact information by police forces and forensic laboratories about the prevention of accidental contamination of samples and appropriate physical constraint measures in the DNA laboratories should overcome its major shortcoming, i.e. false positive results.

Acknowledgements
This work was supported by the grant Concerted Actions from the Belgian Government and by the Interuniversity Network for Fundamental Research sponsored by the Belgian Government (1987–1991). We thank Sabine Hermans, Peggy Servranckx and Kathy Merckx for their excellent technical assistance, Staf Doucet for synthesizing the oligonucleotides and Karel Rondeau for the photographic work.

References

Baird, M., Balazs, I., Giusti, A., Miyazaki, L., Nicholas, L., Wexler, K., Kanter, E., Glassberg, J., Allen, F., Rubinstein, P., and Sussman, L. (1986). Allele frequency distribution of two highly polymorphic DNA sequences in three ethnic groups and its application to the determination of paternity. *Am. J. Hum. Genet.* 39: 489–501.

Balazs, I., Baird, M., Clyne, M., and Meade, E. (1989) Human population genetic studies of five hypervariable DNA loci. *Am. J. Hum. Genet.* 44: 182–190.

Boerwinkle, E., Xiong, W., Fourest, E., and Chan, L. (1989) Rapid typing of tandemly repeated hypervariable loci by the polymerase chain reaction: Application to the apolipoprotein B 3' hypervariable region. *Proc. Natl. Acad. Sci. USA* 86: 212–216.

Bowcock, A. M., Ray, A., Erlich, H., and Sehgal, P. B. (1989) Rapid detection and sequencing of alleles in the 3' flanking region of the interleukin-6 gene. *Nucleic Acids Res.* 17: 6855–6864.

388

Bronstein, I., Voyta, J. C., Lazzari, K. G., Murphy, O., Edwards, B., and Kricka, L. J. (1990) Rapid and sensitive detection of DNA in Southern blots with chemiluminescence. *Bio Techniques* 8: 310–314.

Ceccaldi, P. F. (1989) Cent ans d'identification (de l'homme à son génome). *Bull. Acad. Natle. Méd.* 173: 835–843.

Chimera, J. A., Harris, C. R., and Litt, M. (1989) Population genetics of the highly polymorphic locus D16S7 and its use in paternity evaluation. *Am. J. Hum. Genet.* 45: 926–931.

Chimini, G., Mattei, M.-G., Passage, E., Nguyen, C., Boretto, J., Mattei, J.-F., and Jordan, B.R. (1989) *In situ* hybridization and pulsed-field gel analysis define two major minisatellite loci: 1q23 for minisatellite 33.6 and 7q35-q36 for minisatellite 33.15. *Genomics* 5: 316–324.

Cooper, D. N., Smith, B. A., Cooke, H. J., Niemann, S., and Schmidtke, J. (1985) An estimate of unique DNA sequence heterozygosity in the human genome. *Hum. Genet.* 69: 201–205.

Dallas, J. F. (1988) Detection of DNA "fingerprints" of cultivated rice by hybridization with a human minisatellite DNA probe. *Proc. Natl. Acad. Sci. USA* 85: 6831–6835.

Decorte, R., Hilliker, C., Marynen P., and Cassiman, J.-J. (1990a) Rapid and simple detection of minisatellite regions in forensic DNA samples by the polymerase chain reaction combined with a chemiluminescence method. *Trends Genet.* 6: 172–173.

Decorte, R., Cuppens, H., Marynen, P., and Cassiman, J.-J. (1990b) Rapid detection of hypervariable regions by the polymerase chain reaction technique. *DNA Cell Biol.* 9: 461–469.

Eckert, K. A., and Kunkel, T. A. (1990) High fidelity DNA synthesis by the *Thermus aquaticus* DNA polymerase. *Nucleic Acids Res.* 18: 3739–3744.

Economou, E. P., Bergen, A. W., Warren, A. C., and Antonarakis, S. E. (1990) The polydeoxyadenylate tract of *Alu* repetitive elements is polymorphic in the human genome. *Proc. Natl. Acad. Sci. USA* 87: 2951–2954.

Gasparini, P., Savoia, A., Pignatti, P. F., Dallapiccola, B., and Novelli, G. (1989) Amplification of DNA from epithelial cells in urine. *N. Eng. J. Med.* 320: 809.

Gasparini, P., Trabetti, E., Savoia, A., Marigo, M. and Pignatti, P. F. (1990) Frequency distribution of the alleles of several variable number of tandem repeat DNA polymorphisms in the Italian population. *Hum. Hered.* 40: 61–68.

Gelfand, D. H. (1989) *Taq* DNA polymerase. In: Erlich, H.A. (ed), PCR Technology. Principles and Applications for DNA Amplification. Stockton Press, New York, pp. 17–22.

Georges, M., Lequarré, A. S., Castelli, M., Hanset, R., and Vassart, G. (1988) DNA fingerprinting in domestic animals using four different minisatellite probes. *Cytogenet. Cell Genet.* 47: 127–131.

Gill, P., Jeffreys, A. J., and Werrett, D. J. (1985) Forensic applications of DNA "fingerprints". *Nature* 318: 577–579.

Gill, P., Lygo, J. E., Fowler, S. J., and Werrett, D. J. (1987) An evaluation of DNA fingerprinting for forensic purposes. *Electrophoresis* 8: 38–44.

Gill, P., Sullivan, K., and Werrett, D. J. (1990) The analysis of hypervariable DNA profiles: problems associated with the objective determination of the probability of a match. *Hum. Genet.* 85: 75–79.

Giusti, A., Baird, M., Pasquale, S., Balazs, I., and Glassberg, J. (1986) Application of deoxyribonucleic acid (DNA) polymorphisms to the analysis of DNA recovered from sperm. *J. Forens. Sci.* 31: 409–417.

Gjertson, D. W., Mickey, M. R., Hopfield, J., Takenouchi, T., and Terasaki, P. I. (1988) Calculation of probability of paternity using DNA sequences. *Am. J. Hum. Genet.* 43: 860–869.

Higuchi, R., von Beroldingen, C. H., Sensabaugh, G. F., and Erlich, H. A. (1988) DNA typing from single hairs. *Nature* 332: 543–546.

Hillel, J., Plotzky, Y., Haberfeld, A., Lavi, U., Cahaner, A., and Jeffreys, A. J. (1989) DNA fingerprints of poultry. *Anim. Genet.* 20: 25–35.

Horn, G. T., Richards, B., and Klinger, K. W. (1989) Amplification of a highly polymorphic VNTR segment by the polymerase chain reaction. *Nucleic Acids Res.* 17: 2140.

Impraim, C. C., Saiki, R. K., Erlich, H. A., and Teplitz, R. L. (1987) Analysis of DNA extracted from formalin-fixed, paraffin-embedded tissues by enzymatic amplification and hybridization with sequence-specific oligonucleotides. *Biochem. Biophys. Res. Commun.* 142: 710–716.

Jeffreys, A. J., Wilson, V., and Thein, S. L. (1985a) Hypervariable "minisatellite" regions in human DNA. *Nature* 314: 67–73.

Jeffreys, A. J., Wilson, V., and Thein, S. L. (1985b) Individual-specific "fingerprints" of human DNA. *Nature* 316: 76–79.

Jeffreys, A. J., and Morton, D. B. (1987) DNA fingerprints of dogs and cats. *Anim. Genet.* 18: 1–15.

Jeffreys, A. J., Wilson, V., Neumann, R., and Keyte, J. (1988) Amplification of human minisatellites by the polymerase chain reaction: towards DNA fingerprinting of single cells. *Nucleic Acids Res.* 16: 10953–10971.

Jeffreys, A. J., Neumann, R., and Wilson, V. (1990) Repeat unit sequence variation in minisatellites: a novel source of DNA polymorphism for studying variation and mutation by single molecule analysis. *Cell* 60: 473–485.

Kanter, E., Baird, M., Shaler, R., and Balazs, I. (1986) Analysis of restriction fragment length polymorphisms in deoxyribonucleic acid (DNA) recovered from dried bloodstains. *J. Forens. Sci.* 31: 403–408.

Keohavong, P., and Thilly, W. G. (1989) Fidelity of DNA polymerases in DNA amplification. *Proc. Natl. Acad. Sci. USA* 86: 9253–9257.

Kwok, S., and Higuchi, R. (1989) Avoiding false positives with PCR. *Nature* 339: 237–238.

Lander, E. S. (1989) DNA fingerprinting on trial. *Nature* 339: 501–505.

Lanzillo, J. L. (1990) Preparation of digoxigenin-labelled probes by the polymerase chain reaction. *BioTechniques* 8: 621–622.

Lawyer, F. C., Stoffel, S., Saiki, R. K., Myambo, K., Drummond, R., and Gelfand, D. H. (1989) Isolation, characterization, and expression in *Escherichia coli* of the DNA polymerase gene from *Thermus aquaticus*. *J. Biol. Chem.* 264: 6427–6437.

Lench, N., Stanier, P., and Williamson, R. (1988) Simple non-invasive method to obtain DNA for gene analysis. *The Lancet* i: 1356–1358.

Lewin, R. (1989) Limits to DNA fingerprinting. *Science* 243: 1549–1551.

Li, H., Gyllensten, U. B., Cui, X., Saiki, R. K., Erlich, H. A., and Arnheim, N. (1988) Amplification and analysis of DNA sequences in single human sperm and diploid cells. *Nature* 335: 414–417.

Litt, M., and Luty, J. A. (1989) A hypervariable microsatellite revealed by *in vitro* amplification of a dinucleotide repeat within the cardiac muscle actin gene. *Am. J. Hum. Genet.* 44: 397–401.

Lo, Y.-M. D., Mehal, W. Z., and Fleming, K. A. (1988) Rapid production of vector-free biotinylated probes using the polymerase chain reaction. *Nucleic Acids. Res.* 16: 8719.

Ludwig, E. H., Friedl, W., and McCarthy, B. J. (1989) High-resolution analysis of a hypervariable region in the human apolipoprotein B gene. *Am. J. Hum. Genet.* 45: 458–464.

McNally, L., Shaler, R. C., Baird, M., Balazs, I., De Forest, P., and Kobilinsky, L. (1989a) Evaluation of deoxyribonucleic acid (DNA) isolated from human bloodstains exposed to ultraviolet light, heat, humidity, and soil contamination. *J. Forens. Sci.* 34: 1059–1069.

McNally, L., Shaler, R. C., Baird, M., Balazs, I., Kobilinsky, L., and De Forest, P. (1989b) The effects of environment and substrata on deoxyribonucleic acid (DNA): the use of casework samples from New York City. *J. Forens. Sci.* 34: 1070–1077.

Mullis, K., Faloona, F., Scharf, S., Saiki, R., Horn, G., and Erlich, H. (1986) Specific enzymatic amplification of DNA *in vitro*: The polymerase chain reaction. Cold Spring Harbor Symp. Quant. Biol. 51: 263–273.

Nakamura, Y., Leppert, M., O'Connell, P., Wolff, R., Holm, T., Culver, M., Martin, C., Fujimoto, E., Hoff, M., Kumlin, E., and White, R. (1987) Variable number of tandem repeat (VNTR) markers for human gene mapping. *Science* 235: 1616–1622.

Norman, C. (1989) Maine case deals blow to DNA fingerprinting. *Science* 246: 1556–1558.

Odelberg, S. J., Plaetke, R., Eldridge, J. R., Ballard, L., O'Connell, P., Nakamura, Y., Leppert, M., Lalouel, J.-M., and White, R. (1989) Characterization of eight VNTR loci by agarose gel electrophoresis. *Genomics* 5: 915–924.

Pääbo, S. (1989) Ancient DNA: Extraction, characterization, molecular cloning, and enzymatic amplification. *Proc. Natl. Acad. Sci. USA* 86: 1939–1943.

Royle, N. J., Clarkson, R. E., Wong, Z., and Jeffreys, A. J. (1988) Clustering of hypervariable minisatellites in the proterminal regions of human autosomes. *Genomics* 3: 352–360.

Saiki, R. K., Gelfand, D. H., Stoffel, S., Scharf, S. J., Higuchi, R., Horn, G. T., Mullis, K. B., and Erlich, H. A. (1988) Primer-directed enzymatic amplification of DNA with a thermostable DNA polymerase. *Science* 239: 487–491.

Saiki, R. K., Walsh, P. S., Levenson, C. H., and Erlich, H. A. (1989) Genetic analysis of amplified DNA with immobilized sequence-specific oligonucleotide probes. *Proc. Natl. Acad. Sci. USA* 86: 6230–6234.

Sensabaugh, G. F. (1981) Uses of polymorphic red cell enzymes in forensic science. *Clin. Haematol.* 10: 185–207.

Southern, E. M. (1975) Detection of specific sequences among DNA fragments separated by gel electrophoresis. *J. Mol. Biol.* 98: 503–517.

Vassart, G., Georges, M., Monsieur, R., Brocas, H., Lequarré, A.S., and Christophe, D. (1987) A sequence in M13 phage detects hypervariable minisatellites in human and animal DNA. *Science* 235: 683–684.

Weber, J. L., and May, P. E. (1989) Abundant class of human DNA polymorphisms which can be typed using the polymerase chain reaction. *Am. J. Hum. Genet.* 44: 388–396.

Williams, C., Weber, L., Williamson, R., and Hjelm, M. (1988) Guthrie spots for DNA based carrier testing in cystic fibrosis. *The Lancet* ii: 693.

Wong, Z., Wilson, V., Patel, I., Povey, S., and Jeffreys, A. J. (1987) Characterization of a panel of highly variable minisatellites cloned from human DNA. *Ann. Hum. Genet.* 51: 269–288.

Wong, Z., Royle, N. J., and Jeffreys, A. J. (1990) A novel human DNA polymorphism resulting from transfer of DNA from chromosome 6 to chromosome 16. *Genomics* 7, 222–234.

Wu, S., Seino, S., and Bell, G. I. (1990) Human collagen, type II, alpha 1, (COL2A1) gene: VNTR polymorphism detected by gene amplification. *Nucleic Acids Res.* 18: 3102.

Zimran, A., Glass, C., Thorpe, V. S., and Beutler, E. (1989) Analysis of "color PCR" by automatic DNA sequencer. *Nucleic Acids Res.* 17: 7538.

DNA Fingerprinting: Approaches and Applications
ed. by T. Burke, G. Dolf, A. J. Jeffreys & R. Wolff
© 1991 Birkhäuser Verlag Basel/Switzerland

Genetic Typing Using Automated Electrophoresis and Fluorescence Detection

J. Robertson, J. Ziegle, M. Kronick, D. Madden*, and B. Budowle

Forensic Science Research and Training Section, FBI Laboratory, FBI Academy, Quantico, Virginia, USA
*Applied Biosystems, Inc., 850 Lincoln Center Dr., Foster City, CA 94404, USA

Summary
Multi-color fluorescence detection systems offer unique advantages when compared to single label detection methods for DNA typing, genetic disease testing, population fingerprinting, and DNA mapping. Internal controls are easily used and identified by different color dye labels. Multiple independent samples or multiple analyses of the same sample are run in each lane of a gel. Precision of size assignment and quantification are improved. Here, we will review a variety of methods used to analyze DNA and present the advantages of the multi-color fluorescence dye approach. An automated and quantitative DNA typing assay for human identification is shown. This method is an improvement over previous manual techniques and uses multi-color fluorescence labeling, electrophoresis and real-time detection methodology.

Introduction

Electrophoresis has been used widely to separate and identify DNA fragments. Applications of this technique include DNA typing, population fingerprinting, genetic mapping, and genetic disease testing. Many methodologies are available to detect the electrophoretically separated DNA fragments. One may incorporate a radioative label, stain with ethidium bromide or silver, view fluorescence under ultraviolet light, label with fluorescent dyes or couple enzymes for a colorimetric assay. Many techniques are used to estimate DNA fragment size and to quantify the amount of DNA in a gel band. This is especially useful when DNA size is correlated or directly translatable to allele type, and DNA concentration to gene dosage. Each labeling and detection system has its own unique advantages and disadvantages for particular applications.

Here we will review the unique advantages of a multi-color fluorescent dye approach to genetic analysis. Human identity testing and a genetic disease test for Duchenne muscular dystrophy will be used to illustrate the failings of current systems and the advantages of the new approach.

Human Identity Testing

DNA based identity tests rely on the genetic differences among individuals. Variable number of tandem repeat (VNTR) loci currently are the most informative genetic markers for characterization of people. These markers, like genes, follow standard Mendelian inheritance. VNTR alleles differ by size or the number of tandem repeat sequences. Traditionally, the VNTR alleles are identified and detected using Southern blotting protocols and restriction enzyme digestion of sequences bordering the VNTR. Due to sensitivity requirements of the assay, detection methods which rely on radioactive labels are almost universal.

The use of polymerase chain reaction (PCR) recently has made it possible to analyze the VNTR loci without the need for either Southern blotting or isotopic labeling. PCR products separated by electrophoresis are detected by staining with ethidium bromide or silver (Allen et al., 1989). The PCR method is simpler and much faster than Southern blotting. Both methods, however, are subject to band shift artifacts, lack of adequate controls, and error prone manual manipulations. These problems lower precision and accuracy of the tests and reduce the ability to assign correct alleles to an individual. This is especially problematic in forensic applications where allele identity typically is used either to exclude or include an individual from a population of humans who could have deposited biological material at the scene of a crime or to relate the individual to a putative family member. The issue of adequate controls and standards has been strongly stressed in the recent report, "Genetic Witness: Forensic Uses of DNA Tests" from the Office of Technology Assessment in the Congress of the US (OTA-BA-438, 1990). It is obvious that a more precise and quantitative approach to the analysis is needed.

Four years ago researchers achieved a more analytical approach to DNA sequence analysis using Sanger reactions with four different color fluorescent dyes (Smith et al., 1986). Since each of four primers can be labeled with a different dye, the reaction products from four different sequence reactions can be identified in a single gel lane. Each sequence reaction can be terminated with a different base. Later Carrano et al. showed that the same four fluorescent dyes could be used to map restriction fragments of cosmids with higher sizing precision and higher throughput (Carrano et al., 1989). With this method an internal size standard tagged with one dye type is used for automatic sizing of fragments tagged with another color dye. Three DNA restriction digests are each tagged with a different dye type and run within the same gel lane as the size standard. The use of an internal standard circumvents the problems of band shift artifacts, provides an internal control for gel lane to gel lane comparison of data, and also offers a means to automate the analysis. More recently, this same technology has been

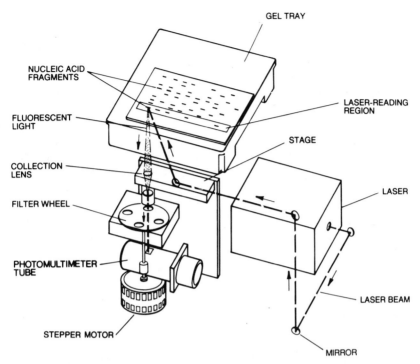

Figure 1. Real-time fluorescence detection of four fluorophores during electrophoresis. Fluorophores labeled to DNA fragments are excited when the fragments pass by a fixed laser scanning window. The fluorescence emission is sequentially passed through four band filters each designed to pass the dyes maximum emission. The signal is digitized and stored on a computer hard disk for later analysis.

used for designing genetic disease tests incorporating dye labeled primers in PCR amplifications (Zimran *et al.*, 1989).

Here, we show an example of this technology with the analysis of PCR generated VNTR alleles. An automated, electrophoresis and real-time detection fluorescence scanner is used. The fluorescence scanner system incorporates a multi-line argon ion laser which scans across the gel to excite fluorescence from passing dye labeled PCR products (Fig. 1). Four different color dyes, fluorescein and rhodamine derivatives, are excited at the same wavelength of light, and the emission signal passed sequentially through four different band filters specific for each dyes' maximum emission (Fig. 2). The signal is collected and analyzed to assign fragment size and quantify fluorescence. Fragment size is automatically extrapolated from the internal size standard curve.

Fluorescently tagged PCR primers are used to label DNA fragments *in situ*. The length and relative quantity of PCR products from VNTR alleles are automatically determined by comparison with in-lane size standards labeled with a fluorescent tag different from that used on the

394

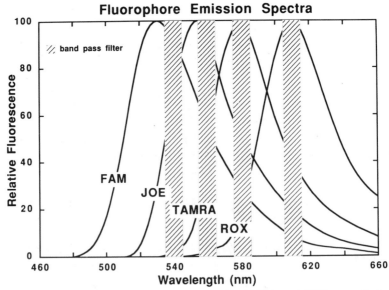

Figure 2. Emission spectra of four dyes, FAM, JOE, TAMRA, and ROX, respecitvely. All dyes are either fluorescein or rhodamine derivatives. Dyes are excited by 40 mW 488 and 514 nm dual line argon ion laser.

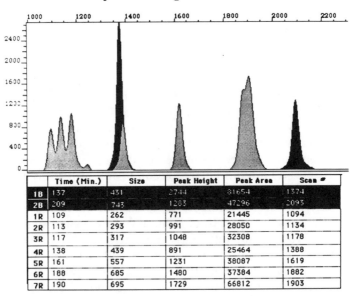

Figure 3. Electrophoretogram representing the peak data from real-time fluorescence detection of fluorescently labeled (Fam dye) PCR products. PCR products are from the amplification of the VNTR locus, pMCT118. The internal-lane size standard (pBR322 digested with AluI), labeled with ROX dye, is shown in light grey. The PCR products are shown in black. Automatic size determination of the PCR procucts shows two fragment bands at 431 and 743 bp, respectively.

PCR primers. The use of in-lane standards result in high precision size assignments that are typically 98% or greater.

A study of a small population ($n = 50$) humans was made. The PCR products of the VNTR locus pMCT118 (Kasai *et al.*, 1990) were analyzed by both silver staining and real-time fluorescence detection. The allele sizes differ by 16 base pairs. The presence of an internal size standard is crucial because of the close proximity of the fragment bands within an agarose gel and the risk of assigning the wrong allele (size) to a particular fragment band. The data show that the automated multi-color fluorescence method correlates with the silver-staining manual

Table 1. Distribution of frequency of alleles (size) of the pMCT118 locus by both the fluorescence and silver staining methodologies. The study is of 50 Caucasian individuals

Allele Size (bp)	Frequency	
	Fluorescence Method	Staining Method
432	23	23
447	2	2
461	2	2
479	5	5
494	5	5
510	3	3
524	27	27
539	7	7
554	1	1
582	8	8
598	5	5
614	1	1
628	7	7
643	0	0
659	1	1
719	1	1
743	1	1
779	1	1

method and also can be used for identity testing. Figure 3 shows an electrophoretogram scan of peak data obtained from the multi-color dye analysis. All 50 DNAs showed automatic typing results equivalent to the manual staining technique. Table 1 shows the distribution of allele frequency of the data.

The laser system provides higher detection sensitivity than standard methods. Typically, between 2 and 200 nanograms of DNA can be used for the PCR analysis. Only 1% of the PCR products of a 25 PCR cycle reaction is used. This is approximately 200 times more sensitive than standard staining techniques.

Genetic Disease Testing

Before the advent of PCR, genotyping for genetic disease status was accomplished by Southern blotting procedures and Restriction Fragment Length Polymorphisms (RFLP) analysis. These analyses were used to identify normal, affected, and carrier status among family members and the unborn. Typical tests used radioactive probes and X-ray film development, and required one to two weeks for analysis. PCR changed many of the approaches to genetic disease assessment.

One new approach is the multiplex PCR assay for Duchenne muscular dystrophy (DMD), an x-linked disease (Chamberlain et al., 1989). In this test, exonic deletions correlated to disease state in 60% of the affected population are identified by the strategic placement of PCR oligonucleotide primers. Nine primer sets designed for nine exonic deletion sites are "multiplexed" in a single reaction tube. The primers also are designed to produce products of differing fragment sizes. Each fragment band, or lack thereof, can be identified when the "multiplex" reaction is analyzed by agarose gel electrophoresis and ethidium bromide staining. This assay circumvents the standard Southern blotting procedure for identifying affected males. Female carrier status or gene dosage, however, is usually not possible with this method due to the poor quantitation of the PCR assay and the qualitative visual assessment of stained fragments. Southern blotting is often used to quantify PCR product in each band, but this technique uses radioactivity and is laborious and time-consuming. Use of fluorescence tagged primers in the assay and the use of sensitive laser based fluorescence detection provides a means to circumvent Southern blotting procedures to determine gene dosage.

The fluorescence multiplex PCR assay uses different color fluorescent dyes to label the nine oligonucleotide primers. In addition, two other primer sets are added. One is designed as a control for reaction conditions and presence of DNA. (This primer set was designed for the human beta globin gene.) The other primer set is used for internal sample identification and to prevent sample mix-up and affirm familial relationships. (This primer set was designed for the VNTR locus, pMCT118.) An internal size standard was used to assign DNA fragment size.

This assay is more reliable for assigning fragment size, and provides better quantification for gene dosage than indirect detection of fragments by Southern blotting/densitometric protocols. Figure 4 shows, in tabular form, the results of a small quantitative study on DMD gene dosage.

Another creative approach to genetic disease testing is *competitive oligonucleotide priming* PCR (Ballabio *et al.*, 1990). This technique uses two primers whose sequences varies by one base pair. One primer represents the "wild type" sequence. The other primer represents the mutation sequence. Three primers are used in the PCR reaction. The two variant primers compete for a site that is complementary to only one of them. Base mutations are identified by a radioactive label

band size in base pairs

sample	196	268	331	360	388	416	459	506	547
Normal									
female 1	1.00	1.05	1.05	1.01	1.11	1.10	1.07	1.12	1.07
2	1.00	1.02	1.00	0.96	1.02	1.01	1.01	1.12	0.97
3	1.00	0.99	1.03	1.01	1.07	0.99	1.07	1.00	1.01
4	1.00	0.97	1.01	1.00	1.00	0.99	0.99	0.90	0.97
5	1.00	0.95	0.90	1.00	0.80	0.91	0.87	0.90	0.97
Carrier									
female 1	1.00	1.05	0.89	0.91	1.27	1.17	1.12	1.36	**0.55**
2	1.00	**0.53**	1.09	1.05	1.19	1.16	1.07	1.12	0.97
3	1.00	0.97	1.04	1.02	**0.46**	1.01	0.94	**0.36**	0.93
4	1.00	0.87	0.85	0.92	**0.38**	0.85	0.83	**0.24**	0.79
5	1.00	0.96	0.95	0.98	0.88	0.96	0.91	**0.48**	**0.46**
Affected									
male 1	1.00	1.06	1.09	1.05	1.25	1.14	1.11	1.51	**0.00**
2	1.00	1.07	1.04	1.03	1.24	1.03	1.06	1.48	**0.00**

Figure 4. Quantitative assessment of gene dosage using the fluorescence multiplex PCR assay. Data are ratios of peak areas of sample PCR product bands to mean peak area of peak bands obtained from female normal individuals. Amplification conditions for the PCR product band, 506 bp, is not optimized. Deletions appear as 0. Normal results appear close to 1, and carrier results appear close to 0.5.

attached to one of the primers and then the reaction is repeated using a radioactive tag on the other primer. Thus, both homozygous and heterozygous individuals are identified.

The multi-fluorophore approach greatly simplifies the procedure. The two competitive primers are differentiated by two different color fluorescent dye labels and then used in the same PCR. Overlapping fragment bands from heterozygous individuals are easily identified by the system (Chehab et al., 1989; 1990).

Conclusions

A variety of DNA fragment detection methods are available to both geneticists and molecular biologists. Choice of a particular system is usually motivated by the level of quantitative analysis needed, the simplicity of the test desired, and the economics of cost, labor, and time. Improved detection procedures, such as the multi-fluorophore approaches, allow experiments which require increased sensitivity, accuracy, and precision in the determination of DNA fragment size and concentration. Multi-color procedures also provide means to "multiplex" sample analyses and to design novel DNA assay methods such as the competitive oligonucleotide priming PCR. Detection methods that utilize more than a single type of label offer new avenues for experimental design and are only limited by the creativity brought to the applications.

References

Allen, R. C., Graves, G., and Budowle, B. (1989) Polymerase chain reaction amplification products separated on rehydratable polyacrylamide gels and stained with silver. *Biotechniques* 7(7): 736–744.

Ballabio, A. *et al.* (1990) PCR test for cystic fibrosis detection. *Nature.* 343: 220.

Carrano, A. V. *et al.* (1989) A high-resolution fluorescence based, semi-automated method for DNA fingerprinting. *Genomics* 4: 129–136.

Chehab, F. F., and Kan, Y. W. (1989) Detection of specific DNA sequences by fluorescence amplification: a color complementation assay. Proc. Natl. Acad. Sci. 86: 9178–9182.

Cheehab, F. F., and Kan, Y. W. (1990) Detection of sickle cell anaemia mutation by color DNA amplification. *Lancet* 335; 15–17.

Chamberlain, J. S. *et al.* (1988) Deletion screening of the Duchenne muscular dystrophy locus via multiplex DNA amplification. *Nucleic Acids Res* 16: 11141.

Kasai, K. *et al.* (1990) Amplification of a VNTR locus (pMCT118) by the polymerase chain reaction (PCR) and its application to forensic Science. *J. of Forensic Science* 35(5): 1196–1200.

Smith, L. M., Sanders, J. Z. *et al.* (1986) Fluorescence detection in automated DNA sequence analysis. *Nature* 321: 674–679.

U.S. Congress Office of Technology Assessment, Genetic Witness: Forensic Uses of DNA Tests, OTA-BA-438 (Washington, DC: U.S. Government Printing Office, July 1990).

Zimran, A. *et al.* (1989) Analysis of "color PCR" by automatic DNA sequencer. *Nucleic Acids Res.* 17/18: 7538.

Subject Index

Page numbers refer to the beginning of the chapter.